Praise for *Crisis Preparedness Handbook*

STEP-BY-STEP GUIDE

"The survival instinct has been part of us since the first humans walked the planet. Unfortunately, crisis preparedness has not. My own journey to prepare has been a hodgepodge of fits and starts. The *Crisis Preparedness Handbook* has changed that for me. It has provided me a logical, step-by-step guide on what to think about, how to prepare for, and how to survive any—usually unforeseen—crisis."

—LEW MASON, KIRKLAND WA
Managing Broker, Windemere Yarrow Bay

TIMING COULD NOT BE BETTER

"*Crisis Preparedness Handbook* delivers a comprehensive how-to guide for home storage and physical survival in all types of natural and man-made disasters.

"Chapters are filled filled with well-researched content, worksheets, and getting-started tips. The systematic approach truly makes the process of preparing manageable instead of overwhelming.

"This comprehensive guide moves well beyond water and food storage to include information on how to manage a medical or dental emergency, how to select and store survival firearms and other survival weapons, and how to plan for a terrorist attack or nuclear crisis.

"The timing of this book could not be better. In this turbulent world fraught with pandemics, natural disasters, and riots, it is a must-have resource for every family!"

—JANET ANDERSON
*Distinguished Professor of Nutrition, Dietetics, and Food Sciences,
Utah State University*

SAVES YOU COUNTLESS HOURS OF RESEARCH

"As a former CERT coordinator, local emergency-preparation specialist, and avid gardener, I have found *Crisis Preparedness Handbook* to be invaluable. It provides just the right amount of detail in a single book, especially in those areas where detail is important. It will save me countless hours of research. I highly recommend this book for anyone at any level of interest in preparing for an emergency for the family, group, or community."

—KELLY STRONG, CACHE COUNTY, UT
Former CERT Coordinator

BEST PREPAREDNESS INFORMATION IN A SINGLE VOLUME

"You face many choices in trying to accomplish the arduous task of planning, executing, acquiring, storing, and utilizing all the aspects of essential preparedness products, equipment, and supplies. *Crisis Preparedness Handbook* will help you make decisions without having to learn from your mistakes—saving you time, money, and frustration.

"This book is divided into sections and subsections, with accurate information detailing what to do, how to do it, and how to utilize all the "parts and pieces" of a serious preparedness inventory. Patricia is thorough, smart, and organized in sharing how to implement and integrate your preparedness goals step-by-step. Her timely information will accelerate your preparedness efforts by distilling them into easily digested, bite-size chunks.

"All things considered, *Crisis Preparedness Handbook* provides the most practical and best preparedness information in a single volume for new preppers, experienced survivalists, and for all individuals, families, or groups practicing preparedness."

—JAMES TALMAGE STEVENS
Author, Making the Best of Basics

GIVES YOU THE NECESSARY STEPS TO PREPARE

"Preparedness planning requires commitment and knowledge. *Crisis Preparedness Handbook* provides an organized and extensive review of the necessary steps to get a household prepared for almost any crisis. In this new edition, author Patricia Spigarelli Aston includes some of her own experiences that will inspire you to get busy and do something to ensure you are prepared for the next crisis."

—VON T. MENDENHALL
Professor of Food Science Emeriti, Utah State University

FILLED WITH PRACTICAL IDEAS—A REFERENCE GUIDE

"I have been using the *Crisis Preparedness Handbook* since 1984. The knowledge I gained from it helped me through many emergency-preparedness quandaries when I was a young mother, such as how to store food, water, and clothing; where to locate emergency essentials; and how to prepare our large family for various challenges.

"Now that we are empty nesters, this new, updated version of the *Crisis Preparedness Handbook* is helping me reevaluate our preparedness for emergencies, our food-storage methods, and how to feel prepared for the trials the two of us may encounter.

"I like the way the topics are organized in this current and relevant update. The topics are organized so each is easy to find, and each section suggests five things I can work on today! Filled with practical ideas, this book can be used as a reference guide so I can look up whatever I need when I need it."

—ROSEMARY LIND, ALPINE, UT
Community volunteer and dedicated mother

TEACHES YOU NOT JUST WHAT, BUT WHY AND HOW TO DO IT

"No one can afford to be without the knowledge in *Crisis Preparedness Handbook*. This new revised and updated edition is pertinent to what all of us are going through, and no responsible member of society should go another day without having this information readily available so they can start applying its knowledge.

"Clearly, Aston takes preparedness seriously and teaches you not just what to do, but why and how to do it. This handbook has all the information of a master-course textbook without the yawn-inspiring boredom of academic books. And a quick look through the table of contents shows that this book is more than a bunch of regurgitated facts from a broad Google search.

"The information in this handbook is invaluable whether you live in a dormitory with roommates, a sprawling estate with a bunker, or anything in between. Including plans for both long- and short-term scenarios, *Crisis Preparedness Handbook* includes both an easy-to-use quick reference guide as well as the foundation for new useful skills and knowledge.

"By helping me avoid buying the wrong things over and over, this book will save me hundreds of dollars—dollars I can spend to improve my crisis plan and level of preparedness."

—JASEN CHANDLER, OUTDOOR ENTHUSIAST
Director of Sales and Market Development, Custom Installation and Design

HELPS YOU AT ALL LEVELS OF PREPAREDNESS

"*Crisis Preparedness Handbook* is a great resource for everyone, not just preppers. It helps you target your preparedness actions by breaking preparedness down into palatable work portions using a risk-based-approach worksheet for rating your likelihood of experiencing a disaster. The book is not designed to be read cover-to-cover; instead, it uses a self-guided approach. Simply select the areas that interest you, and then do as much or as little as you'd like using the levels of preparedness outlined in this guide. This book underscores the truth that something is always better than nothing when it comes to emergency preparedness."

—CARL FARLEY, PREPAREDNESS SPECIALIST
MS in Public Health, Certified Industrial Hygienist, Certified Safety Professional

CRISIS
PREPAREDNESS
HANDBOOK

A Comprehensive Guide to
Home Storage and Physical Survival

3RD EDITION

Patricia Spigarelli Aston
Jack A. Spigarelli

Cross-Current
PUBLISHING

NORTH LOGAN, UTAH

Cross-Current Publishing
North Logan, Utah
cross-current.com
Printed in the United States of America

ISBN, print edition   978-0-936348-01-8
ISBN, EPUB edition   978-0-936348-07-0

First Edition, 1984 © Jack A. Spigarelli
Second Edition, 2002 © Jack A. Spigarelli

Cover and interior design by Daniel Ruesch Design
danielruesch.net

Disclaimers:

The purpose of this book is to offer information to help persons who want to prepare for possible future crises. It is sold with the understanding that every effort was made to provide the most current and accurate information. Any errors or omissions are unintentional. Any use or misuse of the information contained herein are solely the responsibility of the user.

This book is independently authored and published and no sponsorship or affiliation with any trademarked product mentioned or pictured within is claimed or suggested.

Contents

Introduction

SECTION 1
Planning for a Crisis

SECTION 2
Storing Water for a Crisis

Storing Food for a Crisis

SECTION 4

Growing and Preserving Food for a Crisis

SECTION 5
Preparing Your Shelter for a Crisis

SECTION 6
Communications, Transportation, and Protection for a Crisis

Resources

Acknowledgements

My Support Team

Craig Aston ~ My husband, Craig, has been patient, encouraging, and wise. I couldn't have asked for a more supportive partner.

Steve Spigarelli ~ My son Steve was my go-to guy whenever I had a technical issue or needed a fresh opinion or advice about what I was writing.

Adam Spigarelli ~ My son Adam was the first to read an entire draft of the book and offer insights and revisions. He often encouraged me with "Mom, you have to get this book out there!"

Daniel Ruesch ~ A wonderful graphic designer, Dan was with me every step of the way as he formatted the text, designed the cover, and created the original graphics.

Expert Advice

Michele Preisendorf ~ Michele edited the manuscript, helping me say things just right.

Sam Arnold-Boyd ~ Sam expertly created the index with all its complexities.

Robert Aston ~ My stepson Rob patiently shared his perspective and expert advice as he taught me about survival vehicles and firearms.

Jesse Washburn, MD ~ My friend Jesse provided feedback on the medical chapter, ensuring everything was accurate.

Brent Gray, DMD ~ Dr. Gray gave me pointers on how to deal with dental emergencies.

Special Thanks

Thea Spigarelli ~ One of my talented teenage granddaughters, Thea photographed and edited the various images of food storage containers in the book.

Mara Spigarelli ~ Another one of my talented teenage granddaughters, Mara searched the text looking for typos and anything that didn't make sense.

The Story Behind This Book

MY FIRST HUSBAND, Jack Spigarelli, wrote the first edition of *Crisis Preparedness Handbook* (CPH) in the early '80s, when we were facing 14 percent mortgage rates, soaring gasoline prices, and near double-digit inflation. Nuclear war with the Soviet Union threatened, unrest in the Middle East was on the rise, and the government had lost its credibility following the Vietnam War and Watergate.

As the father of our young family, Jack felt responsible and, I think, vulnerable. So he decided to prepare us for the uncertainties and crises he felt certain lay ahead. His planning became an almost evangelical passion he felt compelled to share with others.

With a vision of what he wanted to accomplish, Jack left his real-world professional job and began the process of researching and writing CPH. I was with him the whole way. I don't know how we did it. With four small children, it wasn't feasible for me to work outside the home, but we were somehow able to make our savings of six months stretch into more than a year. Then we juggled finances and borrowed and begged, and a few friends helped us out. Our dedication to self-reliance and the preparedness lifestyle helped us get through that difficult financial period. For a while, Jack put writing the book on hold and got a job, but compelled to finish it, he came back to it after a few months.

As his chief sounding board, editor, adviser, and critic, I read his drafts, questioned his assertions, and added my two cents. But my biggest job was testing the survival equipment Jack recommended—water filters, emergency lighting, grain mills, outdoor gear. I became a bit of an expert on using stored wheat and powdered milk, learned to dehydrate fruits and vegetables, and experimented until I could grow a productive vegetable garden.

After an incredible amount of sacrifice, CPH was completed in 1984. Back then it was extremely difficult to market a self-published book. With our meager budget, we placed a few ads in national survival magazines. CPH received favorable reviews from leaders in the tiny survival industry and managed to sell over five hundred copies. But that wasn't enough to sustain our family, so Jack again looked to other pursuits, though his passion for preparedness was never far from his heart and mind.

Nobody Is Immune

Fast-forward to the tragedy of September 11, 2001. Forever etched in our minds, it heightened everyone's concern about the safety of our world. Jack decided it was time to offer his survival expertise in an updated CPH. He spent about a year revising and getting it ready to market. By this time, Amazon was becoming the seller of books and a self-published book could more easily make its presence known on the market.

Second edition of best-selling *Crisis Preparedness Handbook*, first published in 1984

In 2005, Jack started getting orders from Amazon and selling to bookstores, as well as through our website. Then, on May 5, the unthinkable happened. Jack died from an aortic aneurysm. Crisis and survival became real for me and my family. The sorrow, emotional trauma, and upheaval were almost unbearable. But somehow we were able to manage, not only because of the people who came forward, but largely because I had been raised to be resilient and Jack and I had lived a self-reliant lifestyle. One of the ways I coped was by continuing to work as a schoolteacher. It was a relief to focus on my students. I will ever be grateful for those who compassionately listened and cheered me on.

In time I began reflecting on Jack's dream. The July 2005 bombings in London, the devastation of Hurricane Katrina, and the economic upheaval of 2008 increased demand for the book. With a heightened awareness of how fragile our world is, people wanted to know how they could be prepared. Over the next five years, nearly forty thousand copies of *Crisis Preparedness Handbook* were sold!

I felt his book would benefit from a fresh editing and decided to update the content and make the text more reader-friendly. In 2015, I began this task in earnest.

Being Prepared Brings Peace

As I thought about what our readers really needed, I discovered that my voice and thoughts and intentions were different from Jack's. Before long I was doing more than making a few revisions. I was rewriting and, more importantly, repurposing the book. Jack's intent was to help people prepare for what he considered an inevitable breakdown of society. My complementary goal is to help you get through the kind of crises you see happening on the news—huge events when they happen to us personally but which, with the right preparation can lessen the suffering until we again find normalcy in our lives.

It's highly likely that everyone will have to endure crises at some point. I've survived a hurricane; come-to-a-standstill snowstorms; a house fire; minor earthquakes; how-will-we-pay-the-mortgage struggles; and, most profoundly, Jack's death. I have seen in-laws deal with three feet of floodwater in their home and a dear friend and her family evacuate during a wildfire. And most recently, we've experienced a worldwide pandemic. While, understandably, there has been uncertainty, sorrow, loneliness, anxiety, and panic, I want you to know you can feel the peace and confidence that comes with being prepared.

This updated third version of *Crisis Preparedness Handbook* offers a broader vision of what it means to be prepared. Frankly, there is enough division and polarization, animosity and intolerance in our world in 2020 to make a societal breakdown conceivable, but CPH teaches you how to be prepared for temporary disasters and personal crises so that instead of being life-threatening, many crises may be reduced to minor inconveniences.

—*Patricia Spigarelli Aston*

How to Use This Book

Because *Crisis Preparedness Handbook* has a lot of information, it can feel intimidating. You might wonder, *Where do I start? Do I have to do **all** this to help prepare my family? How will I do it?* **Stop!** The last thing I want you to feel is overwhelmed. This *is* doable. I will show you how.

This Is a Handbook

The first thing to remember is that this is a *handbook*—a guide, a resource. You don't have to read it cover to cover or do every single thing in all forty-four chapters. Be selective. Choose what's relevant to you. Then, in six months, focus on something new.

Some of you may read *Crisis Preparedness Handbook* cover to cover. If so, you'll have a good foundation for every aspect of preparedness and something you can turn to when ready.

If you haven't done so, please read the story behind why I decided to essentially rewrite *Crisis Preparedness Handbook* on page xiii. Doing so will give you insight into the purpose of this book.

How Do I Get Started?

Begin with the basics: water, food, shelter. Think of one thing you want to accomplish in each of these areas. For example, you might want to store enough water to last two weeks. Chapters 6 and 7 help you figure out how to do that. Or think about what you currently have on hand—do you have foods you can easily put together for meals that don't require electricity? And could that be augmented with freeze-dried meals, say, enough to last a few days to two weeks? Chapters 14 and 15 show you how. As for shelter, are you wondering what you can do to be ready for natural disasters and to secure your home? Look for solutions in chapter 34. What if you become stranded in your vehicle? In chapter 41, you will find suggestions for survival on the road.

"Five Things You Can Do Now"

How do you know what to do first? Most chapters include five things you can do right now. These are simple things, like keeping a small reserve of cash on hand for when ATMs don't work, or stashing a pair of shoes at work in case you have to walk home, or teaching family members how to shut off water and gas mains. You get the idea. If you regularly choose one or two tasks, the security and peace of mind that comes with being prepared will be yours.

Use the Table of Contents

The forty-four chapters are arranged in six main sections. Where the first five chapters outline the general principles of becoming prepared, it's a good idea to read them in their entirety. They will help you determine your unique preparedness goals.

Chapter 5 includes the "Rule of Three": we can survive for three minutes without air, three hours without adequate shelter, three days without water, and three weeks without food. The rest of the book is organized around these principles.

Text Features

Tables

Throughout this book, you'll find tables that make important information more accessible. They may outline the steps in a process, list pros and cons, or recommend items for you to store. They are numbered consecutively within the chapters for easy reference.

Quick Looks and Quick Checks

"Quick Looks" highlight and condense key ideas and make it easier to find important information. "Quick Checks" help you evaluate your progress or give you criteria for making comparisons.

Worksheets

The worksheets in *Crisis Preparedness Handbook,* free on our website, CrisisPreparedness.com, will help you organize and plan. Right away, chapter 2's "Rating the Likelihood of Experiencing a Disaster" worksheet opens your eyes to what you're likely to face. Chapter 4 has worksheets for making goals, setting priorities, and planning your preparedness journey. In "Storing Food for a Crisis," you'll find worksheets for evaluating and organizing your food-storage options. Use these worksheets right in the book or download them from our website. You'll also notice lined pages throughout the book for notetaking.

Personally Speaking

Interspersed throughout the text are passages called "Personally Speaking," where I share insights and stories about the things I teach in the book. While writing this book, I felt the need to connect with you on a more personal level so that preparedness might feel more "real"—and more doable. I hope these vignettes give you added perspective.

Resources and Index

As you read through the chapters, you'll see examples of unique products to help you prepare. The appendix includes a quick and easy reference guide for these products. It's organized by chapter and includes descriptions and website information. You'll also find recommended books and other reference materials. The detailed index at the back of the book will be indispensable for helping you find information about whatever preparedness topic you choose.

Website

Be sure to check out CrisisPreparedness.com, where we offer free PDF downloads of all the worksheets in the book. Also, you will find the latest updates about preparedness and reviews of unique products that will help you make decisions about what to buy.

SECTION 1

Planning for a Crisis

Security in an Uncertain World

We live in a world of unprecedented abundance and convenience. We adjust a thermostat if we are hot or cold. We think nothing of making a quick trip to the store if we need a few groceries. We can microwave dinner and have it on the table in minutes. We change clothes with the seasons and launder them in our high-efficiency washing machines. Modern transportation connects us to every part of the world. Communication has been revolutionized with smartphones and an endless number of apps. A button or voice command turns up the lights or brings us nonstop entertainment and twenty-four seven world news.

Vulnerability in Our Interdependent World

Modern technology influences every aspect of our lives, and we've become more reliant on it while at the same time further removed from the basics of self-reliance.

What we call survival today is what our great-grandparents called everyday life.

Specialized and Dependent

Our conveniences are due mainly to specialization. Though each of us is skilled in our own narrow field, we know increasingly less about all the rest, deepening our dependence on extensive interconnected systems for even the most basic of necessities.

For example, we depend on a small number of farmers to produce our food. They, in turn, depend on necessities such as seed, machinery, fuel, fertilizers, pesticides, favorable weather, adequate water, and labor. We count on a vast system of trucks, railroads, container ships, processing plants, and supermarkets to transport and distribute our food, often from thousands of miles away, sometimes even continents away. And that system relies on having enough fuel and on an orderly credit system.

We expect utility companies to deliver our electricity, natural gas, water, garbage collection, and communication without interruption. We assume manufacturers will supply our clothing, household goods, appliances, and vehicles. We rely on doctors, dentists, hospitals,

---

FIVE THINGS YOU CAN DO NOW

1 Identify the most common natural disasters in your geographic region.

2 Read about or watch realistic video clips about natural disasters to gain perspective.

3 List ways you are dependent on others for basic needs and services.

4 Talk about your preparedness concerns with friends and family members.

5 Look through *Crisis Preparedness Handbook* and note chapters that are most relevant to you.

---

and pharmaceutical companies for health services. We expect oil companies to supply our gasoline and police and firemen to protect us. And to pay for all this and more, we demand a sound banking system and a reliable source of income.

One Catastrophe Away

Given our deep dependency on these complex, vast, interdependent systems, we are only one or two catastrophic events away from massive disruption and breakdown. We have seen even relatively minor incidents interrupt the flow of food and other essential goods and services. What happens when it's large scale?

Not If But When—Making the Case for Being Prepared

If both man-made and natural events can create crises that impact local areas with intense disruption and major inconveniences, imagine what global catastrophes such as financial failure, civil unrest, political upheaval, and cataclysmic natural events could do.

Even spared such crises, we all know people who have been devastated by accident, illness, divorce, death, or temporary loss of income.

Imagine "If"

Think about your current situation. Imagine trying to live with no outside services for just one week. How would you cope without power for cooking, refrigeration, or lights? What would you do if you couldn't flush the toilets or turn on a faucet to get a drink of water? Without immediate access to doctors and hospitals, how would you care for the sick or injured? If the store shelves were emptied, what would you feed your family? In stifling heat or freezing cold, how could you cool or warm your home?

Become Self-Reliant

The purpose of *Crisis Preparedness Handbook* is to help you confidently and calmly prepare for and face any crises that may come your way. You must become your own advocate. The knowledge gained from this book can help you do just that.

The Ferocity of Nature: Natural Disasters

Natural disasters are part of life. America's Atlantic and Gulf Coasts are frequently hit with hurricane superstorms. The Midwest suffers from killer tornados, blizzards, and floods. And the western states cope with devastating wildfires and drought and the threat of earthquakes, tsunamis, and even volcanic eruptions.

Table 1.1 gives you an idea of the natural disasters common to your area. As you can see, no region is exempt. Severe storms and flooding, followed by hurricanes and tornadoes, are the most common disasters for which presidential disaster declarations are made. When an area has been declared a federal disaster area, the federal government may offer financial assistance to both individuals and businesses, but it's critical to be prepared just in case.

Table 1.1
Presidential Disaster Declarations
1964 – 2013

FEMA Region	States	Total Disasters	Number of Most Frequent Disasters		Other Disasters (Fewer Than 10)
I	CT, ME, MA, NH, RI, VT	156	Severe Storm Flood Hurricane	68 44 21	Snow, Coastal Storm, Severe Ice Storm, Fire, Tornado, Freezing
II	NJ, NY, PR, VI	136	Severe Storm Flood Hurricane Snow	51 37 27 11	Severe Drought, Tornado, Severe Ice Storm, Fire, Earthquake
III	DE, DC, MD, PA, VA, WV	182	Flood Severe Storm Hurricane Snow	59 58 23 18	Tornado, Drought, Earthquake, Coastal Storm
IV	AL, FL, GA, KY, MS, NC, SC, TN	346	Severe Storm Flood Hurricane Tornado	157 66 65 32	Severe Ice Storm, Coastal Storm, Freezing, Fire
V	IL, IN, MI, MN, OH, WI	241	Severe Storm Flood Tornado	119 78 32	Snow, Severe Ice Storm
VI	AR, LA, NM, OK, TX	278	Severe Storm Flood Tornado Hurricane	96 94 34 27	Severe Ice Storm, Fire, Snow, Coastal Storm
VII	IA, KS, MO, NE	188	Severe Storm Flood Tornado	106 54 16	Severe Ice Storm, Snow
VIII	CO, MT, ND, SD, UT, WY	141	Severe Storm Flood	59 62	Tornado, Snow, Fire, Severe Ice Storm
IX	AZ, CA, HI, NV, Pacific Islands	190	Flood Typhoon Severe Storm Fire Earthquake	57 49 30 20 16	Hurricane, Drought, Tsunami, Freezing, Volcano
X	AK, ID, OR, WA	118	Flood Severe Storm	51 43	Fire, Earthquake, Severe Ice Storm, Snow, Tsunami

Data from "Presidential Disaster Declarations December 24, 1964 to December 31, 2013"
https://gis.fema.gov/maps/FEMA_Presidential_Disaster_Declarations_1964_2013.pdf

Pandemics

Each year brings new varieties of life-threatening diseases with potential pandemic results. Modern air travel makes the risk of widespread influenza, acute respiratory infections, and epidemics or pandemics (like Ebola hemorrhagic fever and COVID-19) alarmingly possible.

Climate Change

According to the NASA website, climate change will likely impact us in a variety of ways. There will be more droughts and heat waves, rainfall patterns will be more extreme, and hurricanes will be more intense. Scientists of climate change warn of potential crop failure and worldwide famine.

Astronomical Events

A coronal mass ejection (CME), or solar superstorm, could have devastating effects on the power grid and high-tech communications, impacting every aspect of society. A cataclysmic event, such as an asteroid collision, could drastically change life on earth.

Man-Caused Crises

Terrorism

The horrific acts of terror we have seen in the last two decades have forced Americans to rethink their complacency. Terrorism plagues even civilized peoples, cyberterror and EMP (electromagnetic pulse) attacks are weapons of choice for modern-day terrorists. The resulting unprecedented destruction and social unrest show us how vulnerable we are.

The threads that hold the fabric of society together are very fragile and our condition can become perilous at any time.

Economic Upheaval

Just over ten years ago, we witnessed the largest economic downturn since the Great Depression. We have also observed the economic failure of well-established European nations, chaos in Africa and South America, and threats to financial stability in Asia. And the financial instability of world markets and local economies threatens us with accelerated inflation, unemployment, depression, and even the collapse of our monetary system.

Social Unrest

We have also seen the political polarization of America in the last ten years. Fueled by a media machine always on the lookout for the next controversial outrage, antagonism between the different factions runs high. Could these tensions produce civil unrest? In some areas, protesting has already evolved into rioting and confrontations driven by a mob mentality. Could these kinds of disturbances escalate into widespread societal collapse?

War and Nuclear Threat

Consider the threat of biological, chemical, or even nuclear weapons in the hands of groups with radical ideologies. Political instability and civil wars can spill over into neighboring countries, with conflicts escalating from regional to international.

Prepare Now

It Makes Sense to Prepare

It's likely you have made some basic preparations. Maybe you have put aside savings, stockpiled several cases of water, or stored emergency gear in your car for winter travel.

However, it is human nature to avoid contemplating potential unpleasantries. They can be disturbing to think about. Some even suggest it is senseless or irresponsible, even hysterical, to consider these potential crises. Most of us know we should be doing something, but we tend to procrastinate. We carry on and hope that nothing bad will happen. If we have never experienced a disaster, it is difficult to imagine the impact it could have on us.

We might ignore the warning signs and minimize the potential consequences. But we feel a nagging uneasiness. It may be foolish and dangerous to us and our family to dismiss the need to prepare. It's better to acknowledge how vulnerable society is and to be alert to how fast a crisis can strike and how severe it can become. It is smart to recognize our potential vulnerability, wise to do something about it.

PERSONALLY SPEAKING

In the spring of 2020, as I was making the final edits to this book, COVID-19 was wreaking havoc across the globe. The subsequent economic upheaval, confusion, inconsistency, and uncertainty spurred on by a politically-induced media frenzy underscored the need to be personally responsible for one's self and the well-being of one's family.

Although it would be difficult to find someone who wasn't impacted by the pandemic, those who were prepared with a supply of food and other essentials as well as a savings for several months could relax a little and make rational choices. When my daughter in Arizona called me at the end of February and told me shoppers at Costco were frantically grabbing up toilet paper and cases of water, I admit I had to fight against the panic I felt. It reinforced, even for me, that the possibilities described in this chapter are very real.

If you were caught unprepared for this crisis, let it be a wake-up call. I'm glad I can help you with your preparedness plans. After assessing our own preparedness, my husband, Craig, and I are currently refining our own preparedness goals.

The Time Is Now

Once we are aware of the potential problems, we must act to decrease their effects. There is an urgency in getting prepared because the future is always uncertain. We must take responsibility and begin to systematically prepare so that when we are faced with a crisis, we can ease the possible trauma and improve our chances of survival. I hope you feel the urgency! The time is now. Let's begin!

2 Personal Commitment

Preparing for a crisis is a major commitment. Although it can be a challenge, it can be simplified by acquiring knowledge and developing a methodical plan.

Acquire Knowledge and Create a Plan

Being ready for a major crisis takes substantial thought and effort. It also requires space and some expense. You may need to make significant changes in lifestyle. The process can be overwhelming, and you may hinder your progress if you try to tackle it all at once, but with a little organization and taking it one step at a time, you will succeed in creating a plan that works for you.

Becoming prepared begins by building a solid foundation of knowledge.

Start by building a solid foundation of knowledge. You will be making many decisions, and sufficient and accurate information will help you make good ones. Next, following the guidelines in this book, devise a systematic plan broken down into manageable steps. Since your decisions will be based on your needs, expect to implement them following your own timeline.

Develop a Personal Vision

How you prepare will depend on your circumstances. Take some time to think about which crises are most likely to happen to you.

The vision you develop will guide your preparations. As you learn more, you may alter your vision. But it is important to decide where you will focus. Your vision will give you a foundation for evaluating the recommendations in this handbook as well as in other sources, and it will help you with knowing how to allocate your time and resources.

Learn from Others

As you develop your vision, consider others' ideas. There are many perspectives on preparedness and survival, and you can learn something from each of them.

FIVE THINGS YOU CAN DO NOW

1 Brainstorm and make a list of the crises you or your family could face.

2 Use worksheet 2.1 to rate the likelihood of various crises.

3 Decide what the three most likely crises you could face in the next five years might be.

4 Check out several preparedness websites and blogs that interest you.

5 Discuss your vision with your family and friends.

Become acquainted with the varying philosophies of survival and preparedness specialists by visiting their websites and reading their articles and books. Consider and evaluate them as you form and articulate your own personal vision for preparedness.

Trust Your Thinking

It's good to discuss your ideas with trusted friends and family members, but remember, it's your future at risk, so don't simply accept someone else's point of view. Think it through. Do you agree with their point of view? If you don't, why not?

Also, don't ignore a possible crisis just because it makes you uncomfortable. It is up to you to create the most likely scenarios you believe you could face.

You don't know exactly what crises you will face or when you might face them. Things no one has anticipated or predicted can turn your world upside down. Have a vision but be flexible with that vision.

Cultivate a Self-Reliant Mindset

Being prepared is a mindset. For the self-reliant person, it is a thoughtful and intentional way of living. In our current society, we spend a lot of money paying other people to do things we should probably do for ourselves. When you live in a self-reliant way, preparing for crises is not something additional or separate—it is just part of life and will naturally be a priority as you allocate your resources.

For example, you may want a heat source that makes you less dependent on utilities. You might consider purchasing an efficient woodburning stove now so that procuring firewood and starting up the stove on a cold day becomes second nature. And you may be surprised at the peace of mind it gives you.

Or, if you desire the wholesomeness and pleasure of eating homemade whole-wheat bread, it would be important to purchase a grain mill and a good bread mixer. Making bread will become routine, and not only will you enjoy delicious, nourishing bread, you will also pay for your grain mill and bread mixer with what you save.

Likewise, if you like the goodness of homemade jams and preserves, you may want to plant a small berry patch or a few fruit trees. You will be amazed at the amount of fruit you get from just a few plants or trees and the satisfaction it gives you.

These examples reflect a way of living—simple, self-reliant routines folded into your life. As you contemplate becoming better equipped for crises, consider ways you can incorporate preparedness into your daily living.

> **Preparedness is part of the self-reliant mindset.**

Include Your Family

Most likely, one of the reasons you are preparing is because you care about your family and their physical well-being. Be sure to include your children in this effort. Self-reliance is a big part of being prepared for the unknown, and children can learn skills and plan for their own safety. When children learn to take responsibility and to problem-solve, they are on the road to becoming self-reliant.

PERSONALLY SPEAKING

 Knowing how to work and stick with something is part of the preparedness skill set. Like many parents, we tried to teach our kids how to work. When they were young, I valued giving them a wholesome breakfast. When they were older and I returned to work outside our home, we felt it was important to continue that practice. The problem was, I no longer had the time. So each fall as we organized for the new school year, we gave that responsibility to one of our kids. Jack supervised, and within a couple of weeks, breakfast was running smoothly. Each day of the week had its own menu, and the kids became proficient at cooking oatmeal, scrambled eggs, and French toast. At the time, I didn't realize that this daily responsibility was not only helping them feel at home in the kitchen; it was also giving them confidence and a self-reliant work ethic.

It was natural for our kids to start earning their own money as they got older. Their primary job, of course, was being a student. But they learned to balance school, studying, extracurricular activities, and social life with their jobs. They learned the value of work and continued to develop self-reliance.

I know several families who have a family business their children are a big part of and where those children learn invaluable skills. I have seen successful honeybee operations, lawn-maintenance companies, piano studios, Etsy shops, shaved-ice stands, chicken and rabbit farming, and 4-H livestock-raising projects. Our children built a window-washing company they passed down from sibling to sibling to help pay for college.

Anticipate Consequences

Begin narrowing down the list of crises to the potential crises you might face. Consider the possibilities and examine various perceived threats. Which are most threatening? How would they affect you and your family? What can you do now to lessen their impact?

Do you envision a disaster where your home is at risk? Will you have to leave your home quickly, or will you be able to stay and shelter there? How will you know if or when it is time to leave? What information will you trust?

Chapter 1 contained a general rundown of possible disasters and crises. You probably already started evaluating which you should prepare for. In this chapter, you will continue to make it personal and anticipate which crises you are more likely to face. Use worksheet 2.1, "Rating Your Likelihood of Experiencing a Crisis or Disaster," on page 15 to evaluate and rate the likelihood of certain crises or disasters in your circumstances. Rank them on a scale of one to five, then begin thinking about the preparations you'll need to make. Your responses to this worksheet will help guide your planning in chapter 4. Worksheets can also be found as downloadable PDF files on our website, CrisisPreparedness.com.

PERSONALLY SPEAKING

 As you consider what preparations you need to make, contemplate unexpected events. For example, in the Northwest, where I grew up, hurricanes were unheard of, but on October 12, 1962, our family experienced the fury of a hurricane, technically a typhoon, as it unexpectedly raged up the Oregon and Washington coasts with sustained winds of over 100 miles per hour and gusts over 150 miles per hour. Known as the Columbus Day Storm, it veered inland and wiped out the power grid, causing outages that lasted weeks. It blew down over fifteen *billion* board feet of lumber, knocked out radio and TV transmissions, collapsed barns, damaged over 55,000 homes in Oregon alone, and killed 46 people. Luckily, my parents were resourceful, and we survived with minimal inconvenience. But I will never forget the alarm and uncertainty I felt as we watched the big maple trees in our backyard being fiercely blown about at the height of the storm.

Not twenty years later, from the vantage point of their fishing boat on the Columbia River, my parents watched in awe as Mount St. Helens erupted and plumes of ash filled the sky. They lived southwest of the mountain and were spared the destruction because the prevailing winds carried the ash and debris east, where others, unfortunately, felt the devastation. Fifty-seven people died in the catastrophe. The power of nature reminds us that we need to be ready for anything.

Personal Crises

Perhaps your vision includes being better prepared for individual hardships such as injury, illness, or unemployment. What is the likelihood of these things happening? How can you make such events less traumatic? You may want to view preparedness as another form of insurance for when your family goes through rough times.

Natural Disasters

Do you foresee any natural disasters? Which are most likely to occur where you live? Do you anticipate a winter storm, hurricane, tornado, or earthquake? How long will the disaster last? Will timely help be likely? Who will you be able to count on for support?

Do you need to fortify your home, build a storm cellar, or put in an emergency power source for heat, light, and refrigeration? If your area is subject to prolonged drought, should you consider alternative water sources?

Man-Caused Crises

How do you see any man-caused crises unfolding? How widespread will they be? Do you anticipate monetary instability, price controls, shortages, or rationing? What about civil disturbances, mob rioting, or even a complete breakdown of law and order? Do you see continuing terrorist attacks causing more turmoil?

Will essential services such as transportation, communications, police and fire protection, and utilities be disrupted? Will food production and distribution be interrupted? Will the lack of medical services and proper sanitation lead to health crises and epidemics? Will looting and violence become widespread? If these crises are what you anticipate, you will want to learn what you can do to improve your chances of surviving them.

How Long Should I Prepare For?

Preparing for a Year

Crises come in different magnitudes and durations. Naturally, if you prepare for the worst case, you'll be prepared for anything less; thus, it's wise to prepare for at least a year where possible. In many parts of the world, if the growing season is disrupted and the food crop can't be harvested, it's at least a year before substantial replacement food can be grown.

> **Preparing for a year can see you through crop failure, prolonged unemployment, or civil disruptions and give you time to adjust to new circumstances.**

Additionally, a year of preparation can see you through prolonged unemployment or civil disruptions and give you time to adjust to new circumstances. It also gives you the potential luxury of sharing with others. Thus, most of the recommendations in this book are based on preparing for one year.

Preparing for a Shorter Amount of Time

That said, preparing for a year is a huge undertaking! You will need to decide if that does, in fact, match up with your vision. Your preparation will depend on your situation and your resources. Consider preparing for the duration of a specific crisis. It might be a month, a week, or just three days. Start with what you can manage, then work up to longer periods. In most instances, the advice in this book can be adapted to fit shorter periods.

How Much Will It Cost?

Costs can vary significantly according to perceived needs and desires. Much will depend on what you already have. Maybe you're starting from nothing. If you feel comfortable with just the basics, the cost should not be too overwhelming. On the other hand, an upscale retreat in the mountains with state-of-the-art, off-grid technology could run into the millions.

There are many ways to minimize costs. I include budget tips throughout the book wherever possible. Also, keep in mind that many preparedness items are useful in other situations.

Common Sense Thriftiness

Again, think of preparedness as insurance or a type of investment. If you regularly save or invest a portion of your income, consider doing so in the form of tangible preparedness items.

Make the most of the money you have. Budget for preparedness. Be proactive about your choices. Compare products and find the best values. For

---

QUICK CHECK

Finding Money to Prepare for a Crisis

- ✔ Are we carefully budgeting?
- ✔ Can we organize a family fund-raising project?
- ✔ Can we supplement with a second income?
- ✔ Can we upgrade employment?
- ✔ Do we have unused or luxury items we can sell?
- ✔ Can we limit our purchases to only necessities?
- ✔ Can we eat out less and reduce the use of prepared foods?
- ✔ Can we limit expensive vacations?

example, a family of four will need about 1,500 pounds of wheat for a basic year's supply based on the 7-Plus Basic Plan described in chapter 16. That can cost as little as $800 or as much as $2,000 and more, depending on how it is packaged and where you purchase it.

Look for savings in store ads, special promotions, liquidations, and closeouts. Also, remember that purchasing quality items often pays off.

Be wary of all-inclusive preparedness modules. Although they may seem convenient, they will not be tailored to your specific needs. Plus, they are expensive and may include inferior products. If you want to make an educated decision about food-storage packages, read "All-Inclusive Menu Plans" on page 100 in chapter 14.

> **In the end, you must decide how important it is to be prepared for possible crises. If you are committed, you will prepare.**

Replace Convenience and Processed Foods with Basic Foods

When it comes to stockpiling, basic foods cost less than typical supermarket convenience foods. By replacing convenience and processed foods with more basic foods, you will save money you can use to build up your preparedness reserves.

Chapter 15 helps you plan menus using foods you can store, and chapter 16 shows you how to stockpile basic foods. Chapter 18 lists and describes basic storage foods.

If cutting expenses is not enough, consider how you can increase your income. Do you have marketable skills you can use to earn a second income? Take a hard look at your possessions. Do you have luxury items that go largely unused? Would it serve you better if you sold them or traded down?

Simplify

One thing you can do to free up resources for preparedness and survival is to simplify your standard of living. It may seem contradictory to suggest getting rid of your stuff and at the same time advocate stocking up. But you will be doing it with an intentional mindset.

Take a serious look at your spending habits. Cut out unnecessary possessions and replace them with potentially lifesaving supplies and tools.

A Worksheet to Help You

Worksheet 2.1 will help you to rate your likelihood of experiencing various disasters. For a downloadable PDF file, go to CrisisPreparedness.com.

Worksheet 2.1
Rating Your Likelihood of Experiencing a Crisis or Disaster

Directions: Rate the likelihood of each disaster/crisis from 1 to 5, with 5 being most likely.

Crisis or Disaster	1	2	3	4	5
Personal injury, illness, unemployment					
Hurricane					
Tornado					
Severe storm					
Ice/snowstorm					
Severe cold					
Severe heat					
Flood					
Flash flood					
Landslide					
Earthquake					
Tsunami					
Volcano					
Wildfire					
Drought					
Solar superstorm					
Cataclysmic astronomical event					
Hazardous-material spill					
Nuclear accident					
Disease pandemic					
Economic disruption					
Civil unrest					
Terrorist attack					
Infrastructure cyberattack					
War					
Nuclear attack					
EMP attack					

NOTES

A Framework for Total Preparedness

Total preparedness means being able to provide for your needs as much as possible in any given circumstance. It consists of three areas: personal preparation, material provisions, and financial resources. This chapter looks at each area and helps you determine what type of planning you need to do.

Personal Preparation

The argument can be made that personal preparation is the most important aspect of being ready for the unforeseeable. It includes your physical and mental health, your mindset and attitude, and your knowledge, skills, and experience. It can be the difference between living as a survivor or perishing as a victim.

Physical Health

Survival both during and after a crisis may depend on your physical health. This is just one more reason to live a healthy lifestyle as recommended by health and fitness experts. It is not a new idea but a good time to reflect upon what you might need to change.

Build up your general physical condition and stamina now by developing good health practices. Diseases that currently do not pose a public health threat may become a concern in future situations. Be sure to get your diphtheria immunizations and tetanus boosters every five to ten years.

Mental Health

Perhaps more than any other thing, survival depends on your state of mind. Major crises create stress for everyone, and how you react can make all the difference.

QUICK CHECK

Living a Healthy Lifestyle

✔ Am I eating a healthy diet and in my correct weight range?
✔ Do I exercise regularly?
✔ Do I get adequate rest?
✔ Do I get regular checkups?
✔ Am I current on immunizations, especially tetanus, diphtheria, and pertussis?
✔ Is all dental work current?
✔ Do I have regular vision and hearing screenings?
✔ Am I caught up on any elective surgeries?
✔ Am I careful not to be dependent on drugs and medications?

FIVE THINGS YOU CAN DO NOW

1 Use the Quick Check "Living a Healthy Lifestyle" to evaluate areas of strength and weakness.

2 Make a list of skills you would like to learn.

3 Purchase a book about a survival skill you would like to learn.

4 Read a fiction or biographical book about personal survival.

5 Acquire a small reserve of cash in small denominations and make sure it is secure.

Some react with excessive fear that quickly turns into panic. They may be completely paralyzed by their overwhelming sense of vulnerability and weakness, their actions ineffective and even dangerous. Others become deeply depressed and apathetic and give up.

Faced with the same difficulties, there are those who can cope with an inner strength that comes from being physically and mentally prepared. They are emotionally stable and function with a clear head and a calm presence. A crisis is still a challenge for them but is less threatening.

How to Prepare Mentally

Preparing mentally is a focused process that expands your perspective and increases your adaptability. You can increase your mental capacity by seeking new experiences, putting yourself in different situations, and challenging yourself to do hard things.

Being prepared requires exercising your mind by expanding your perspective and increasing your adaptability.

Acclimate yourself to the possibilities by reading survival experiences. Accept the fact that conditions can abruptly change for the worst and be aware of what to expect. Use worry constructively by planning out exactly what actions you will take in various crises. This will not only increase your confidence but will reduce your anxiety, confusion, and disorientation.

Learn how to reduce tension through physical activity and relaxation. And lastly, a sense of humor builds resiliency.

PERSONALLY SPEAKING

How can you test your mental resolve and adaptability? Go camping or backpacking. From experience, I have observed that no matter how well you think you've prepared for a camping or backpacking trip, there's always something that tests your resourcefulness. Whether you contrive a net bag for drying dishes, improvise for forgotten tent-poles, or find a safe place to hunker down during a thunderstorm, you're adding to your mental resolve and adaptability. Thinking of alternative cooking methods, sanitation demands, and ways to keep warm are all survival skills in practice and foster the feeling of success that comes from "having survived the trip."

Not long ago, one of my sons took two of his boys, ages ten and twelve, on a little backpacking trip. There were some complications with the four-wheel-drive vehicle, a few food and water issues, and a bit of whining, but the biggest frustration my son faced was teaching his boys proper sanitation in the wild! Nothing gets taken for granted when you are camping. Though challenging at the time, these experiences are the best training for a crisis.

Fictional and Biographical Survival Books

Survival fiction and biographical stories can give you a good idea of what it might be like in a crisis. Check out the resource "Fictional and Biographical Survival Books" on page 423 for recommendations and reviews of books that will give you an opportunity to reflect on how fictional characters and real people improvise, adapt, and survive in desperate situations.

Spiritual Anchors

Finally, do not ignore the strength of spiritual anchors. Without hope for the future, all is lost. Consider who you are and what you stand for. Reflect upon your moral values, what you believe, and your relationship with others and with God.

Knowledge and Skills

The more knowledge and skills you acquire, the better equipped you will be to respond with confidence to whatever situation you may encounter. You will be able to calmly handle problems rather than panic or give up.

> **The more varied knowledge and skills a person has, the better he or she will be able to cope with whatever lies ahead.**

Unlike supplies and equipment, knowledge and skills cannot be purchased and stored on a shelf. They are acquired over time through study and practice as you learn to adapt to your needs and circumstances. And once you have mastered them, they cannot be taken away.

Perhaps the least expensive and most convenient way of learning a skill is from a good book or trusted media source. Another sure way to learn something new is to become an apprentice to a practitioner. You can also learn skills by taking a class or seminar through a university, community college, university extension, adult education program, or special-interest group. And countless how-to videos are available on YouTube and other video channels, although you will need to evaluate the accuracy and reliability of the information.

Dovetailing Skills

Learning a skill is just the beginning. Practice is necessary to build familiarity and efficiency. It can be easier to practice and more enjoyable if you dovetail survival skills with other practical pursuits, recreational hobbies, and family activities. When you become serious about preparedness, developing these skills is very motivational.

In table 3.1, you will find a list of skills, though not exhaustive, to get you thinking. Use it to create a list of the skills you would like to acquire. Divide that list into two categories—essential and desirable.

PERSONALLY SPEAKING

 I have learned from experience that you absolutely must practice the skills you want to develop. A good how-to book or YouTube video can be a great help, but you cannot appreciate the nuances of a skill until you try it. The first garden Jack and I grew when we lived on the Front Range of Colorado can only be described as pitiful. But we persisted year after year, acquired more knowledge, tested new ideas, and learned from our mistakes. Gradually, our gardens improved and even flourished. But it was certainly a skill that took time to acquire and only developed through practice and patience, trial and error.

Table 3.1 Preparedness Skills to Develop	
Homesteading Skills	**Home Arts Skills**
· Cooking using food storage · Cooking using unconventional methods · Growing a garden, orchard, and field crops · Making cheese, butter, and yogurt · Preserving food by dehydrating, canning, smoking, curing, and pickling · Raising chickens, rabbits, goats, and other animals	· Designing, sewing, remodeling, and mending clothing · Making and repairing shoes and emergency footwear · Making soaps and candles · Quilting, knitting, crocheting, spinning, weaving, and dyeing wool · Tanning and crafting with leather
Home Repair and Maintenance Skills	**Machinery Repair Skills**
· Building with rammed earth and other construction techniques · Carpentry, woodworking, and painting · Plumbing, electrical wiring, and masonry · Surveying and clearing land, drilling wells, and damming streams	· Basic auto mechanics and repair skills · Doing tune-ups and maintenance · Fixing appliances, machines, and small engines · Metalworking, welding, casting, and blacksmithing
Communication Skills	**Medical Skills**
· Bartering · Radio and communications · Understanding police and fire codes and CB lingo · Using a ham radio	· Certifying in CPR, EMT, or as a paramedic · Emergency childbirth · Practical nursing skills or CNA · Taking beginning and advanced Red Cross first-aid classes
Wilderness and Primitive Survival Skills	**Outdoor Recreation and Sports Skills**
· Cooking without pots and pans · Finding suitable drinking water · Foraging for edible plants and snaring wild game · Knot tying · Making emergency shelters · Making primitive tools · Map reading and navigation · Starting fires without matches · Weather prediction	· Archery and bow hunting · Camping and backpacking · Canoeing, rowing, sailing, and swimming · Field dressing and preparing game for eating · Handling weapons and marksmanship · Hunting, fishing, trapping, and tracking · Mountaineering, snowshoeing, and downhill and cross-country skiing · Practical defensive shooting · Self-defense and martial arts

©Patricia Spigarelli-Aston

Material Provisions Preparation

Material provisions include food, supplies, tools, and equipment, but they can also include other preparations, like making a place for a garden and orchard, acquiring animals, securing a reliable source of water, building a greenhouse or root cellar, and modifying your home for a woodstove.

Dedicate a Notebook for Your To-Do Lists

The chapters in this book address the materials and supplies and important skills you need to be prepared. As you read each chapter, your family's needs will become clear. Dedicate a notebook or computer file to making a list of provisions and the amounts you need.

Build a Good Reference Library

Knowledge is one of your best investments. Any search for books about self-reliance, survival, or preparedness skills will give you a lot of titles. Look for books that offer a philosophy and overall preparedness strategy you agree with, then seek out supplemental books that teach specific preparedness skills. Although electronic books and YouTube can be good resources, they may not be accessible in a survival situation.

QUICK LOOK

Material Provisions To-Do List

- Set up a notebook or computer file to organize preparedness goals
- Inventory supplies, tools, and equipment on-hand
- Make a list of "can't-live-without" nonfood supplies
- Begin a list of provisions to acquire
- Begin a list of tools and equipment to acquire
- Begin or add to a preparedness library
- Acquire items for bartering

Plan Adequately

Planning is an essential preparedness skill. It is easy to underestimate your needs and overestimate your provisions. Erring on the side of having more than you think you need will give you the ability to share or barter. Also, plan to rotate your supplies, using the oldest first and replenishing things as they get used.

Inventory Your Provisions

Use the items mentioned in this book to start your inventory. Augment the inventory by adding consumable items you would find it difficult to live without. Add to that list any items that could be damaged or wear out and need replacing.

Plan for Backup Systems

Have backup systems so that if one fails, you have an alternative. For example, store food as well as grow a garden and raise animals. Stored food is secure, but it will eventually run out. On the other hand, gardens **Always have a Plan B.** and animals are renewable but vulnerable to drought, disease, and vandals. Using both increases your odds. Likewise, have more than one source for water, cooking, heat, and defense.

Financial Preparation

Financial Independence

Need to fine-tune your finances? Begin by following a written budget and paying off any consumer debt. Build an emergency fund with enough cash to cover basic expenses for three to six months and systematically save for the future. Have an up-to-date estate plan and appropriate term-life insurance, particularly if you have dependents. Tried-and-true financial principles can help you gain financial independence.

A Supply of Cash

Most crises will be regional and relatively short-lived, and many will involve a power outage. ATMs and credit-card systems will be down, and electronic transfers may not be possible, so keep a good supply of cash—several hundred dollars in denominations of twenties, tens, fives, and ones as well as rolls of quarters—on hand.

Alternative Livelihood

Unless you already perform a basic service or produce an essential good, it would be wise to develop an alternate livelihood that would remain in demand in a more basic economy. Perhaps it could be a current hobby.

Gold and Silver

While gold and silver can be transporters of wealth to a more normal economy, their intrinsic value is low in a basic society. Also, for this same reason, be wary of diamonds, other gemstones, rare coins, and collectibles. However, small silver coins may be a useful exchange.

Barter

Paper money has value only in a working economy that accepts it and may be worthless in some future scenarios. Your storage of "real" goods for barter can be crucial, so be sure to diversify and keep a balance of provisions. Consider stockpiling small items that are inexpensive now but would be highly valued in a time of scarcity. It makes sense to store those items your family normally uses. See table 3.2 for items to barter.

In dangerous times bartering can have risks. Be careful to keep a low profile and especially don't flaunt your wealth. Mix it up. Don't always barter in the same place or offer the same things to barter.

Table 3.2 Items for Barter	
Personal Items	**Multi-Purpose Items**
· Bar soap, deodorant, shampoo · Condoms · Diapers, baby wipes · Feminine-hygiene products · First-aid supplies · Lip balm, lotion · Over-the-counter medicines · Pain killers, antibiotics, prescription drugs · Personal wipes · Razor blades, disposable razors · Reading glasses · Toilet paper, sanitary supplies · Toothpaste, dental floss	· Aluminum foil · Ammunition-12-gauge shotgun shells · Ammunition-.22 long rifle ammunition, 9mm, common calibers · Detergent · Duct tape · Garbage bags · Gloves—work, gardening, rubber · Hard plastic containers · Paracord 550 · Resealable bags · Tarps, sheeting · Wire
Tools and Equipment	**Energy**
· Cloth (canvas, denim, wool) · Fishing poles, nets, hooks, etc. · Garden tools · Hatchets, saws, hand tools · Knives, multipurpose tools, utility knives · Scissors · Sewing needles, sewing supplies, thread	· Batteries, solar and conventional · Emergency candles · Flashlights · Fuel: gasoline, propane, kerosene, diesel, Coleman fuel, firewood · Hand warmers · Matches, steel wool, lighters, fire starters · Solar-battery chargers
Food and Water	
· Alcohol (vodka) · Basic foods · Coffee and tea · Cooking oil · Garden seeds, open-pollinated · Leavenings (baking powder, baking soda, yeast) · Salt	· Spices (pepper, cinnamon, chili powder) · Sugar · Treats · Vinegar · Water bottles · Water filters, purifiers

NOTES

Your Master Action Plan

Family Emergency Assessment

As a family, make a checklist of all the things you think are important in responding to an emergency. This is highly individual, so your checklist will be unique. What are the possible scenarios? What resources do you have available? What are your basic shelter, water, food, and communication resources? What things do you need to improve upon? Use worksheet 4.1, "Family Emergency Assessment," on page 28 to assess your family's level of preparedness. Worksheets can also be found as downloadable PDF files on our website, CrisisPreparedness.com.

Determining Your Priorities

The first step in creating a Master Action Plan is to assess needs. Since personal and financial needs vary widely, you should determine your priorities based on the discussions in the previous chapters.

Begin by developing a list of goals for personal preparedness, then make a separate list of goals for acquiring supplies. Finally, list any financial goals that will help you be prepared. Use this chapter and subsequent chapters to create these lists. Next, prioritize the listed items.

Table 4.1 on page 26 outlines some possible priorities for material supplies. Use worksheet 4.2, "Master Action Plan," on page 29 to help you organize your priorities in all three areas. It can also be found on our website, CrisisPreparedness.com.

A Realistic Timetable

Once you determine your priorities, you'll need to set specific goals. Realistically, your timetable will be a compromise between urgency and available resources. With your timetable as a guide, set a target date for each of the items on your lists. Use worksheet

---

FIVE THINGS YOU CAN DO NOW TO BE READY FOR AN EMERGENCY

1 Complete worksheet 4.1, "Family Emergency Assessment," to evaluate your preparedness level.

2 Identify three priorities for you family preparation. Use worksheet 4.2, "Master Action Plan."

3 Determine a primary meeting place and an alternative meeting place for your family in case of an emergency.

4 Make a "walk home" plan in case transportation is halted.

5 Plan a "surprise" three-day family drill.

4.3, "Inventory Planning Checklist," on page 30 or from our website. List the amount needed, target date, storage location, and check off when achieved.

Purchase items to complete the inventory planning checklist consistent with your needs and means. Using the amount allowed by your budget, build a master shopping list showing your planned month-by-month purchases. Use worksheet 4.4 on page 31 or from our website.

Now you are ready to implement your plan.

Table 4.1
Example: Possible Priorities for Material Provisions

PRIORITY 1: Most Critical Items to Sustain Life for a Short-Term Crisis

- Water for two weeks (fourteen gallons/person) and water purifier
- Portable food and essential supplies for seventy-two hours
- Medications and first-aid basics
- Minimal sanitation needs for two weeks
- A way to keep warm in freezing conditions
- Sensible shoes and clothing for getting home or to a shelter

PRIORITY 2: Essential Items to Sustain Life during a Three-to-Six-Month Crisis

- Basic food plan for three to six months with recipes
- Garden seeds for one season
- An electric grain mill or quality manual grain mill
- A plan for obtaining, storing, and purifying water
- Basic sanitation for sustained needs
- Durable, sturdy, warm clothing
- Bedding to keep warm without additional heat
- Method/s to cook and heat, along with fuel for three to six months
- Basic survival library

PRIORITY 3: Storage Necessary for a More Normal Lifestyle for up to One Year

- An advanced food plan for one year
- Additional clothing, bedding, medical supplies, and fuel
- Equipment, tools, and supplies for long-term sustainability
- A self-sufficient homestead

©Patricia Spigarelli-Aston

Three Levels of Preparedness

If you have not already established your vision for preparedness, this is a good time to give it some serious thought. Your vision will likely include one or more levels of preparedness.

Level 1—Minor Disruptions

The first level deals with relatively minor disruptions—things that benefit from minimal yet thoughtful preparation. They are small-scale events, local and isolated, lasting only a short time. They need immediate action and may affect just your family. For example, this level of preparedness could be as simple as having gear in your car in case you're stranded in a blizzard. Or it might be isolated to one community with the likelihood of outside help rallying quickly. Either way, you are on your own for a while and want to be as self-reliant as possible.

Level 2—Significant Emergencies or Crises

The second level of preparation is for longer-lasting, regional or national emergencies where transportation and utilities may be significantly disrupted or other circumstances such as pandemics require you to shelter at home for weeks or months. In some circumstances it may include evacuating your home. This level of preparedness also includes temporary personal economic crises, like income loss.

Level 3—Large-Scale Catastrophes

The third level of preparation is for large-scale catastrophic crises. This could include financial upheaval or societal breakdown. Some call it Global Pandemic, Armageddon, Peak Oil Collapse, SHTF, TEOTWAWKI, or Apocalypse. No matter what name you give it, it comes from a healthy distrust of the status quo and the fragile nature of our interdependent society. It also comes from a strong inner conviction that you determine your destiny and you are prepared for whatever comes.

Putting Your Plan into Practice

Keep Track and Update

With time, your stockpile will grow and your preparations will take shape. It is important to rotate your supplies. Use a simple inventory control system to keep things up-to-date and fresh. Establish a regular schedule for monitoring your inventory, and update your lists as situations change. Refer to "Inventory Control" on page 72 for more information about tracking inventory.

Trial Run

Practicing for an emergency can help you identify preparedness strengths and weaknesses. Develop a plan with your family, considering the potential disasters you may face.

You might want to stage a three-day trial run to help identify any weak spots in your preparations. This test lasts seventy-two hours and consists of surviving as though a crisis has deprived you of all outside assistance and services. How will you manage without electricity? What precautions will you need to take if you have open flames? How will you properly vent alternative fuels? What about water and sanitation?

Worksheets to Help You

Use worksheets 4.1, 4.2, 4.3, and 4.4 to create and execute your Master Action Plan. For downloadable PDF files, go to CrisisPreparedness.com.

Worksheet 4.1 Family Emergency Assessment				
Possible Disasters Our Family Might Face	Resources We Need to Be Safe	What We Already Have in Place	What We Need to Improve	✓ When Completed
Disaster Possibility #1:	Shelter: Water: Food: Communication:			
Disaster Possibility #2:	Shelter: Water: Food: Communication:			
Disaster Possibility #3:	Shelter: Water: Food: Communication:			
Disaster Possibility #4:	Shelter: Water: Food: Communication:			

Worksheet 4.2		
Master Action Plan		
Goals for Personal Preparedness	Goals for Acquiring Material Provisions	Goals for Financial Preparedness
Priority 1	Priority 1	Priority 1
Priority 2	Priority 2	Priority 2
Priority 3	Priority 3	Priority 3

©*Patricia Spigarelli-Aston. For a downloadable PDF file, go to CrisisPreparedness.com.*

Worksheet 4.3 Inventory Planning Checklist				
Item Needed	Amount Needed	Completion Target Date	Storage Location	✓ When Completed

Worksheet 4.4			
Master Shopping List for 20____			
Items to Purchase for Each Month	Allocated Budget	Items to Purchase for Each Month	Allocated Budget
January		July	
February		August	
March		September	
April		October	
May		November	
June		December	

©*Patricia Spigarelli-Aston. For a downloadable PDF file, go to CrisisPreparedness.com.*

NOTES

The Rule of Three

You might be thinking, *Where do I start with my preparations? What should I do first?* Where you begin depends on how prepared you already are and which crises you anticipate. It also depends on circumstances like which part of the country you live in, the climate there, and whether you live in an urban or rural setting.

Perhaps you have heard of the survival "Rule of Three." The callout box on this page describes these three rules.

Most of this book is dedicated to teaching you how to prepare to have proper shelter and enough water and food to survive a potential crisis. Naturally, a clean air supply is also critical.

Three Minutes without Air

Our brains stop working, our hearts stop beating, and all other organs shut down without oxygen.

Think about possible scenarios where you could be left without oxygen. Most happen suddenly and leave you with only seconds to figure out a way to survive. Table 5.1 on page 34 describes the common emergencies that threaten breathing.

CPR

CPR training is a survival skill worth acquiring as it can dramatically improve a person's chance of survival during cardiac arrest. If your blood stops flowing, you will suffer brain death within four to six minutes. CPR is *first* aid, and, if performed properly, it effectively keeps blood flowing and provides oxygen to the brain. The goal is to keep you alive until more advanced medical help arrives. Everyone should learn CPR. For some circumstances, you may want to consider gas masks (see chapter 44 for information and recommendations).

> **QUICK LOOK**
>
> ### The Rule of Three
>
> You can survive for
> **three minutes**
> without air.
>
> You can survive for
> **three hours**
> without protection or shelter
> in a harsh environment.
>
> You can survive for
> **three days**
> without water.
>
> You can survive for
> **three weeks**
> without food.

FIVE THINGS YOU CAN DO NOW

1 Make sure smoke detectors and carbon-monoxide detectors are installed and in working order.

2 Take a CPR class.

3 Stock your car with supplies for surviving a winter storm.

4 Fill a small backpack with emergency equipment to keep at your workplace.

5 Keep an extra pair of comfortable shoes in your car or workplace in case you must walk to safety.

Table 5.1
Responding to Emergencies Where Breathing Is Threatened

Crises That Threaten Breathing	Potential Causes	Possible Actions and Remedies
Choking	· Food · Illness · Asthma	· Knowledge of how to administer Heimlich maneuver · Supply of medications for asthma relief
Anaphylactic shock	· Severe allergies	· EpiPens · Antidotes for known allergies
Carbon-monoxide poisoning	· Faulty furnace · Improperly vented heating and cooking appliances, such as outdoor charcoal grills, space heaters, and generators · Automobile exhaust pipes plugged with ice or snow	· Carbon-monoxide detector and alarm · Properly ventilate cooking and heating appliances. · Clear automobile exhaust piles when stranded in a winter storm.
Smoke inhalation	· House fires	· Smoke detectors · Practice fire-escape plan with children.
Poisonous gas	· Accidental chemical spills · Terrorist activity · Chemical warfare	· Hold your breath and move as quickly as possible away from the source. · Filter air with a makeshift mask. · Move to higher ground since many poisonous gases are heavier than air. · Invest in a gas mask.
Smothering	· Being confined in a small place during an earthquake or building collapse · Being trapped in an avalanche	· Get close to an outer wall or a large piece of furniture where a triangle of life can protect you from the debris and give you an air pocket. · Learn and follow avalanche protocol. Wear an avalanche beacon.
Drowning	· Flash floods, flooding from rising rivers, hurricanes, and tsunamis	· Try to hold on to something that floats and keep feet pointed downstream. · Follow tsunami safety guidelines.

Three Hours without Shelter

After breathing, shelter is the most important requirement for survival. Exposure to extreme conditions, whether hot or cold, can put you in a life-or-death situation. This principle of preparedness is often overlooked, but with a little preparation, you can prevent a tragedy. If you will be traveling in cold conditions, make sure you have an emergency kit in your car as well as a way to keep warm. In hot climates, be sure you have plenty of water and a way to create shade.

Respect the Forces of Nature

According to a story in Backcountry Chronicles, a couple from Atlanta went for a hike on a beautiful September day in the High Uinta Mountains of Utah. They were advised to take extra clothing with them in case the weather changed, but they did not. And the weather changed. An unexpected, fast-moving early winter snowstorm hit the area. Because they did not let anyone know their exact plans, nobody knew they were missing until they failed to arrive in Atlanta five days later as scheduled. Their bodies were discovered the following spring.

This story illustrates the importance of shelter. Shelter includes those items that protect you from the elements. Clothing that keeps you warm and dry, appropriate footwear, and matches and fire-starting tools are the minimum basics for emergency shelter. In this case, the people in the story did not take simple preparations seriously.

Reasonable Precautions

If you head out for an adventure, make sure you let someone know where you are going, even if it is just for an afternoon hike. And it is always a good idea to carry a PLB (personal locator beacon) when in the wilderness. Make sure you have the minimum outdoor survival equipment you need to stay protected in whatever environment you will be in.

Land Shark Instant Survival Shelter & Stealth Bag is a micro-thin layer of aluminized film that will protect you against exposure to the elements. It is heat retaining, windproof, waterproof, and built to reduce punctures and tears. It can keep you alive in extreme conditions on both land and in water.

PERSONALLY SPEAKING

 My husband moved to Minneapolis in 1983. He was not used to the notorious Midwest blizzards and was talking with some acquaintances about how he could prepare for them. One man scoffed, "We really have you nervous about the winters out here, don't we!" That November, right after Thanksgiving, this man and his family drove his mother to her home just an hour and a half southwest of Minneapolis. As they made their way back to Minneapolis, the region was hit with a powerful, record-breaking blizzard that trapped thousands across the Midwest. Unfortunately, this family slid off the highway, which then closed. The family was not rescued, and they perished.

Lessons learned? First, complacency can be dangerous. Respect the forces of nature. Second, expect the unexpected and prepare for it. Third, have multiple options. And fourth, let people know your plans, and check in periodically if possible.

Everyday Precautions

If you commute to and from work, be sure to have emergency kits at your workplace and in your car. If it becomes necessary to leave your workplace, you should have appropriate shoes and outer clothing so you can make it home or to a shelter.

For more detailed information about preparing your shelter, refer to chapters 33 through 39.

Three Days without Water

You cannot live much longer than three days without water. Specifics about water preparedness are discussed in chapters 6 through 9, but for now, think about how much water you might have on hand for an emergency. At the least, you need a minimum of two quarts per person per day just for drinking. You will need more for cooking and sanitation. Acquiring and storing water is something you can do right away.

Three Weeks without Food

Although you can survive without food for three weeks, it is far better to have the food you need to endure a crisis. There are many things to consider when you begin to store food. See chapters 10 through 17 for more information about storing food.

SECTION 2

Storing Water
for a Crisis

Emergency Water Supply

Water is one of your most critical preparedness needs. We take water for granted, but without it, a person will suffer exhaustion, dehydration, cramps, heatstroke, and illness. Death will occur within four to ten days and even earlier in severe conditions. For this reason, when it comes to stockpiling provisions, I recommend starting with water.

Besides for drinking, you'll need water for food preparation, especially if you are relying on dehydrated foods. It is also needed for basic hygiene and bare-minimum sanitation.

Water Supply Interruptions

As water sources are susceptible to pollution, it is not uncommon for whole communities to have their water supply compromised. But the greater danger may come from water infected with disease, inadvertently contaminated with chemicals, or intentionally poisoned by terrorists. Water may also be cut off due to drought, hurricanes, floods, landslides, earthquakes, power outages, and labor disputes.

There are two parts to the solution. First, have your own water reserves. Second, have at least one way to purify or filter additional water as you need it.

Have at least one way to purify or filter additional water.

How Much Water Should You Store?

How Much Water Do You Use?

According to the United States Geological Survey, Americans use an average of 80 to 100 gallons of water per day. This includes water used for manufacturing and farming. On the other hand, the average African uses about five gallons per day. Obviously, Americans can get by with much less water in a crisis. But even so, using just five gallons of water per day requires a little over 1,800 gallons per year per person. To give you a comparison, a standard bathtub holds about 50 gallons of water.

FIVE THINGS YOU CAN DO NOW

1 Learn how to shut off the main water valve to your home.

2 Calculate how much water your family needs for two weeks.

3 Calculate how many bottles of water your family needs for two weeks.

4 Purchase several cases of bottled water the next time you go to a grocery store.

5 Clean the toilet tanks so that the water in them is fresh.

Water is bulky, heavy, and hard to contain. For most people, it is impractical to store an emergency supply of 1,800 gallons per person for an entire year.

PERSONALLY SPEAKING

Curious to know how much water our family used, I looked at the water bills for the winter months when we were not watering lawns or gardens, only to discover we were using fifteen to twenty gallons of water per person per day! In addition to drinking water, that included handwashing, teeth brushing, daily showers, toilet flushing, dishwashing, and laundry. Even though we were below the American average, we were way above the recommended emergency amounts!

Your water bill will probably tell you how much you use a month in thousand-gallon units. Divide the total number by thirty to figure the average daily amount your family uses, then divide that by the number in your family to find out each person's average daily use.

Recommended Storage Amounts

The Red Cross recommends you store a minimum of one gallon per person per day for two weeks, or fourteen gallons per person. The actual amount needed for that period will depend on a person's age, physical condition and activity level, type of food consumed, and the environment. If you plan to use dehydrated food, you will need two to five gallons more per person. Also, store water for your pets.

> Store a minimum of one gallon of water per person per day for two weeks, or 14 gallons per person.

How Many Bottles of Water is That?

Typical water bottles hold 500 ml, or 16.9 ounces, which is a little more than a pint. There are 8 pints in a gallon. If you need 14 gallons per person, that totals 112 bottles of water per person. Since there are usually 24 bottles of water in a case, that would be a little more than 4½ cases of water per person. Use Worksheet 6.1 on page 42 to calculate how many containers of water you need with different sizes of water containers.

Emergency Reserves in Your Home Water System

The Water in Your Pipes, Tanks, and Water Heaters

Remember that you already have water in your home in the pipes, toilet tanks, and water heater. If you prepare ahead, you will be able to make this water available when needed.

You should know how to shut off external water pipes. Keep a wrench or valve-shut-off tool in a handy location so you can close your main valve if there's a warning that the water system might be compromised. This will prevent unsafe water from entering your home as well as keep the water in your home contained.

Another safeguard is to install an antisiphon valve on the water inlet to keep the water from flowing back out of your house if the water pressure drops substantially.

Water in Pipes

To collect the water from your pipes, open the highest water valve in the house, then drain the water from the pipes by opening lower faucets.

Water in Toilet Tanks

The water in your toilet tanks is also a source of fresh water—if the tank is kept free of scale, rust, and organic growth by occasionally cleaning with bleach.

Water in Water Heaters

The water in your water heater can be used if its inlet valve is closed before any contamination takes place. To prevent damage to the water heater and yourself, turn off the gas or electricity to it prior to emptying it. The water can be drained from a self-vented tank by simply opening the bottom valve. If the tank is not self-vented, air must be let into the system from the top valve. To guarantee contaminant-free water, regularly drain your water heater to prevent the accumulation of rust and sediment at the bottom of the tank.

Quick Storage

One quick way to store water before a severe storm, such as a hurricane, arrives is to use existing various household and collapsible, five-gallon storage containers.

The bathtub is a large water-holding vessel. However, it may not be clean or may have chemical residue, and once filled, the water in it can become contaminated. A solution to this problem is the WaterBOB—an inexpensive, expandable container shaped to fit bathtubs that holds up to one hundred gallons of water and comes with a pump.

Adding Extra Water to Your Home

One way to increase your water supply is to add one or more extra tanks to your home water system. If you have the room, the ideal place for these extra tank(s) is next to your water heater, where the water is continually cycled through and always available.

Remember, however, that any contamination of the regular water supply will contaminate these additional in-system tanks as well. If you anticipate water contamination or a shortage, simply turn off the main water valve. The tank should have an antisiphon valve on its main inlet, as well as a secondary outlet, such as a hand pump.

Rainwater Harvesting

How useful rainwater collection is depends on the amount of rain your area receives and the type of rainwater-collection system you have. One inch of rain falling on a two-thousand-square-foot roof will provide 1,250 gallons of water. Since you will need a way to collect the water, I recommend RainXchange. Their containers can be an attractive addition to the outside of your home.

Although the rainwater itself is free of impurities, the surfaces it is collected on such as roofs and gutters may not be. To be a safe water source, rainwater should be filtered. You can use non-potable rainwater for laundry or gardening. Also, check the rainwater collection regulations in your state.

Worksheet 6.1		
Determining How Much Water Your Family Needs for Two Weeks	
(Recommendation: one gallon per person per day)	

Total number in household: _____	14 x number in household = total gallons: _____	Total gallons x 128* = _____ total ounces *(*128 ounces in a gallon)*
How many 0.5-liter (16.9-ounce) water bottles?	Total ounces ÷ 16.9 = _____ (total number of bottles)	
How many 2-liter (67.6 ounces) bottles?	Total ounces ÷ 67.6 = _____ (total number of 2-liter bottles)	
How many 32-ounce canning jars?	Total ounces ÷ 32 = _____ (total number of jars)	
How many three-gallon containers?	Total gallons ÷ 3 = _____ (total number of three-gallon containers)	
How many three-and-a-half-gallon containers?	Total gallons ÷ 3.5 = _____ (total number of three-and-a-half-gallon containers)	
How many five-gallon containers?	Total gallons ÷ 5 = _____ (total number of five-gallon containers)	
How many seven-gallon containers?	Total gallons ÷ 7 = _____ (total number of seven-gallon containers)	

©*Patricia Spigarelli-Aston*

Storing Water

Water must be stored where it will not be accidentally contaminated or freeze and break the container. If stored properly, it can potentially last indefinitely. As a precaution, consider sampling your water once or twice a year. If the water has an odor, looks dark or cloudy, or tastes bad, it should be replaced.

Water will taste flat if it has lost the air normally trapped in it, but that can be reversed by vigorously shaking the container or by pouring the water back and forth between two containers. A kitchen whisk or blender also works.

Bottled Water

Next to tap water, bottled water is the easiest to obtain and most reliable source of water available. It is portable, clean, and economical. Make it your first acquisition in water storage.

However, disposable bottles begin to break down as soon as they are manufactured. Compared to other types of containers, bottled water has a short shelf life—about two years, but more if stored in a cool, dark place away from contaminants. Bottled water stored in your car will be exposed to heat and should be recycled frequently.

Some people fear the chemicals in plastic are harmful, whereas others argue that the traces of plastic are insignificant, especially if you are only using it for emergency purposes. Plastic bottles are also not as sustainable as more durable reusable containers. Nevertheless, they are a practical and convenient way to store water.

As calculated in the previous chapter, it takes about 4½ cases of 24 half-liter bottles to provide 14 gallons of water—the recommended amount for one person for two weeks. Use Worksheet 6.1 on page 42 to calculate how many containers of water you need for your family.

FIVE THINGS YOU CAN DO NOW

1 Determine how many gallons of water your family should store for emergencies.

2 Decide on what kind of permanent water containers you will use to store your water.

3 Calculate how many containers you will need to store the desired amount of water for your family.

4 Begin purchasing containers for your water storage.

5 Systematically begin stockpiling emergency water.

Water Storage Containers

Water can be stored in either portable or stationary containers, but be sure they are FDA approved as safe for human use. Many containers are made from durable, high-density, moderately opaque polyethylene (HDPE) and come in a variety of shapes and sizes. However, if they have previously been used to store food products, they will slowly leech the chemical residues of those products into the water. Containers can also be made of glass, fiberglass, or enamel-lined metal.

As you decide which size containers work best for you, keep in mind that a gallon of water weighs eight pounds. A 3½-gallon container filled with water will weigh about 28 pounds, a 5-gallon container about 40.

Storing Water in Small, Portable Containers

There are many options for storing smaller quantities of water. Look for containers that stack easily and feature comfortable handles, tight-fitting gasket-type lids, and pouring spouts. In table 7.1 you will find several good options for small, portable water-storage containers. Other containers for storing water include canning jars, glass jugs, and soda-pop bottles. Clearly label and date all containers.

> Store water in both small and large containers for more flexibility.

Containers to Avoid

Never use a container that has previously held fuel, poisons, or other toxic materials. Bleach containers are not safe to store water meant for human consumption. Plastic milk jugs aren't a good option either because the seams are not durable and there can be milk residue and bacteria left in the container. Also, the water in waterbeds should not be used. The material waterbeds are made of is not FDA approved for use as a safe water container. Plus, the treatment chemicals are not safe, nor can the residues be entirely removed with filters.

PERSONALLY SPEAKING

 About ten years ago, we purchased several five-gallon containers to add more water to our emergency supply. We thoroughly rinsed them out, filled them with tap water, tightly closed the lids, and put them in the storage area under our front steps. Recently, we were curious about the quality of the water and decided to try it. I'll admit I was slightly skeptical. After all, the water had been stored for almost ten years. I let my husband try it first. He opened the spigot and poured the water into a transparent glass. The water was clear—no cloudiness, nothing floating in it, no impurities of any kind. Next, he sniffed it—no odor. Then he tasted it.

He smiled. "It tastes like water!" I poured myself a glass. He was right; it tasted like water, just as if it had come fresh from the tap. My conclusion was that putting clean tap water into a clean water container is a safe and practical way to store it for an emergency. I still think best practice is to periodically recycle the water or condition it with a chemical like chlorine dioxide, but if you fail to do that, you'll still likely have a safe water supply.

Table 7.1 Portable Water Storage Containers		
Water Containers	**Features**	**Advantages**
Hedpak		
	· 5 gallon, semirigid · Sturdy HDPE container · 3 containers are needed for two weeks for one person	· Built-in handles · Spigot available · Vent lid for easy pouring · Can carry for short distances
Front Runner Jerry Can		
	· 5.3 gallons · Heavy-duty, puncture-resistant · Food-grade BPA plastic · Plastic tap can screw out to connect hoses or taps	· Protected tap at bottom · Can be strapped to vehicles · Vertical shape makes it easier to carry
Aqua-Tainer		
	· 4-gallon and 7-gallon sizes · Sturdy HDPE container	· Ergonomic handle · Hideaway spigot · Stackable
WaterBricks		
	· 3.5 gallon · Sturdy HDPE, stackable · Easy-grip handle · 10-brick stacks	· Maximum efficient storage for small spaces · Individual containers for grab-and-go use · Large opening for easy access
Fold-a-Carry		
	· 2.5 gallon and 5 gallon · Polyethylene plastic · Fold-down handle · Integrated spigot	· Collapsible and flexible · Flexible at very cold temperatures · Foldable for storage · Less expensive than others

©*Patricia Spigarelli-Aston*

Storing Water in Large, Stationary Containers

Storing Water in 55-Gallon Drums

Large drums are useful for more immobile, permanent water storage. The typical 55 -gallon blue drum made from BPA-free plastic provides minimum emergency water for a family of four for about two weeks.

The durable blue material is nearly opaque and may be used for storing water in garages, sheds, or outdoors. To compensate for expansion from freezing, do not fill the barrel completely. Also, cover barrels that are outdoors, if possible, to minimize UV deterioration and possible algae growth.

Vertical Storage

For easy filling and emptying, consider a system that stores two or more barrels vertically, such as Titan Ready USA's Hydrant Water Storage System (pictured). It houses two to four 55-gallon water barrels that connect and stack vertically.

Accessories

You will likely need a food-grade hose for filling the container and a hand pump to help you access the water. Look for a hand pump with several sections so that it can fit a variety of container heights. A siphoning hose and siphon pump are also helpful for removing and recycling the water.

> **QUICK LOOK**
>
> ### Accessories to Help with Storing Water
>
> - Food-grade Hose
> - Hand pump or
> - Siphoning hose with pump
> - Water treatment chemicals

Storing Water in Water Tanks

Larger water tanks range in size from one hundred to three hundred gallons or more. Look for tanks that are UV resistant. A blue tank universally means fresh water. However, blue tanks as well as white tanks are still susceptible to algae growth because light can penetrate these containers. You may want to consider a black or dark-green tank if it will be stored outdoors, but be sure it is food-grade.

A tank with an elliptical shape will more easily fit through a doorway. Those with both a faucet and a drain make recycling water convenient and easy. A keyed faucet and drain will prevent accidental flooding. Sure Water, Inc. and WaterPrepared both specialize in large water-storage tanks. (The one pictured is Sure Water's 260-gallon water tank.)

Storing Water in Larger Water-Storage Systems

Greater quantities of water can be stored in large, underground tanks, cisterns, and reservoirs. Some people believe they can use the water in a swimming pool, but depending on the chemicals used, it may not be usable even if filtered. At the very least, the water will be subject to contamination and pollution and will need to be purified or filtered.

Check with farm-equipment suppliers that sell large water tanks, especially if you have animals that will need water.

Pumps and Siphons

As stated previously, a hand pump makes large containers more usable. Big tanks merit a larger rotary pump or a high-volume, high-pressure pump. Be sure to include a food-grade siphon for water recycling.

Storing Water in Flexible Water Bags

Water bags made from flexible, durable, food-grade materials are another alternative for water storage. EmergencyZone offers a system of five-gallon Mylar bags contained within protective, rigid cardboard boxes (pictured) that have cutouts for handles to make them portable. The system uses the chemical Aquamira to maintain freshness. (See the section "Conditioning Your Water Storage" on page 48.)

Aquaflex's flexible Aquatank II emergency water-storage tank bag, made of heavy-duty, food-grade, polyurethane-coated nylon (pictured), has outstanding tensile, tear, impact, and abrasion resistance. Its garden-hose fittings make it easy to fill (you will need a hose safe for drinking water). The 30-gallon size measures 3 feet by 3 feet. The containers also come in 15- to 300-gallon sizes.

Maintaining Your Water Supply

Cleaning and Refreshing Water Containers

Rinse new barrels or tanks with a diluted solution of regular, unscented household bleach (5.5 percent to 6 percent concentration sodium hypochlorite) and water. Swish it around the container, then rinse with clean tap water until no odor remains.

Though barrels of water can be a challenge to clean, the good news is, if you originally filled the barrels with fresh tap water, there is a good chance the water is safe to drink even after five years, although it may be less palatable. However, if you want to be assured the water is fresh and pure, plan to recycle it about once a year.

If a barrel or tank is fitted with spigots, remove the water that way. If not, use a pump or siphon to remove the water. You will need a food-grade hose that is about two and a half times the depth of your container. Refresh the container with the diluted solution of household bleach described previously.

Flushing Out 55-Gallon Drums

Where it can be a hassle to worry about recycling the water you have stored in 55-gallon drums, Emergency Water Corp's EZ Drum Cycler (pictured) makes it simple.

Begin by pouring one cup of undiluted bleach into the 55-gallon drum of water. Let it sit for twenty-four hours to kill any bacteria or slime growing in the barrel. Attach the EZ Cycling System valves to the water barrel, connect drinking-water hoses, and flush out the old water, replacing it with fresh water. This takes about a half hour.

Power washers are also effective for cleaning the insides of barrels or tanks that have more serious contamination or grime.

Conditioning Your Water Storage

If you intend to store water in containers for a long time, be sure to condition it to prevent the growth of organisms. If you plan to regularly recycle the water with fresh tap water, conditioning is not necessary.

Bleach

You will find a thorough discussion about chemicals for purifying water in chapter 8. These same chemicals can also be used to condition your water for storage. If you are using clean tap water, however, simply use eight drops of bleach per gallon of clear water or one-half teaspoon per five-gallon container. Let it stand for about twenty minutes. If you can still smell chlorine, it can be stored; if not, repeat.

Bleach loses potency with age and should be rotated yearly, kept tightly capped, and stored in a cool location.

Sterilizing with Heat

For longer storage, you may want to bottle and heat process the water. Begin by filling clean canning jars to within a half-inch of the top. Either pressure can them for five minutes at ten psi or boil them in a water bath—twenty minutes for a quart, twenty-five for a half gallon. For each one thousand feet above sea level, increase the time for quarts by one minute and half-gallons by two. The water will keep longer than just putting it in clean jars.

Commercial Water Treatments

Liquid chlorine dioxide is an excellent way to condition water you intend to store. There are several commercial water preservers and refreshers that take the guesswork out of water purification. Aquamira is a liquid dioxide water treatment designed specifically for conditioning stored water and requires only a teaspoon to purify water five gallons of water. H2O ResQ is another treatment that prevents the recurring growth of bio-film in water-storage tanks and containers.

Purifying Water

Purification is one of the two main ways to obtain clean water. Water filtration, as discussed in the next chapter, is the other. The purpose of water purification is to kill any of the pathogens that cause disease. Iodine and chlorine are the two main chemicals that kill pathogens. Heat and UV are also effective.

Water Contamination

Protozoa, bacteria, and viruses can cause serious and life-threatening diseases. Giardia, cryptosporidiosis, cholera, salmonella, typhoid, infectious hepatitis, and dysentery parasites are a few water-borne diseases. Water contaminated by sewage is especially unsafe and may also contain harmful parasites, chemicals, pesticides, poisons, heavy metals, and radioactive particles. Likewise, it may contain algae, sediment and silt particles, or have a bad odor or taste.

Pollutants cannot always be detected. All water from unknown sources or of uncertain purity should be purified or filtered prior to use for drinking, food preparation, or personal hygiene.

Although there are several methods for purifying water, they work with varying degrees of effectiveness depending on the type of pathogen.

Protozoa, Bacteria, and Viruses

It is helpful to understand the relative sizes of viruses, bacteria, and protozoa. Think of it this way—if viruses were the size of a BB, bacteria would be the size of a golf ball and protozoa a beach ball.

Viruses and bacteria are easier to kill with purification. But because of their smaller size, they are more difficult to filter. On the other hand, relatively larger protozoa are easier to filter but more difficult to kill with purification.

Table 8.1 on page 50 summarizes common disease-causing organisms transmitted by water. Almost all are found in water contaminated with sewage. Giardia can be in

---

FIVE THINGS YOU CAN DO NOW

1 Become knowledgeable about potential harmful water-borne pathogens.

2 Identify potential water sources other than your present culinary water.

3 Learn how to use ultraviolet (UV) energy to purify water and purchase a UV device.

4 Determine how much disinfectant you will need for purifying water for two weeks.

5 Purchase chlorine dioxide tablets or another disinfectant chemical.

---

water used by domestic animals or wildlife. E. coli is often transmitted through infected, undercooked meat but can be transmitted by water as well. Salmonella typhi and cholera are most commonly found in the water sources of developing countries.

Table 8.1 Common Disease-Causing Organisms Transmitted by Water and Their Symptoms	
Viruses 0.004 – 0.1 microns	Protozoan Cysts 8 – 12 microns
· Norovirus: diarrhea, vomiting · Hepatitis A (HAV): jaundice of skin and eyes, dark urine, loss of appetite, vomiting, fever, stomach pain	· Giardia lamblia: diarrhea, abdominal cramps, gas, malaise, chills, vomiting, headache · Cryptosporidium: watery diarrhea, abdominal cramping, nausea, headaches · Entamoeba histolytica: amoebic dysentery*
Bacteria 0.2 – 4 microns	
· Campylobacters (several kinds): most common cause of diarrhea worldwide, also abdominal pain, fever, headache, nausea, vomiting · Escherichia coli (E. coli): bloody diarrhea and abdominal cramps	· Salmonella typhi (Typhoid Fever): high fever, aches, weakness, stomach pains, loss of appetite · Vibrio cholera: acute diarrhea, vomiting, lower-leg pain, dehydration, shock
*Dysentery is a generic term for the body's reaction to an intestinal infection caused by bacteria or protozoa. Profuse bloody diarrhea, fever, abdominal pain, and weight loss are usual symptoms.	

©Patricia Spigarelli-Aston

Clean Water

As clean water is extremely important, you should carefully consider the methods for purifying water and determine which are most suitable for your potential crisis. Remember that it is a good idea to have backup systems, so consider several methods.

Water Source Selection

You should assume that your water is contaminated if the culinary water system has been compromised. Make every effort to select a water source that is free from sewage or fecal-related bacteria and toxic chemicals. Use groundwater (an underground source) if possible. If you are forced to use surface water, avoid sources containing unknown floating material or that have a dark color or odor.

Straining

Before boiling, distilling, purifying, or filtering water, it is best to remove any suspended particles. This can be done by allowing them to settle to the bottom of a container or by straining through several layers of coffee filters, paper towels, or a clean porous cloth.

Boiling Water

Boiling is the single most effective way to kill pathogens in water. Boiling water for one full minute will kill all bacteria, but other harmful microorganisms may require up to three minutes. Increase boiling time by one minute for each thousand feet above sea level. You will need to double the time if your water is cloudy. Boiling has several limitations. You need a heat-proof container, a heat source, and adequate fuel. It also takes time to cool it. Aerating can improve the flat taste of boiled water.

Ultraviolet Radiation

Sunlight

Sometimes referred to as SODIS (solar water disinfection), sunlight UV-A can be used to disinfect water. Use a clear plastic bottle, preferably PET plastic, such as a typical soda or water bottle, no bigger than two liters so the UV rays can penetrate it. Place the bottle in direct sunlight for at least six hours. Cloudy water cuts down UV effectiveness because UV rays cannot kill viruses or bacteria hidden behind particles. Sunlight can kill many potential disease-causing organisms but will not purify chemical contamination.

Ultraviolet Light Devices

Ultraviolet light devices kill all microorganisms. However, like sunlight, they do not work well in murky water. It is necessary to strain or prefilter water that is not clear. UV devices are lightweight and require batteries. SteriPEN Classic 3 (pictured), now part of the Swiss Katadyn Group, makes a line of handheld UV water purifiers for personal use.

QUICK LOOK
How to Disinfect with Sunlight
1. Prefilter or strain water to eliminate murkiness.
2. Fill a bottle with clear water.
3. Allow the bottle to lie flat in direct sunlight for six hours.
4. Increase UV focus by setting the bottle on a reflective surface, such as a shiny metal tray or aluminum foil.
5. Turn the bottle occasionally.

Chlorine-Based Disinfectant Chemicals

As the principle chemicals used for water purification, chlorine and iodine are both highly effective against viruses and bacteria, though they are not as effective against protozoa. And neither removes sediment, radioactivity, or chemical contaminants. They tend to give the water a chemical odor and taste, but treated water can be made more palatable by aerating and exposing the container to air or by filtering with carbon.

With either iodine or chlorine, the exact amount of chemical and length of time required for complete disinfection depends on the temperature of the water, its alkalinity, the amount of suspended solids, and the quantity and type of dissolved organic compounds. For best results, you should warm the water to 60°F (15°C) and strain it of larger impurities.

Chlorine Dioxide

Chlorine dioxide, chemically different from regular chlorine bleach, is the most effective chlorine compound for killing viruses and bacteria, as well as giardia, and works in about thirty minutes. It is only moderately effective in killing cryptosporidium, however, and can take as long as four hours to do so. Chlorine dioxide comes in both liquid and tablet form and includes Aquamira (pictured), Katadyn Micropur, and Portable Aqua Chlorine Dioxide Tablets.

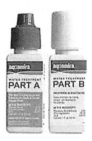

Household Bleach

Sodium hypochlorite is the active ingredient in household bleach and the most common chlorine compound. Look for unscented with a 5.5 percent

Bottled bleach loses its effectiveness and should be rotated at least annually.

to 8.25 percent concentration. Household bleach loses its effectiveness over time, and its ability to purify is reduced if the water has a high concentration of microorganisms. If you intend to store a gallon of bleach for emergency water purification, it is important to rotate it at least annually. Safely dispose of expired bleach by pouring it down the drain with plenty of water.

Granular Calcium Hypochlorite

> **QUICK LOOK**
>
> ### How to Make a Granular Bleach Solution*
>
> 1. Dissolve 1/4 ounce or about one heaping teaspoon of high-test granular calcium hypochlorite in two gallons of water to make a stock chlorine solution.
> 2. To disinfect water, use one part stock solution to one hundred parts water. That's about two and a half tablespoons of solution per gallon of water.
> 3. If chlorine taste is too strong, aerate by pouring water from one clean container to another and let it stand for a few hours before use.
>
> *Based on EPA recommendations.

Calcium hypochlorite is a granular form of bleach commonly used to sanitize swimming pools and purify drinking water. A one-pound bag of calcium hypochlorite will make 128 gallons of bleach solution. That is enough to purify 12,800 gallons of water!

Since this granular solution has a short shelf life, it should be used right away. If properly stored, it can last about five years. Calcium hypochlorite is a highly reactive chemical that should be stored in a cool, dry place, away from metal, and in an airtight, noncorrosive container, such as a glass jar with a plastic lid. Be sure to select a brand that contains no other active additives and is safe for drinking water. Follow the instructions on the label for safe usage.

> **QUICK LOOK**
>
> ### How to Purify Water with Household Bleach
>
> 1. Allow water to settle and then strain to remove any debris.
> 2. Pour the clarified water into a clean container and add eight drops of bleach per gallon.
> 3. Allow treated water to stand for thirty minutes. It should have a slight chlorine odor.
> 4. If there is no chlorine odor, repeat the treatment and let stand for fifteen minutes before drinking.
> 5. If chlorine taste is too strong, let water stand overnight or pour from one clean container to another to allow excess chlorine to dissipate.

Other Chlorine Compounds

NaDCC (Sodium Dichloroisocyanurate or Troclosene Sodium)

This chlorine disinfectant is used worldwide to treat water during emergencies or when available water supplies are unreliable. It is highly effective against viruses, bacteria, and giardia but not cryptosporidium. Aquatabs and Oasis Water Purification Tablets are sources of NaDCC. Both come in effervescent tablet form and take about thirty minutes to purify water at room temperature, sixty minutes if the water is cold.

Halazone Tablets

Halazone tablets, another chlorine-based product, were used in the past to purify water. They were effective but have been largely replaced by other water treatments. Their primary limitation is a short life once exposed to air.

Iodine-Based Disinfectant Chemicals

Though iodine will kill bacteria and many viruses, according to the Center for Disease Control and Prevention, it will not kill cryptosporidium and has only a low to moderate effect on giardia.

Iodine treatments come in three forms: tincture of iodine, crystals, and tablets. Iodine is more dependable than chlorine because it is less affected by heavy organic pollution and alkalinity. It is also more effective over a wider temperature range. But, contrary to some claims, iodine treatments will affect the taste of water. Ascorbic acid (vitamin C) tablets can help neutralize and diminish the iodine taste, but add them only after the water has been purified.

Remember that iodine is an emergency water purifier. Like chlorine, iodine is a chemical. Do not plan to use it for a long time. Also, be aware that some people are allergic to iodine.

QUICK LOOK
How to Make an Iodine Solution
1. Measure 4–8 grams of iodine crystals into a small dark bottle with a leak-proof cap.
2. Add water to the bottle.
3. Shake vigorously for 60 seconds.
4. Let the crystals sink to bottom of the bottle.
5. Use the saturated solution to purify the water.
6. Use 12.5 ml per quart at room temperature and let water stand for 30 minutes.
7. Use 20 ml per quart if water is 40°F (4°C).

Iodine Crystals

There are two ways to use iodine crystals, the most convenient the commercial product Polar Pure Water Treatment. It is ready to use—simply add drops of the solution to water and follow product directions. It has a built-in thermometer to help you determine how much iodine is needed. One bottle treats up to two thousand quarts of water. Or you can make your own saturated solution using iodine crystals (USP-grade sublimed iodine). Iodine crystals have an indefinite shelf life and are not affected by temperature, but keep the bottle tightly sealed.

Iodine Tablets

Convenient and easy to use, iodine tablets (tetraglycine hydroperiodide, or TGHP) are sold under several brand names (see page 424 in resources). They may be sold alone or with taste-neutralizing vitamin C tablets. The recommended dose is two tablets per quart of water. Wait time is thirty minutes. The shelf life is about four years but longer if stored in cool, dark, dry conditions. Tablets change from gray or dark brown to yellow or light green as they lose potency.

Tincture of Iodine

Tincture of iodine 2 percent solution can also be used to purify water. The recommended dosage is five drops to one quart of room-temperature water. Wait time is thirty minutes.

PERSONALLY SPEAKING

Iodine and chlorine are both poisons that kill any organisms that may be hiding in water. I will admit the skull and crossbones on tincture-of-iodine bottles and the cautions listed on bleach containers have made me a little nervous about using either chemical. But I have learned that all you must do is follow directions and use the correct concentrations. The chemicals will have killed all the organisms in about thirty minutes. There is usually a slight, unappetizing, telltale odor, but if you can wait and let the water stand overnight or agitate it, the excess chemicals will dissipate into the air and the water will taste better. Sometimes we have masked the off-taste with lemon-flavored drink mix.

Ozone Purification, Flocculants, and Copper-Silver Ions

Ozone

Ozone is an effective treatment and is used in municipal systems around the world. Roving Blue makes a personal pocket ozone water purifier called the O-Pen (pictured). It is placed in a container of water, where it creates ozone, which effectively kills viruses, bacteria, and protozoa. It has rechargeable batteries and comes with a USB cord. Also, an ozone purifier can be installed in a home water system, using electricity to operate.

Flocculants

Flocculants and coagulants are often used to help clean the turbid water in municipal water systems. They work by pulling solids into clumps. This process also works on an individual scale. P&G Purifier of Water (PuR) packets combine flocculation, coagulation, and disinfection to remove dirt, cysts, and pollutants and to kill bacteria and viruses. It works in about thirty minutes. A single packet cleans 2.5 gallons, or ten liters, of water.

Copper-Silver Ions

H20ResQ is a point-of-use copper and silver ions water treatment specifically for the prevention of the recurring growth of biofilm slime that may develop on the surface of water in storage containers. Scientific research shows that most bacteria are in this biofilm. This purifier is used in combination with a light concentration of sodium hypochlorite (bleach).

Filtering Water

The purpose of water filtration is to remove impurities, both organic and inorganic. Unlike purifying, filtering does not leave organic remains or a chemical taste.

Distilling Water

Distilling produces safe, clean water and adds redundancy to your water preparedness. The American Red Cross recommends it as one option for obtaining fresh water in a disaster. During distillation, water is boiled and changed into steam. As the steam cools, it condenses, forming fresh water, which is then collected.

While distillation effectively decontaminates water, it requires heat and produces results slowly. Hard water can clog distillers and necessitate frequent cleaning. The effort, inconvenience, time, and cost of equipment and fuel can make distilling a challenge.

Plus, if your water is contaminated with chemicals like chlorinated hydrocarbons or other organic chemicals that have a boiling point lower than water, distilling will not be effective. To remedy this problem, these chemicals can be dissipated and minimized before distilling by boiling the water for several minutes before attaching the still.

Survival Still

The nonelectric Survival Still was originally created to help people in Third World countries. This carefully designed stainless-steel still can be used with several different heat sources but requires monitoring and produces fresh water slowly.

It requires two household pots to boil and cool the water during distillation. It can produce from one to two quarts of fresh water per hour depending on the temperature of the heat source and how cool you can keep the upper pot, which causes the condensation. It can be purchased separately or with pots.

FIVE THINGS YOU CAN DO NOW

1 Determine what kinds of pathogens are most likely to affect your drinking water during a crisis.

2 Determine what kind(s) of water filters your family would ideally use.

3 Purchase an inexpensive personal water filter for each member of your family.

4 Obtain and store prefiltering materials, such as cloth, nylon mesh, coffee filters, or paper towels.

5 Purchase a group filter best suited to your family's needs.

Water Filtration Basics

What Should You Look for in a Water Filter?

A good microfiltration system will remove any suspended microscopic organic matter. Most filters will remove bacteria and protozoan cysts, but only a few will remove viruses, which are much smaller.

A good system will also produce a reasonably fast stream of water should you need pure water quickly. Various filters may use a pump, squeeze-action, or gravity-fed system. Some use more than one method.

Consider whether the filter can be cleaned and reused or if it is disposable, and whether it will filter enough water to meet your needs before the filter clogs. Some filters also have a charcoal component or other method that removes bad tastes and odors.

Viruses harmful to humans are not generally found in water supplies in North America, but in a crisis, if your water is contaminated with human waste, there is the potential for viral contamination. As you look for a filter, consider any likely pathogens you'd need to remove.

> **QUICK CHECK**
>
> ## What to Look for in a Filter?
>
> ✓ Does it remove bacteria and protozoa?
> ✓ Do you need one that removes viruses?
> ✓ Does it improve taste and odor?
> ✓ How easy is it to use?
> ✓ Does it have an adequate flow?
> ✓ Does it filter quickly enough?
> ✓ Does it pump, squeeze, or use gravity?
> ✓ Can the filter be cleaned and reused or replaced inexpensively?
> ✓ Are the components sturdy?
> ✓ What is its cost versus value?

A Two-Step Solution

If you anticipate you not needing to filter viruses, choose a personal filter that targets bacteria and protozoa. As a backup, have sodium chloride tablets or a UV light to kill viruses if the need arises.

Prefilter Water

Since filters are susceptible to clogging, if you want to extend the life of your filter, prefilter water to remove any visible debris. Use a clean cloth, nylon stocking, fabric, paper towels, or coffee filters.

Some filters come with a prefiltering mechanism that saves you time and hassle. However, this option usually increases the cost of the filter.

Hollow-Fiber Filters

Fiber filters, also called hollow-fiber membranes, use state-of-the-art technology developed by kidney-dialysis research. They are extremely effective in increasing the filtering surface area to allow a high-flow rate while filtering out microscopic pathogens. A big advantage of these filters is that they are cleaned by backwashing. Clean them with distilled or fresh water as soon as possible after use, and be careful not to let fiber filters freeze as it will damage them.

Ceramic Microfilters

The most effective ceramic filters have a submicron pore size, and depending on that pore size, they can filter out cysts, bacteria, and even viruses. They can also remove radioactive solids, asbestos, and fiberglass fibers but not dissolved minerals, toxic industrial chemicals, or salts, nor will they purify brackish water or desalinate water.

Activated Carbon Filters

Carbon filters improve the taste and odor of water. Their porous, honeycombed internal structure is highly adsorptive, which means molecules and ions naturally adhere to it. The fresher they are and the more carbon they contain, the more effective.

They remove chlorine, chloroform, hydrogen sulfide, THMs (trihalomethanes, which are chlorine by-products), and other hydrocarbons. Carbon filters also remove or reduce concentrations of organic chemicals, including pesticides, herbicides, industrial chemicals, dissolved organics, heavy metals, and some trace minerals.

Bacteria and viruses are only partially removed by carbon media alone, and carbon filtration is not effective against fluoride, nitrates, salts, or asbestos fibers.

Types of Water Filters

Water-filtration technology has improved greatly in the last ten years and will likely continue to improve. While there are many filters that will do a good job, you should consider three types of water filters for preparedness: personal, small-group, and freestanding.

Personal Water Filters

Personal water filters are small and portable. Although they are often designed for recreational or international use, they are also ideal for emergency preparedness and emergency-evacuation kits. They are available as straws, squeeze tubes, bottles, and compact filters. The ones pictured are the Water Basics Redline Filter Bottle and the Sawyer Squeeze Water Filter.

Ideally, every person in the family should have their own personal filter.

Since they are often point-of-use filters, it is important to take precautions against contaminating the filter parts that hold the clean water or that go in your mouth.

Every person in the family old enough to understand how to keep from contaminating the filter should have their own personal filter.

Personal filters are quite cost-effective. Most filters remove bacteria and protozoa. Those that filter out viruses are usually more expensive and have a slower flow. Some filters have a dual-filtering system that uses charcoal to eliminate some chemicals and improve odor and taste.

Filters vary in how fast they filter and how much they can filter before clogging. If filters need to be cleaned, you'll want to know how difficult and how effective that cleaning is. Generally, ceramic filters are cleaned by scrubbing and microfiber filters are cleaned by backwashing. A few filters are disposable or have cartridges that need to be replaced.

Small-Group (Family) Portable Water Filters

You should have at least one small-group portable filter and preferably a backup unit. There are quite a few excellent filters in this category. They may use a pump-action system or be gravity-fed and sometimes both. Be sure to also have backup replacement filter cartridges if needed.

What to Look for in a Family Filter

Look for a filter that will produce enough water for your family's needs. How easy is it to use, and how quickly will it filter water? Is it adaptable for a variety of filtering needs? Can you filter into a range of containers? Does it require filter replacements? Can the filter be easily and effectively cleaned? Does it have a prefilter that removes sediment and larger particulates? The filter pictured is the highly rated MSR Guardian Purifier.

You should have at least one small-group portable filter.

Freestanding Home Water Filters

Finally, you may want to have a large freestanding filtration system you can use in an extended crisis or off-grid situation. These filters are not very portable and are independent of the water system in your home. They usually fit on a countertop.

What to Look for in a Freestanding Filter

Look for a filter that both filters and purifies, ideally without chemicals. It should filter nonpotable water, e.g., streams, lakes, rainwater, eliminating living and nonliving materials, a wide range of inorganic substances, and viruses, bacteria, and protozoa.

Consider a gravity-fed drip system that minimizes the amount of work it takes to obtain filtered water and one that filters quickly enough for your needs. Look for a filter that has a long life and does not need frequent replacing. Think about not just the cost of the unit but cost per gallon of water filtered. The freestanding filter pictured is the Big Berkey Water Filter.

SECTION 3

Storing Food for a Crisis

10 Getting Started with Your Food Storage Plan

If you have not already done so, consider how comprehensive you want your food storage plan to be. What are your perceived needs? In other words, what are you storing food for? What level of commitment can you dedicate to preparedness and food stockpiling? Would you like to have backup food storage but do not have a lot of extra resources? Is acquiring a reserve supply of food something you can work at over time? This chapter will help you decide what level of preparedness will fit your needs.

Three Levels of Food Storage Preparedness

How and why you prepare is very personal. It's helpful to think about preparedness in three levels, or stages. If your concern is about temporary disruptions or possible short-lived personal challenges, and if you want your meal plans to include foods you already use, you might select the Everyday Level of preparedness.

If you are interested in having up to a year's supply of simple, basic foods, take a look at the Basic Level. If you feel an urgency to be fully prepared to the extent possible for any major natural disaster or breakdown of society and are strongly committed, consider the Advanced Level.

Everyday Level

The Everyday Level of food storage preparation focuses on short-term needs. It is balanced between foods you use daily and a few basic foods you store and continually replenish. It includes fruits, vegetables, eggs, dairy products, and meats and may include basic foods, like beans and grains. Use the food storage plans in chapter 15 to help achieve this level of preparedness.

Basic Level

The Basic Level focuses on essential storage foods and is minimal but adequate. This is a good place to begin if you want to keep it simple or if budget is a concern. It can be done inexpensively by purchasing foods in bulk and preparing them for storage yourself. The

---

FIVE THINGS YOU CAN DO NOW

1 Decide which type of food storage plan will fit your needs–Everyday, Basic, or Advanced.

2 Use a food storage calculator. Evaluate the results.

3 Make a list of the foods you already use that are suitable for storing.

4 Use the Quick Check to evaluate your existing food storage.

5 Decide on one food you know you want to have in your storage and purchase enough of it for one month.

7-Plus Basic Plan described in chapter 16 is a Basic Level plan. It can be combined with the Everyday Level and upgraded as your resources allow.

Advanced Level

The Advanced Level of preparedness starts with intensive stockpiling. Besides basic foods, it includes storing canned and freeze-dried meats, along with canned, dehydrated, or freeze-dried fruits and vegetables to approximate fresh food. Chapter 17 gives you detailed support for this level of preparedness.

You sustain this level of preparedness by supplementing stored foods with lifestyle modifications and self-reliance skills. This can include growing and preparing much of your own food, the cost depending on the lifestyle changes you adopt. Your overall living costs may actually decrease. Later chapters will discuss the attitudes, skills, and equipment necessary for making lifestyle changes.

Chapters 15, 16, and 17 will teach you how to create a comprehensive food storage plan, show you how to customize that plan, and give you the flexibility to meet your individual requirements.

PERSONALLY SPEAKING

 My ideal plan includes all three levels of food storage and takes place in phases. To begin with, we stockpile foods we normally eat. This food is in my pantry and freezer. It is convenient to use and will see my family through about three months. It means I am not panicking if there is a local or regional disruption and that I can usually figure out something to fix for dinner without running to the store or relying on a box of mac and cheese.

I look at the second phase as food insurance. This is where we add to our three months of regular food and begin gathering basic foods that will keep us alive in the event of a long-lasting crisis. Certain foods, like whole grains, beans, sugar, and powdered milk, last a long time if stored properly. I keep them in the storage room in my basement. During this phase, it is important to acquire specialized cooking equipment and develop the skills to use these foods.

It is not realistic for most of us to store a whole year's worth of the kinds of foods we eat daily. The drive-through at fast-food restaurants and pizza delivery are hard to duplicate. And it is just as challenging to reproduce the garden-fresh produce and fresh eggs, meat, and cheese we enjoy in our normal diet. Yet in a lengthy crisis, we want to be able to take care of our families. That's where phase three comes in. I like it because it helps me plan to match, as close as possible, the kinds of foods we normally eat. It includes the basic foods we stored in phase two, plus things like fruits, vegetables, and meats. We also grow a garden and fruit trees to add fresh foods. This phase requires a total preparedness mindset along with accompanying skills and lifestyle changes. Though it takes time and considerable effort to achieve this level of preparedness, to the degree we can accomplish this phase, we will enjoy a feeling of self-reliance and peace of mind.

A Personalized Plan

Because no two individuals or families are identical and situations vary considerably, the best plan is the one that fits your unique goals, needs, and circumstances. There are no set rules and no easy, one-size-fits-all method. You will want your plan to be adequate, balanced, and consist of foods your family enjoys. Use worksheet 10.1 on page 66 to help you evaluate your food storage plan. The worksheet is also found as a downloadable PDF on our website, CrisisPreparedness.com.

Planning from a List

Many preparedness experts will provide lists of essential food items to store. These lists usually consist of basic, economical foods well-suited to long-term storage. They can be a good starting place and may sometimes be improved upon with only minor changes or additions.

A list generated by someone else will not likely fit your specific needs.

Although it is tempting to simply choose a list and be done with it, someone else's list of specific items is not likely to fit your family's needs exactly, plus it can be confusing to even choose a list. Recommended items are often inconsistent from one expert to the next, and the suggested amounts of the same items often vary.

Food Storage Calculators

If you've explored food storage options, you've come across preparedness sites with food storage calculators that require you to enter the number of people you plan to feed and the number of months you want the food to last. These calculators then determine the amounts of specific foods you'll need.

A basic list of generic storage foods is built into the calculator. The types and amounts are based on the expertise and opinions of the calculators' creators. Using a basic list, the calculator multiplies the number of people by the number of pounds of each specific food per person. Some break it down by age to compensate for different caloric needs, but the results are only as reliable as the assumptions and the lists of foods programmed into them. Using a calculator has the same shortcomings as using a list.

> **QUICK CHECK**
>
> ### Evaluating a Food Storage List
>
> ✓ Does it fit your family's situation?
> ✓ Does it include all important nutrients, including oils?
> ✓ Does it account for higher nutritional demands during a crisis?
> ✓ Is there enough variety to prevent menu fatigue?
> ✓ Does it include foods that your family normally eats and likes?
> ✓ Does it require specialized equipment?
> ✓ Does it require specialized cooking skills?

Another kind of calculator is found on commercial food storage sites that sell dehydrated meal packages. These calculators are designed to sell their products and are not very useful unless you intend to purchase all your food from that company, which is not recommended.

7-Plus Basic Plan Food Storage Calculator

We have designed a calculator based on the 7-Plus Basic Plan, which takes into account all the variables discussed in the 7-Plus Basic Plan section of chapter 12. It is available on our website, CrisisPreparedness.com.

Other Considerations

Table 10.1 gives you some additional things to consider as you create your food storage plan.

How and Where to Buy Food Storage

Acquire Preparedness Items Systematically

There's no need to buy items quickly or all at once. Simply add small amounts of foods to your stockpile on a regular basis. Use worksheet 4.4, "Master Shopping List," on page 31 to organize your purchases. You may want to acquire a month's supply of each item before increasing the amount of any one item, or you may purchase items when they are on sale. Use shelf life to determine desired rotation and a replacement schedule. Shelf life is discussed in chapter 11.

Table 10.1 Things to Consider as You Design Your Plan	
Plan for extra people.	Sharing with relatives and friends, unforeseen guests, a longer-than-expected crisis, or for barter
Plan for two years in the future.	Needs change over time.
Review and reevaluate at least yearly.	Harvest time might be the best time because of abundant supplies of food.
Plan for unique needs.	Infants, young children, pregnant or nursing women Those with allergies or chronic health problems

©Patricia Spigarelli-Aston

Shop for Quality and Economy

Because you'll seldom find a single source offering a complete selection of the highest quality items at the best prices, it pays to purchase from a variety of sources. Depending on where you live, you may be able to find local sources for food storage. Check with bakers, millers, grain brokers, feed dealers, and grocery wholesalers. Though you may have to repackage them for storage, purchase institutional sizes of dehydrated and other foods at restaurant supply stores, warehouse clubs, and similar outlets. Do not hesitate to ask for quantity pricing. Often, the most economical time to buy food is at harvest, when supply is at its peak. Be aware that closeouts may be offering last year's goods, so check expiration dates.

Online Purchasing

Many preparedness companies do much of their business selling online. They sell dehydrated and freeze-dried foods in cans and in Mylar bags inside buckets. They also sell grains and beans in cans, poly buckets, Mylar bags, and in bulk quantities by the bag. Do not overlook shipping costs, which can substantially increase the total price.

Purchasing through a Co-Op

One way to minimize shipping costs is to join or start a co-op and buy foods with others by the pallet or truckload. You might also consider becoming a dealer or distributor so you can save 20 percent to 50 percent off normal prices.

A Worksheet to Help You

Use worksheet 10.1 on page 66 to help you evaluate your food storage plan. For a downloadable PDF file, go to CrisisPreparedness.com.

✓ Yes	✓ No	**Worksheet 10.1** **Evaluate Your Food Storage Plan**
		Nutritionally Complete
		Are there adequate calories and nutrients?
		Does it consider age, weight, gender, and unique dietary needs?
		Enough Variety to Avoid Food Fatigue
		Are there different flavors, textures, and colors to add interest to meals?
		Does it include foods for treats and special occasions?
		Allows for Individual and Family Preferences
		Does it consider familiar tastes, likes, and dislikes?
		Are our family favorites included?
		Convenience
		Can meals be easily prepared?
		Do we have any specialized equipment that is needed?
		Stores Well
		Do the foods have a reasonable shelf life?
		Are the foods ones we will use and rotate?
		Space and Weight
		Do we have enough space to store it?
		Does it meet our portability and mobility needs?
		Affordability
		Does it fit within our budget?
		Availability
		Are the foods we desire available?
		Can they be easily replenished?

©Patricia Spigarelli-Aston. For a downloadable PDF file go to CrisisPreparedness.com

Principles of Successful Food Storage

It would be a tragic waste of time, money, and effort if you found your stored food unusable when needed. Follow the guidance in chapters 11 through 14 to help prevent this.

Using Your Food Storage

Store What You Eat and Eat What You Store

The ideal preparation is to store the kinds of foods you normally eat. For most, however, there's a considerable gap between what we normally eat and what we store. We either store large quantities of commodities like wheat, rice, corn, beans, and powdered milk, or we store a prepackaged commercial supply put together by a food storage retailer. Either method can be problematic.

> **The first principle of food storage is to store what you eat and eat what you store.**

Unfamiliar Foods

A common belief is that people will eat anything if they are hungry enough. While there are historical and modern accounts of people surviving by eating things they normally would not consider eating, many people cannot do this. Some will go without, even starve, rather than eat unfamiliar or distasteful foods. This is particularly true of the young, ill, and aged. The stress of a crisis tends to increase rejection of strange foods.

Sudden Diet Changes Can Cause Distress

Your digestive system is used to your regular diet. It takes time to adjust to changes. In a crisis or stressful situation, a dramatic change in diet can cause additional stress as well as gastrointestinal problems

Many of the storage foods suggested in this book may be different from your present diet. So how do you add storage food to your diet? What is the solution? First, store foods that are as close to what you normally eat as possible. Second, acquaint your family with the foods you plan to store and help them develop a taste for them.

FIVE THINGS YOU CAN DO NOW

1 Introduce to your family's diet one new food you want to store.

2 Try one new recipe using food storage items.

3 Become acquainted with shelf life. Read labels for shelf life.

4 Place a permanent marker in your food storage area and write dates on the foods you purchase.

5 Place newly purchased foods behind older foods on your shelves.

Introducing Unfamiliar Foods

The ideal situation is to use your food storage in your regular menus. Although you'll still eat foods not in your storage, you should make adjustments to bring the two closer together. Gradually begin introducing unfamiliar storage foods into your meal plans. Begin with the foods most like your normal diet so changes are less noticeable. This is better than attempting to drastically alter cooking and eating habits all at once, especially with children. It also minimizes digestive discomfort and gives you time to develop the necessary cooking skills. Do not store foods in large quantities until you are comfortable eating them.

PERSONALLY SPEAKING

 I wish I had some great secret to successfully introducing new foods, especially storage-friendly foods. I have observed children who'll eat what they are given without complaint and children who'll balk at almost anything. I have watched my adult children, particularly my daughters-in-law, with admiration as they've taught their children how to enjoy a variety of foods. They seem to be most successful when they don't make it a big deal and are just quietly persistent. They serve healthy foods with confidence and a mindset that it's not negotiable. One other pretty significant thing is that their husbands are supportive.

Develop Cooking Skills for Using Storage Food

If you store foods you do not normally use, you'll need to develop cooking skills to prepare acceptable and appetizing meals using them. You may also need specialized tools and equipment to make the task easier.

One of the biggest challenges in using storage foods is to present meals that are interesting and visually appealing. Find and develop recipes for the items you store, especially for those unfamiliar foods, and purchase specialized cookbooks for unique food storage items as you buy them. Use herbs, seasonings, and condiments to make the foods more closely resemble your regular diet.

Food Intolerances and Allergies

Unfortunately, some of the most common storage foods are associated with food intolerances and allergies. People with gluten intolerance or celiac disease cannot tolerate the gluten protein found in wheat, rye, and barley. Intolerance to lactose in milk products and egg allergies also cause problems. Even though you may need to adjust which items you store, the principles are the same.

Table 11.1 Shelf Life De-Mystified	
"Sell by"	Date manufacturer wants store to remove product from shelves to ensure best quality
"Best if used by"	Date manufacturer recommends to consumer for best flavor and quality
"Use by"	Last date recommended for using the product (usually for perishable items, such as meat and dairy)
Closed or coded dates	Tell when the product was manufactured. Deciphering the codes can help you make informed decisions about the longevity of the product.
Life-sustaining shelf life	Sometimes known as "emergency shelf life," this is the actual length of time a food can be stored and still be edible and offer nutritional benefit.

Understanding Shelf Life

What Is Shelf Life?

No food remains entirely static during storage, and all foods eventually become inedible. The length of time a food takes to become inedible is called its storage, or shelf, life. Understanding shelf life will help you make informed decisions about how long to safely store foods. Manufacturers use dates to help you determine shelf life. Table 11.1 on page 68 will help you understand the definitions and limitations of product dating.

Recent Data on Shelf life

Over the last decade or so, research has verified that many storage items, particularly those with low moisture content, can be stored safely and with nutritional integrity for at least thirty years. Research has also shown there can be significant differences based on the quality of the packaging, whether it is maintained in an oxygen-free environment, and other storage conditions. The shelf life of foods packaged for grocery-store shelves is typically shorter. Canned goods have a reasonably long shelf life, while items packaged in boxes and bags have a shorter shelf life. Fresh produce, meat, and dairy have a very short shelf life.

Table 11.2 Ways Foods Can Be Unsuitable for Eating	
Loss of Nutrients	Loss of Palatability
· Loses vitamin potency · Proteins break down · Minerals oxidize	· Rancidity · Crystallization · Colloidal modifications · Oxidation, fading pigments
Spoilage and Contamination	Loss of Functionality
· Bacteria: botulinum toxin, salmonella · Enzymes from fungi, yeasts, and molds · Weevils, mites, rodents · Environmental odors and dust	· Loss of leavening ability · Loss of thickening or whipping ability · Loss of gelling or set-up ability

What Makes Food Unsuitable for Eating?

Although shelf-life figures for foods are easy to find, various sources often disagree about the shelf life of various foods. So how do foods become unsuitable for eating? The ability to predict shelf life is affected by the four variables summarized in table 11.2. You'll want to consider each of these as you make choices about a food's potential shelf life.

Factors That Determine Shelf Life

Remember that a stated shelf life is only an estimate and foods may be safe, palatable, and nutritious beyond these dates. Even those recommended in this book are only based upon the best currently available data modified by personal experience with various conditions and climates. Actual shelf life is dependent on unique conditions, such as packaging,

---

QUICK LOOK

Factors Affecting Shelf Life

- Quality of the product at time of storage
- Processing standards
- Packaging materials
- Storage conditions: heat, light, moisture, oxygen levels

---

temperature, humidity, contamination, etc. Periodic sampling will help you determine if a food has reached its shelf life.

Food Weaknesses

Each component of food—fats, proteins, carbohydrates, vitamins, minerals, and color pigments—undergoes its own form of deterioration. Individual foods have unique physical and chemical properties that give them a unique storage potential. Storage also tends to magnify whatever weakness there is in the quality of the original food. For example, foods high in fat content quickly become rancid, and foods with bright or dark pigments fade.

Processing Methods

Among the variety of methods used to preserve food, each process has a significant influence on product stability and shelf life, affecting nutrient retention, particularly water-soluble and heat-susceptible vitamins. Chapter 14 discusses the pros and cons of different food-preservation methods.

Packaging Methods

Quality of packaging also affects shelf life. For example, a thin plastic bag or paper sack will be less effective in protecting foods than a metal can coated with food-grade enamel and sealed with oxygen absorbers.

Storage Environment

The principal factors affecting shelf life are the conditions in which it is stored. Packaging methods and storage environment are both discussed in chapter 12.

Food Spoilage

Recognizing Food Spoilage

In a crisis, food poisoning is most likely to occur due to a lack of refrigeration. Since medical care will likely be overburdened during a crisis, be extra cautious about potential food spoilage.

The following precautions will help keep your family safe.

QUICK CHECK
Look for These Signs of Spoilage
✓ Broken seal
✓ Bulging can or jar lid
✓ Gas bubbles
✓ Foamy or spurting liquids
✓ Visible molds or slimes
✓ Emulsion separation
✓ Cloudiness
✓ Discoloration
✓ Off odors
✓ Sour or bitter taste
✓ Soft or mushy foods

- After opening an item, carefully examine it for any signs of spoilage before tasting it.

- Smell it. If it smells acidic, cheesy, fermented, musty, putrid, sour, or like rotten eggs, it's probably spoiled. If in doubt, remember that heating, especially in a covered pan, brings out the characteristic smells of spoilage.

- Lastly, taste a small portion of the food. Off flavors do not always indicate spoilage but should be considered suspicious.

Salvaging Spoiled Food

Some foods can be salvaged. Mold can be cut off firm fruits and vegetables, such as apples and cabbage, as well as hard cheeses, dry-cured hams, and hard salami. If mold is found on butter, dairy products, jams and jellies, breads, grains, or soft fruits and vegetables, the product should be discarded. The problem here is that mold grows beneath the surface and may be invisible.

The bacteria that causes botulism can be present even without any signs of spoilage. Always use approved USDA guidelines for home canning. (See page 431 in the resource section.) As an extra precaution, boil home-canned meats and vegetables for ten minutes before tasting. The high acidity of fruits and tomatoes canned with additional acid prevents botulism growth. If you suspect spoilage, destroy or bury the food immediately so it's not accidentally eaten by children or animals.

Radiation will not harm a food unless actual fallout particles come into contact with it. Clean the outside of intact cans before opening them if fallout contamination is suspected. Fruits and vegetables can be peeled and the peeling discarded.

Rotating Your Food Storage

Rotating your food storage ensures that foods are as fresh as possible, are used in your daily diet, and are sampled on a regular basis. It also minimizes loss of food value and flavor as well as chances of loss due to spoilage, contamination, or damage.

Rotating your food storage keeps your supply as fresh as possible.

Keep Easy Access to Food Storage

To make rotating your storage easy you should have quick access to the storage area. One option is to have a pantry close to the kitchen and a secondary storage to replenish the kitchen pantry.

Rotation Plans

In general, you want a rotation plan where you place fresh items at the back and bring the older items to the front. You can develop your own system to fit your circumstances, but it should take into account the approximate shelf life of the foods you store and rotate all of them within the recommended periods.

Each time you purchase or store an item, label and date the container with the month and year. Also mark the replacement dates on the label and separate the foods into different locations according to dates. Systematically organize your storage with rows or sections of shelving, depending on your configuration.

PERSONALLY SPEAKING

We have a dedicated room in our basement lined with shelves for food storage. It is unheated and is the coolest place in the house. Besides food, we store canning and food storage equipment, storage supplies, and boxes of empty canning jars. We also keep a Sharpie pen and stick-on labels handy for labeling the things we store. One side of the room contains large buckets of grains and legumes, as well as containers of water. The other wall is lined with shelves dedicated to foods we regularly use and keep in constant rotation. It is our home "grocery store." We stock up on items when they are in season or on sale. Our menus include the food stored in our storage room. When it turns out we have not been using an item, we donate it to the local food bank while it is still good.

Rotation Systems for Cans

If canned items are a big part of your diet, you may also want to build or buy a rotation system for your food storage pantry. You can find many plans for building your own with a quick internet search. Commercial systems come in a variety of shelf sizes and as entire wall units. These slant-shelf units allow you to insert new cans at the back and retrieve the oldest cans first. Look for a system that allows you to adjust  it to fit different can sizes. Shelf Reliance manufactures a series of free-standing shelving systems and smaller units that sit on existing shelves. FIFO Can Trackers (pictured) is an example of a smaller systems that help you organize and rotate canned foods.

Balancing Quantities

If it becomes necessary to depend entirely on your storage, you may discover another problem. Because a plan does not always match normal usage, some items may be used at a different rate than estimated. This is especially true of convenience foods, "treat" items, or other foods stored in less-than-normal quantities.

For example, if you have chosen to allocate only one-quarter the amount of butter you normally use in a year, and if your family uses it at the normal rate, you'll run out in just three months!

Allocating Food Weekly or Monthly

To keep from using up a certain kind of food, divide the total amount of each food into weekly or monthly allowances and ration that food, planning your menu from that allowance. It's easiest if you regularly transfer the weekly inventory basis to a different location from the rest of your storage. As you use your storage and learn which foods your family prefers, you can increase the amounts of those foods so you don't suffer shortages.

Inventory Control

Once you have accumulated storage, it is important to track what you use and how much you have on hand with an accurate written record you keep current. Table 11.3 summarizes what you should consider for inventory control.

You can use either manual or electronic methods for inventory control, several of which are listed below, but be sure to include the following: item name, amount desired, amount purchased, amount used, purchase date, shelf life, replacement date, item size, and storage location.

The Inventory Planning Checklist introduced in chapter 4 can help you get started with your storage inventory plan. Keep up to date with inventory and rotation by using one of the methods listed in table 11.4.

Table 11.3 Inventory Control Basics	
Goal	How much do I need?
On hand	How much do I now have stored?
Replacement date	What is my rotation schedule for this item?
Location	Where will it be stored

Table 11.4 Manual Inventory Methods	
Checklist Methods	· Laminate or use a plastic sleeve for your Inventory Planning Checklist (chapter 4) and attach it to a clipboard. · Use a washable marking pen to track items on a worksheet.
	· Keep a sheet attached to a clipboard or bulletin board near the storage area. · Write the names of items removed from storage and use the sheet as a shopping list to replenish.
	· Use a dry-erase board. · Write the names of items when removed and erase when replaced.
Card File Method	· Set up a card file using 3 x 5 or 4 x 6 cards, with separate cards for each item or category. · Regularly note the items in inventory and record on each file card any items removed · Periodically check inventory and replenish.
Notebook Method	· Keep inventory sheets in a loose-leaf binder. · Keep track of individual items with separate sheets for each category and lines for each individual item.
Inventory Card Method (While this takes some time to set up it is very visual and each member of the family can be taught how to remove the cards as items are used.)	· Make an inventory card to represent units of items stored and a poster board with a pocket on it for each type of item in your storage. · Whenever an item is used, take the corresponding card and place it in a pocket or envelope to keep until it's time to buy the replacement item. · Take a quick inventory of each item by simply counting the cards. · You may want to color-code these cards and list relevant data for the item on the card, such as size, amount, date purchased, etc.

Interactive Spreadsheets

An interactive electronic spreadsheet is one way to organize the foods you store. Will a predesigned spreadsheet work for you, or can you design your own spreadsheet?

Food Storage Apps

Smartphone apps that help you organize your pantry and refrigerator usually have a function for creating a list and a feature that scans product barcodes so you can track your inventory. Some also give you recipe ideas for the ingredients you have on hand. These features can be useful if you are willing to meticulously scan barcodes and if the foods you want to store do indeed have bar codes.

With the Home Food Storage app, available only for iPhones, you can scan foods, create a food storage shopping list, organize foods into groups, and customize amounts.

Food Storage Conditions and Containers

When it comes time to use your food storage, you want it to be usable. How and where foods are stored greatly affects shelf life. Although deterioration cannot be stopped entirely, proper packaging and storage conditions can minimize it.

Factors in the Food Storage Environment

The ideal storage location is cool, dry, dark, and clean. The best containers maintain an oxygen-free environment and keep out insects and rodents. Each environmental factor not only has its own effect, but when combined with other factors, the rate of deterioration is multiplied. Refer to table 12.1 on page 76 for a summary of how to create an optimal storage environment.

Heat

Heat is a powerful destroyer of food quality. The rate of a simple chemical reaction roughly doubles with each 18° F (10° C) rise in temperature. Because of their biochemical and enzymatic nature, reactions in food may increase at a greater rate. These reactions affect the color, flavor, texture, and nutritional value of food. Heat increases the growth rate of both microbes and insects. And while it can dry out certain foods, it can also generate moisture inside containers, increasing microbial activity.

> To assure quality in stored food, maintain a temperature of less than 70° F.

Moisture

A certain amount of moisture is necessary to maintain quality in most foods, but too much moisture—above about 10 percent—promotes the growth of yeasts, molds, bacteria, and insects. It also increases nonenzymatic browning and the breakdown of fats, provides a medium for chemical reactions, and corrodes containers. Foods properly prepared for

---

FIVE THINGS YOU CAN DO NOW

1 Dejunk, clean, and organize to make room for food storage.

2 Purchase or build shelves for food storage.

3 Purchase several storage containers and a bucket wrench.

4 Calculate how many square feet you will need for the 7-Plus Basic Plan for your family.

5 Install dowels or strips of wood on shelves to contain canned goods during an earthquake

---

long-term storage contain the correct amount of moisture and keep that moisture at a constant level.

Light

Although light inhibits the growth of molds, it is best to shield your storage from both artificial light and sunlight. Ultraviolet light is the primary culprit in damage to flavor and texture. It increases rancidity and destroys vitamins. Highly pigmented foods are especially susceptible to bleaching and discoloration. Shield light-penetrable containers by keeping them in boxes.

Oxygen

Oxidation destroys vitamins, alters pigments, and promotes decomposition. Fruits, vegetables, and foods high in oils and fats are especially susceptible to the negative effects of oxygen. Even small amounts of oxygen will oxidize the fat in nearly all foods and produce rancid flavors and odors.

Oxygen is essential for the growth of fungi, yeasts, molds, and microbes. Also, weevils and other insects that contaminate flour and grains require oxygen to live and reproduce. Oxidation is increased by slicing and grinding foods and by the presence of heat and light.

Odors and Dust

Table 12.1 Creating the Ideal Storage Environment	
Heat	· Maintain a temperature less than 70° F (21° C) and, ideally, close to 40° F (5 ° C). · Avoid freezing to prevent breakage of containers and damage to textures and flavors.
Moisture	· Create a dry storage environment where containers will not corrode. · Use suitable packaging to maintain the right amount of moisture.
Oxygen	· Use airtight containers and fill them to the top. · Process with oxygen absorbers, a vacuum, or an inert atmosphere to reduce the effects of oxygen.
Light	· Lmit exposure to light with opaque containers. · Store in dark closets. · Cover light-penetrable containers with opaque material.
Odors and Dust	· Store food away from strong odors, chemicals, and dust.
Insects and Rodents	· Store food in resistant containers. · Use proper environmental controls.

Some foods, particularly grains, flour, and milk, absorb odors from the environment and should never be stored near substances with strong odors, such as garlic, onions, soaps, and petroleum-based products, like gasoline, oil, kerosene, paint, paint thinners, and pesticides.

Insects and Rodents

To inhibit insect infestation, maintain a cool, oxygen-free and moisture-free environment. Use durable containers to prevent rodents from damaging food storage.

Food Storage Containers

A suitable container will maintain proper moisture and natural food aromas while keeping air, light, dust, foreign odors, insects, and rodents out. Such containers can be made from a variety of materials in different shapes and sizes. Each

A good container keeps moisture and flavor in and additional moisture, air, light, dust, foreign odors, insects, and rodents out.

type of material has its benefits. You'll likely use different containers for different needs. Food storage containers are not difficult to acquire. Several companies that specialize in food storage containers are listed on page 425 in the resource section. Table 12.2 on page 78 compares the advantages and disadvantages of common storage containers.

Polyethylene HDPE Buckets

Rigid, high-density polyethylene (HDPE) food-grade plastic buckets, often called poly buckets, are practical for long-term storage. Always select food-grade plastic containers that are BPA and phthalate free. Look for an SPI resin identification code of 1, 2, 4, or 5 stamped on them.

Round buckets come in 5-, 6-, and 6.5-gallon sizes, ideally with wire handles. Square poly buckets come in 2- and 4-gallon sizes and can be stacked to use space more efficiently.

You can get used buckets from sources such as bakeries, school cafeterias, restaurants, and ice cream stores, but make sure they are FDA-approved, have contained only foods, are in useable condition, and are thoroughly cleaned and dried before use.

Bucket Lids

Lids for round buckets come in either snap-on or screw-on styles. Look for snap-on lids that have gaskets for a tight seal. The screw-on gamma lid pictured has two-parts—a ring with a gasket that fits over a 12" bucket, and a twist top that screws onto it to form an airtight seal. This is especially useful if you plan to open and close the container often. These lids are available in several colors, which makes it nice for organizing your storage by color.

Bucket-Lid Wrench

A bucket-lid wrench makes for easy opening, and a rubber mallet is handy for securing snap-on lids to ensure an airtight fit.

Polyester Foil (Mylar) Bags

Polyester foil storage bags, commonly known by the trade name Mylar, are made from layers of polyester, plastic, and foil. Mylar bags, when used with oxygen absorbers, significantly increase shelf life for long-term food storage because of their moisture- and oxygen-barrier properties.

However, Mylar bags are susceptible to puncturing. One way to prevent punctures is to use a Mylar bag as a liner in a poly bucket or place several smaller bags in a bucket. A heat sealer is needed to seal Mylar bags. This process is described in chapter 13.

Bags made from Mylar foil are the top choice for storing many items because they are moisture and oxygen resistant.

Table 12.2
Advantages and Disadvantages of Common Storage Containers

Container	Advantages	Disadvantages
Polyethylene HDPE Bucket	· Lightweight · Stackable · Resistant to breakage · Will not rust or corrode · Seamless for easy cleaning · Airtight, with resealable lid · Resistant to insects and rodents · Variety of sizes	· Container and contents susceptible to UVA deterioration · Permeable to air over time · Absorbs odors of foods stored · Bulky, less efficient use of space · Moderately expensive
PETE or PET Containers	· Lightweight · Portable · Resistant to breakage · Will not rust or corrode · Long-lasting given right conditions · Resistant to insects and rodents · Variety of shapes and sizes · Free when recycled	· Container and contents susceptible to UVA deterioration · Permeable to air over time · Difficult to thoroughly clean for reuse · Sizes are usually too small for large quantities of food · Moderately expensive if purchased
Polyester Foil (Mylar) Bags	· Lightweight · Portable · Impermeable to air, moisture, and light · Use space efficiently · Flexible and stackable · Long-lasting given right conditions · Inexpensive · Variety of sizes	· Susceptible to holes and tears · Susceptible to rodents · Requires sealing equipment for DIY
Metal Cans	· Strong and unbreakable · Impermeable to air, moisture, and light · Stackable · Long-lasting given right conditions · Resistant to insects and rodents	· Susceptible to rust and corrosion · Relatively heavy · Bulky, less-efficient use of space · Expensive, not reusable
Glass Jars	· Portable · Impermeable to air and moisture · Will not rust or corrode · Easily sterilized · Reusable · Long-lasting given right conditions · Variety of sizes	· Breaks easily · Contents susceptible to UVA damage · Practical only for small quantities · Relatively heavy · Expensive, initially

©Patricia Spigarelli-Aston

PETE or PET Plastic Containers

These plastic containers are identified by the #1 inside the triangle usually found on the bottom of the container. PETE containers come in a wide variety of shapes and sizes and can be purchased with lids with gaskets for an airtight seal. PETE containers are also available for free in the form of soda pop and juice bottles, and snack and condiment jars.

Metal Containers

Metal cans are an ideal storage container as they are excellent barriers to air, moisture, and light. They are available in a wide variety of sizes and shapes and are used for both wet- and dry-pack food processing.

Glass Containers

Glass jars are the primary container for home-canned produce and meats. These jars are impermeable to air and moisture but must be kept in a cool, dark place and protected from breaking. Use quart or half-gallon sizes for storing dry bulk food.

Although jars can be moderately expensive, they're reusable. To save money, purchase them at thrift stores or get them from friends no longer using them. The biggest drawback to jars is that they let in light and break easily, but these two problems can be overcome somewhat by storing the jars in their original cartons. It can also be challenging to seal them properly to keep them airtight for long-term storage.

JarBOX

JarBox is a plastic canning-jar caddy that securely holds twelve jars and comes in both pint and quart sizes. It is a nice organizational tool and is especially useful in areas prone to earthquakes.

Determine the Right Size Container

Food begins to deteriorate once its container is opened. Perishable foods, such as canned meats, require refrigeration or need to be used within hours after opening. The ideal sized container will hold no more than is needed for a meal.

The appropriate container size for dehydrated foods is one where the contents will be used up prior to their open shelf life being exceeded. For example, if an item has an open shelf life of three months and you need twenty pounds for one year, the maximum amount per container would be five pounds. Check your list of food items and determine appropriate container sizes for each.

Table 12.3 Size Considerations	
Advantages of Small Containers	Advantages of Large Containers
· Reduces the possibility of wide-spread contamination · Minimizes contents becoming stale after opening · Can be stored in smaller spaces · Easier to rotate · More portable	· Much more economical · More efficient use of space · Easier to organize · Fewer containers to manage

As you can see in table 12.3 on page 79, smaller containers have quite a few advantages, though large containers are more economical, especially for items not costing much per pound. Wheat, for example, may be purchased in six-gallon poly buckets for about half of what it costs in #10 cans.

Where to Keep Your Food Storage

Food Storage Areas

Your food storage areas should be orderly and readily accessible to accommodate rotation and inventory control. You may also want to spread your storage among several locations to decrease the chance of it being affected all at once by flood, earthquake, theft, or other mishaps.

Ideally, you want two different types of storage areas—a large area with low humidity for storing preserved foods, such as canned and dehydrated foods, grains, legumes, milk, sugar, salt, etc.; and another area with high humidity for storing fresh produce (see chapter 28).

How Much Space?

The amount of space needed depends on how many people you are storing for, how long you are storing for, and the types of food you are storing. This book's 7-Plus Basic Plan (see chapter 16) and similar plans take about fifteen cubic feet per person for a year's supply—imagine a space three feet wide, five feet tall, and one foot deep. The Advanced Plan (see chapter 17) requires about thirty cubic feet. It will be less with air-dried fruits and vegetables and more with freeze-dried foods.

Creating Storage Space

Ideally, you'd dedicate a room or pantry to food storage. While finding space in a small home or apartment can be a challenge, with a little creativity and by reexamining how you are using the space you have, you can probably find space. Here are several ways you can make space:

- Store nonfood items in the garage to make space in the home for food storage.
- Take a good look at how you're using the space you have and rearrange it.
- Dejunk your spaces to make more room for food storage.
- Rearrange how cupboard, pantry, and closet space is used.
- Use the empty spaces behind furniture, on balconies, or under stairways.
- Make furniture, such as bookcases, beds, coffee and end tables, from storage containers.
- Partition a large room with a false wall, room divider, decorator screen, or curtain and use the space for storage.

Remember as you try to find space that food should not be stored at temperatures above 70° F (21° C).

Shelving, Racks, and Bins

You can either buy or build shelves. Most commercial shelves are three feet wide and twelve to eighteen inches deep. Invert shelves to create a lip around the edges. If you build your own, remember to leave two inches of headspace and use vertical supports at least every three feet. A one-inch strip of wood or wire, etc. can be placed at the outer edge for security.

Minimizing Earthquake Damage

During an earthquake, storage items will get thrown around and damaged or destroyed if they are not properly secured. If you live in an earthquake-prone area, you may want to use these simple precautions that can help protect your food storage:

- Use unbreakable containers as often as possible.
- Store glass containers in original cartons and cushion with newspaper or use JarBOX jar-storage containers (see resources).
- Store heavy, large items near the floor to minimize breakage.
- Either do not stack storage, or stabilize the stack against tipping.
- Use sturdy shelving and anchor the top to the wall studs by bolting or tying down securely.
- Enclose cupboards with sliding doors, sturdy latches, or hooks to prevent contents from dumping.
- Install "lips" or guardrails on open shelves with strong cord or wire or nail sturdy dowels a couple of inches above the shelf.
- Install metal shelves upside down to create a small lip.
- Locate storage areas in sturdiest part of the home.

Consider having an outside entrance so you can get to your storage if your home happens to collapse over the storage area.

Pest Control

Cockroaches, silverfish, ants, and rodents ruin billions of dollars of food each year. and contaminate far more than they eat. Although silverfish and ants can be a nuisance if they contaminate food, they are not disease carriers. Cockroaches and rodents, however, are disease carriers. And rats carry more human diseases than any other animal except for mosquitoes. During an extended crisis with poor or nonexistent sanitation and a buildup of refuse, pests can be a major problem. Prepare now to minimize it.

Dealing with Infestations

Should an infestation occur, destroy the contaminated food and clean the shelves. Recommendations of pesticides safe for use around food are constantly changing, so check with your cooperative extension or a reputable pest control company.

QUICK CHECK
Preventing Pests
✓ Inspect your storage area for spaces where pests can enter or hide.
✓ Close off spaces around the foundation, doors, windows, and pipes.
✓ Put screens over air vents.
✓ Eliminate dark, humid, warm spaces.
✓ Use insect- and rodent-proof containers.
✓ Clean up spilled food.
✓ Keep storage areas meticulously clean.
✓ Prevent trash from accumulating.
✓ Wash floors and walls with disinfectant.

Packaging Dry Food Storage

Packaging your own food storage has several advantages. First, it's more economical to purchase grains, beans, sugar, powdered milk, and other food items in bulk. Second, with so many packaging options, you get to choose the size and type of container that's best for your situation. And third, you are assured of the quality of the product when you prepare it firsthand.

> To store grain most successfully, start with a high-quality, clean grain and store it in clean containers.

Commercially Packaged Products

While commercially packaged products are convenient, you should be aware that you are paying a lot for that convenience.

Buying basic commodities in #10 cans costs more than buying them packaged in five- or six-gallon food storage buckets, often called poly buckets. And both cost more than packaging them yourself in Mylar foil bags and five-gallon poly buckets. It's worthwhile to compare prices before purchasing bulk foods as you'll find wide price variations.

Packaging and Storing Food in Poly Buckets

Poly buckets made of high-density polyethylene plastic are a quick way to package foods, like grains and beans, that have low-moisture and low-fat content. This is especially economical if you

> Poly buckets are a good choice if you will be continually rotating the foods stored in them.

are continually using your grains and beans. Foods should be clean and free of insects for best long-term storage results. Buckets should be made of food-grade plastic, and lids should have an airtight, gasket-style seal.

Since light can penetrate these white buckets, they should be stored in a dark place. Air can also penetrate them over time, so eventually the container will deteriorate and food will become

---

FIVE THINGS YOU CAN DO NOW

1 Start saving PETE or PET containers for storing dry foods.

2 Purchase several food-grade plastic buckets with airtight, gasket-style seals.

3 Package 100 pounds of wheat in poly buckets.

4 Purchase a 25- or 50-pound bag of grains, legumes, or oatmeal and repackage it using Mylar bags with oxygen absorbers.

5 Make a "breakfast bucket" with packages of oatmeal, cream of wheat, powdered milk, dried apples, etc.

---

susceptible to oxidation. Determined rodents can chew through the buckets, and they're not guaranteed protection against insect infestation.

Despite these concerns, poly food storage buckets are an inexpensive way to store grains and other dry goods. To successfully store dry bulk foods, simply place the food in a food-grade poly bucket, secure the lid, and store in a cool, dark, dry place.

Clean grain with a low moisture content will store for thirty years. Table 13.1 shows how many pounds of different foods you can put in a five-gallon bucket.

Table 13.1 Pounds of Food per Five-Gallon Bucket	
Food Item	Pounds per Bucket
Grains	30–36
Cracked wheat	22–33
Cornmeal	22–23
Flour	23–25
Rolled oats	15–20
Pasta	16–22
Beans	30–35
Split peas, lentils	30–35
Sugar, granular	29–34

Packaging and Storing Food in Mylar Bags

Mylar bags are an affordable and effective storage option for do-it-yourself food packaging and are almost always used with absorbers. A Mylar storage bag should have a layer of aluminum between the polyester and plastic layers. The thickness of high-quality Mylar bags ranges from 3.4 mm to 7 mm. Avoid the inexpensive, lightweight single-ply products.

Purchase quality Mylar bags from a reputable source. USA Emergency Supply and Pack Fresh USA offer quality Mylar bags and oxygen absorbers. They come in one-pint to six-gallon sizes, but you can customize the size of the bags by cutting a larger bag into smaller pouches for items such as herbs and spices. Mylar bags are also available with resealable zip openings that can be used to close the bag once the heat seal is broken.

To get a good seal on thicker bags, you may need to use an impulse seal such as those offered by American International Electric Sealer Supply, Baytec Containers, and Sorbent Systems. The cost is around $100. Bags can be resealed by the same methods. If the bags

Table 13.2 Using Mylar Bags for Packaging Food Storage	
Step 1	Have food and the appropriate size of bags ready before you open the container of oxygen absorbers.
Step 2	Fill the bag with dry bulk-food product, leaving enough space to seal the package. Add the oxygen absorbers. Keep those not being used in an airtight container until needed.
Step 3	Immediately heat seal the bag. Place the top of the filled bag across a hard surface, such as a metal-edged carpenter's level or a two-by four-inch piece of wood.
Step 4	Use an iron set on high. Firmly press the hot iron against the bag where it rests on the hard surface. If your bags are small, a second option is to use a flat iron.
Step 5	The best seals are at least a half-inch wide. Check seals after several hours to be sure they're good. Seal multiple times if necessary.

are small enough, you can use the heat source on a vacuum sealer, but be sure to turn the vacuum pump off. See table 13.2 on page 84 for directions on how to use Mylar bags for packaging food storage. It works best with two or more people working in assembly-line fashion.

Using Mylar Bags With Food Storage Buckets

Mylar bags and oxygen absorbers used in combination with storage buckets are excellent for long-term storage as they prevent insect infestation and oxidation problems. The buckets give support and protection from puncture. Below are several options to consider:

1. Place a twenty-by-thirty-inch bucket-sized bag in a storage bucket and fill with grain. Seal with O₂ absorbers and store until needed.
2. Place several smaller Mylar bags filled with grain and O₂ absorbers in a storage bucket. Open one bag at a time to limit exposure to oxygen and moisture.
3. Fill several bags of related items, such as breakfast foods. Seal with O₂ absorbers and place in a bucket. This will help with rotation and using the food items evenly.
4. Package all the ingredients for one meal in separate Mylar bags and store them together in one container.

> **QUICK TIP**
>
> ### An Easy Way to Fill Bag
>
> For a bag-lined food storage bucket, fill an identical bucket with the food item and then pour it from that bucket into the bag-lined bucket. Vibrate the can and tamp it down until the food item fits.

Packaging Food in Plastic Containers, Cans, and Jars

PETE or PET Plastic Containers

First test the container to make sure it will hold a seal by placing the sealed, empty container underwater and squeezing to see if air bubbles rise from it.

Clean and dry containers thoroughly, then fill with low-moisture dry bulk foods. Placing an appropriate-sized oxygen absorber in them will help fumigate them and decrease oxidation. Screw the lid on tightly. You may want to seal the lid with tape to help protect the seal. Since these containers will allow air to penetrate over time, they're best for foods you will rotate regularly. They need to be stored in the dark to prevent UVA deterioration.

Cans and Dry Packing

Cans are an excellent dry-pack storage container, and food storage is often sold in #10 cans. You will need a can sealer as well as cans and lids to do it yourself. A sealer is expensive but could pay off if the cost is shared. The All American Senior Flywheel Can Sealer (pictured) and the Ives-Way Manual Can Sealer are two options and cost between $500 and $2,000. (See page 276.) Look for used can sealers. Purchasing empty cans in small quantities is expensive. You may be able to find a local source to avoid shipping costs. House of Cans, Inc. and Wells Can Company sell cans in smaller lots.

Packing dry bulk foods in cans is simple. Just fill the cans to within a quarter inch of the top, then tap to settle the contents and add more to fill. If desired, weigh the contents for consistency. Add an appropriate-sized oxygen absorber (300–500 cc), seal the can, and then label each can immediately, especially if you are canning several varieties of food. Store in a dry, cool place. Table 13.3 shows how many pounds of foods you can put in a #10 can.

Table 13.3 Pounds of Food per #10 Can	
Food Item	Pounds per #10 Can
Grains	5.25–5.5
Cracked wheat	4.0
Cornmeal	4.0
Flour	4.0
Rolled oats	2.4–2.8
Pasta	2.7–3.5
Beans	4.8–5.6
Split peas, lentils	5.5
Sugar, granular	5.6

Jars

Dry bulk food can also be stored in Mason jars or other food jars. Fill the jar with a dry bulk food, then settle the jar and add more product, leaving a small space at the top. Place an oxygen absorber in the jar (50–100 cc for a quart), then place the lid and ring on the jar and label it. Store the jars in a dark, cool place. The downside to jars is that they are expensive, small, and breakable.

Using Oxygen Absorbers, Vacuum Packing, and Dry Ice to Prevent Oxidation

Oxygen Absorbers

Oxygen absorbers have revolutionized home storage. They are an easy-to-use, inexpensive way to create an oxygen-free storage environment and increase the storage life of dehydrated and dried foods.

Oxygen absorbers contain iron powder, which quickly reacts with oxygen to create harmless iron oxide (rust), and are typically found in small packets or sachets. When placed in a sealed environment, the oxygen in the air (21 percent) combines with the iron in the packet, leaving the remaining nitrogen (78 percent) and trace gases found in the air. If used properly, oxygen absorbers will reduce the oxygen level to 0.01 percent.

Oxygen absorbers range in size from 20 cc to 3,000 cc and come sealed in packages of twenty to one hundred individual absorbers. (Those pictured are OxyFree 60 Oxygen Absorbers with OxyEye.)

Once the package seal is broken, the absorbers in it immediately begin to absorb oxygen from the air. Quickly place unused oxygen absorbers in a small airtight glass jar or airtight PETE container, where they can be stored for up to three months, or use a vacuum sealer to repackage the remaining absorbers into smaller, more convenient packages.

Using Oxygen Absorbers Safely

As oxygen absorbers remove oxygen from the storage environment, they improve food quality by reducing oxidation, preventing insect growth, and inhibiting the growth of mold and most bacteria.

To prevent botulism, use only dry, low-moisture food products.

There is one exception. The bacterium Clostridium botulinum, which causes botulism, thrives in a moist, oxygen-reduced environment. For this reason, it is important to use oxygen absorbers only with dry, shelf-stable foods that have a low moisture and oil content. These include whole grains, oatmeal, beans, split peas and lentils, pasta, powdered milk, and dehydrated fruits and vegetables.

How Many Oxygen Absorbers to Use

Oxygen absorbers can be used with a variety of storage containers. The cc measurement stated on an oxygen absorber indicates how many cubic centimeters of oxygen it will absorb. Table 13.4 gives you an idea of how many oxygen absorbers you'll need for various size containers, though it's not harmful to use more than recommended. Use several oxygen absorbers to equal the total amount needed. For example, use ten 200 cc absorbers or four 500 cc absorbers to total 2,000 cc.

Pasta or beans require more oxygen absorbers because there is more air volume in the container. Recommendations vary from source to source, but as a rough guide, use about 100 cc per gallon of dense food and 200 cc for less-dense food. Oxygen absorbers should not be used with sugar or salt because they will cause them to clump.

Table 13.4 How Many Oxygen Absorbers?		
Food Storage Item	Container Size	Amount of Oxygen Absorber
Wheat, rice, oats, small grains	• 5- or 6-gallon bucket • 20" x 30" Mylar bag	1,000–2,000 cc
Pasta, noodles, beans	• 5- or 6-gallon bucket • 20" x 30" Mylar bag	2,000–2,500 cc
Assorted food items	• 1-gallon jar • #10 can • 10" x 16" Mylar bag	300-500 cc
Assorted food items	• 1-quart jar • 6" x 9" Mylar bag	50-100 cc

Vacuum Packing to Prevent Oxidation

Vacuum packing removes the air and thereby the oxygen from containers and is ideal for packaging small quantities of dry goods, such as nuts, as well as dehydrated foods, such as herbs and dried fruits, for short-term storage. It's also ideal for repackaging meats purchased in bulk into smaller quantities to be frozen. Vacuum packing helps prevent freezer burn.

However, vacuum packing is not recommended for long-term storage because the seal produced by typical home-vacuum packers is not reliable. It is also somewhat time and cost prohibitive to vacuum pack large quantities of storage products, such as grains, legumes, and powdered milk.

Also, some precautions and common sense are necessary. Vacuum packaging is not a substitution for the heat processing of home-canned foods. Perishable foods still need to be refrigerated or frozen.

Dry Ice with Poly Buckets

Dry ice is carbon dioxide in its solid form, and when placed in a bucket, it sublimates and the carbon dioxide replaces the air. This method of reducing oxidation is safe, inexpensive, and easy to use. You can expect about 90 percent of the air to be replaced, leaving about 2 percent oxygen—effectively suffocating adult insects and their larvae.

Dry ice is sold in some grocery stores, usually in five-pound chunks. You will need about three ounces of dry ice per five-gallon container. One pound (450 grams) will treat six five-gallon buckets. Although nontoxic, dry ice is very cold (-109.3° F [-78.5° C]) and should be handled with tongs or heavy gloves to avoid cryogenic burns. Transport it in an open or loosely closed container in a ventilated vehicle. Do not place it in a refrigerator, freezer, or closed ice chest because CO_2 gas builds pressure as it sublimates.

Carbon dioxide is heavier than air and tends to remain in the container even after opening. Much of it will stay in the container if you scoop the food out rather than pour it.

Follow the directions in table 13.5 to use dry ice in storing food.

	Table 13.5 Preparing Food Storage with Dry Ice
Step 1	Select a low-humidity day outside or a well-ventilated room. Brush off any water vapor that may have condensed from the atmosphere on the dry ice to prevent additional moisture from being added to the food.
Step 2	Use a hammer to break ice into smaller pieces. Prepare 3 ounces, or about 1/3 cup, of dry ice for a 4 to 6-gallon container (2 cubic centimeters per liter).
Step 3	Place the dry ice in the middle of the container as low as possible. If using plastic containers, first place 2 or 3 inches of grain in the bottom to prevent the intense cold from cracking the bottom of the container.
Step 4	Finish filling the container, then wait until any visible fumes have disappeared. Depending on the container size, this can take thirty minutes or longer.
Step 5	Place the lid on but do not seal until the dry ice is entirely vaporized, or the pressure may cause the container to burst. If any bulging occurs, quickly remove the cover, wait a few minutes to let excess gas escape, and then replace it.

Using Desiccants to Remove Excess Moisture

Desiccants are the little packets commonly found in new shoes and pill bottles. Technically a desiccant adsorbs the moisture (different from absorb), which means water molecules stick to it. Governments and industries use desiccants to protect delicate electronic equipment, communication instruments, and fine guns.

Dehydrated foods, including grains, tend to attract moisture from the air, which can result in spoilage or insect infestation. If the food you are storing has the correct moisture content to begin with and is sealed in an airtight container, no desiccant should be needed. However, under humid conditions, or if there is doubt about the stored foods being dry enough, or if you will be opening the container regularly to use the contents, you may want to use a desiccant to help keep the food dry.

The easiest method is to use commercially prepared desiccants. Do not use desiccants with salt, sugar, or flour because removing their moisture will cause them to lump. If you use both an oxygen absorber and a desiccant, place the desiccant toward the bottom of the container and the oxygen absorber toward the top since the desiccant will inhibit the oxygen absorber.

Reducing Insect Contamination

Although insect-infested food does not sound very appetizing, in a crisis, it may be important to salvage contaminated food.

The need for eliminating insects depends on your situation. If you live in a dry climate, you probably do not need to worry about insects if you start with properly cleaned, low-moisture grains and store them

> **The most effective way to prevent an insect infestation in your storage is to create a reduced-oxygen atmosphere by using oxygen absorbers, vacuum packing, or dry ice.**

in suitable containers. If you live in high-humidity or have previously had an infestation, you'll want to inspect regularly and take appropriate action if needed. Generally, legumes are not as susceptible to insect infestations as grains are.

Reduced-Oxygen Atmosphere

A reduced-oxygen environment will prevent the growth of insects and their larvae in grains, legumes, pasta, and dehydrated fruits and vegetables. As described in the previous section, the use of oxygen absorbers, vacuum packing, and dry ice are effective methods for reducing the oxygen atmosphere.

Freezing

Place the container of grain in the freezer. For containers weighing one to fifteen pounds, two or three days at or below 0° F (-18° C) will kill adult insects. A temperature of -10° F (-23° C) will destroy eggs and larvae. Upon removal, wait twenty-four hours and then dry any condensation prior to storing. Larger containers require more time to ensure the cold has penetrated the food.

Heating

All forms of insect life can be killed with heat. Table 13.6 shows the lengths of time required at different temperatures. Higher temperatures or longer periods of time may affect the quality of flours milled from grains and may reduce germinating abilities.

Table 13.6 Temperature and Time for Eradication	
Internal Temperature	Minutes Required
150° F (66° C)	4 minutes
140° F (60° C)	10 minutes
120° F (49° C)	20 minutes

Small packages may be heated directly. The contents of larger packages should be placed in a shallow pan to a depth no greater than three-fourths of an inch to assure complete penetration of heat. Leave the oven door slightly ajar to avoid overheating and stir the food occasionally at the higher temperatures to keep it from scorching.

To avoid reinfestation, place treated food in insect-proof containers before it cools. Heating can also be used to reduce the moisture content of foods.

Mechanical Removal

Infested grains can also be submerged in a container filled with cold water; the bugs will float to the top, where they can be skimmed off. Then either use the grain quickly or oven-dry it until it is again hard enough.

Types of Food Storage

Foods can be stored in different forms, including fresh, canned, frozen, and dried. This chapter examines the characteristics of each type so you can make informed choices for your preparedness plan.

Keep in mind that individual foods may be good in one form yet unsatisfactory in another. The best plan consists of a combination of food types. Consider your budget limitations, desired convenience, perceived needs, and personal preferences and sample all foods before including them in your storage.

For all types of food, purchasing combination foods, such as casseroles or stews, will almost always be more expensive than buying the foods separately.

Basic Storage Foods

The biggest part of a complete food storage plan are the basics: grains, cereals, flours, pasta, dried legumes, powdered milk, sugar, honey, and salt. You can buy these packaged in cans, Mylar bags, and poly buckets. See chapter 18 for information about purchasing and storing basic foods.

You can also prepare your own by purchasing in bulk quantities and repackaging them for storage. See chapter 13 for information about packaging your own food storage. Companies that specialize in selling grains, legumes, and powdered milk are listed on pages 425-426 in the resource section.

Fresh Foods

A few fresh foods can be stored in root cellars and other cold-storage places to extend storage life after harvest, though these methods only provide a storage life of about six months and work best in northern climates with cool or cold winters. See chapter 28 for information about cold storage.

FIVE THINGS YOU CAN DO NOW

1 Find out where you can purchase grain.

2 Buy six cans of a canned-entrée type food your family enjoys.

3 Try an MRE or a retort food.

4 Buy several freeze-dried entrées for your family to sample.

5 Analyze the ingredient and nutrition labels in a food storage entrée.

Canned Goods

Canned foods have a useful place in food storage. Foods that are thoroughly cooked are usually well-suited to canning. Taste tests show that most people consider canned foods better tasting than their dehydrated counterparts.

All foods begin to lose nutrients soon after harvesting and during preparation.

Nutritional Value

For top nutritional quality, fresh fruits or vegetables should be picked at peak ripeness, properly stored at cool temperatures, and eaten soon after harvest. Loss of nutrition occurs during the preparation of all foods, and even fresh foods suffer loss soon after harvesting. Canned foods are processed shortly after harvest at the peak of their nutritional benefit, and although the processing of canned goods causes them to lose some of their original nutritional value, they are still a healthful choice. A study at UC Davis concluded that "fresh fruits and vegetables usually lose nutrients more rapidly than canned or frozen products." And a study at the University of Michigan compiled the results of over forty nutritional studies, concluding that fresh, canned, and frozen vegetables all had similar nutritional values.

Many canned items have sugar, salt, preservatives, and other food additives added to them during canning. Manufacturers are becoming more sensitive to consumers who want to avoid these additives, so look for such products. Home canning is an excellent alternative as it allows you to monitor the contents. See chapter 29 for information about home canning.

Canned goods are an ideal food source to be purchased and replaced as they are used.

QUICK LOOK

Advantages of Canned Foods

- Familiar to most diets
- Wide variety available
- Inexpensive
- Can be eaten without heating
- No water needed to prepare
- Small containers are easy to transport, share, and trade

Shelf Life

The USDA recommends that for highest quality, you should plan to use high-acid canned fruits and tomato products within eighteen to twenty-four months and low-acid meats and vegetables within two to five years.

If stored in cool, dry conditions, the emergency shelf life of canned foods—the time they can be kept and remain palatable and nutritional—can be significantly longer. Industry and military research suggests that most canned goods stored at 70° F (21° C) or lower stay nutritious and palatable even after five years. One study of ten canned vegetables showed no significant loss of vitamin A after three years of storage and an average of 61 percent of the vitamin C remaining. In fact, after five to six years, there was still 50 percent of the original vitamin C.

Cost and Vulnerability

Canned goods are often less expensive when compared to freeze-dried and dehydrated products. Studies have shown that about 60 percent of commercially canned goods are even cheaper than comparable fresh foods. Canned items are frequently on sale at your local grocery store, and you can save by comparing prices. And save even more when you home-can your own in-season or homegrown produce.

But be aware that cans are vulnerable to corrosion and are susceptible to bursting if they are dropped or fall. The acids in foods such as tomatoes and fruits can cause cans to deteriorate from the inside out.

PERSONALLY SPEAKING

 I often use canned foods as building blocks for my meals and as a quick enhancement to recipes. Canned chicken, beef, tuna fish, and salmon are the foundations for many of my meals. I use canned olives, green chiles, and mushrooms to jazz up my recipes. Canned beans, like pinto, kidney, black, and garbanzo, are great in soups, salads, and spreads. Even if they are not part of your everyday meal plans, it's a good idea to have some quick meals in a can on hand—things like chili, stew, macaroni and cheese, ravioli, and kid-friendly Spaghetti O's. Canned pumpkin and applesauce are great for adding flavor and moisture to sweetbreads and cakes.

Every fall, we can our own sauces, fruits, and vegetables. Spaghetti sauce is our family specialty, and everyone gets involved! I frequently use home-canned tomato sauce and tomatoes in Italian and Mexican dishes. Where I grew up in the Northwest, we were spoiled with home-canned Columbian River salmon, Dungeness crab meat, and razor clams.

Although I prefer fresh or frozen vegetables, I keep cans of green beans, corn, and peas on hand, too. I use my canned foods throughout the year and replenish them at the end of each growing season when the new crops are harvested.

Frozen Foods

Frozen foods are the nearest to fresh foods of all the stored foods and are generally considered superior in appeal and nutritional value to canned or dehydrated foods.

Nutritional Value

Fruits and vegetables chosen for freezing, like those canned, are processed shortly after harvest at peak ripeness, when they are most nutrient-dense. And yet, despite their high quality, frozen foods still suffer loss of water-soluble nutrients, such as Vitamin C, during processing and storage. For example, after six to twelve months, frozen fruits contain only 70 percent of their original vitamin C and vegetables retain only 50 percent. The blanching process necessary to halt enzyme action is responsible for a large portion of this nutrient loss.

Shelf Life

Shelf life is anywhere from three months to beyond one year. Frozen foods may suffer "freezer burn," which is caused by dehydration and oxidation when food comes in contact with the air. Vacuum packaging food can minimize this.

Cost and Vulnerability

Because of the electricity required, frozen foods are costly to store, often tripling the initial price after one year. But the biggest downside is their vulnerability. Frozen food is dependent on refrigeration. Freezers need electricity or gas to run and can suffer mechanical breakdown at any time, making them vulnerable in times of crisis.

One way to avoid this vulnerability is to have an alternative energy source so that outages and shortages have less impact. If you live where you are frequently affected by power outages, you may want to have a backup generator.

According to the NASD, or National Ag Safety Database, you'll need a generator that has four times the wattage used by the appliance it is intended to run so there is enough power to start or cycle the appliance. So if your freezer runs on 400 watts, you'll need a generator that produces at least 1,600. Of course, if you are running more than one appliance, you'll need more wattage.

You might also consider refrigerators and freezers that run on propane. Unique Off Grid Appliances makes a 2.2 cu/ft and a 6.0 cu/ft propane model. See chapter 36 for information about alternative methods of refrigeration. You will also need to store fuel for the generator or freezer.

Retort Foods and Aseptically Packaged Foods

Retort Foods

Retort foods are a recently developed food packaging where no refrigeration is required. Food is placed in a lightweight, flexible pouch made from a durable multilayer plastic and metal laminate. The air is then evacuated, and the pouch is sealed. Next, it is placed in a special pressure cooker called a retort oven, where it is sterilized at temperatures between 240° and 250° F to prevent spoilage and deterioration. They are popularly called MREs (Meal, Ready-to-Eat), and are used as a standard individual field ration for the military. Some commercial foods, including juices, tuna fish, soups, and Asian foods are commonly sold in retort pouches.

Nutritional Value

Because the thin pouch requires less processing than metal cans, the food undergoes less deterioration and has the potential of being superior in quality to even frozen foods. Foods remain moist, and their colors, textures, and flavors compare favorably with canned and frozen foods. They generally contain few preservatives and can be eaten directly from the pouch.

The pouch is lighter than metal cans and requires less time to heat. To eat, place the MRE directly into boiling water for three to five minutes or use a flameless ration heater (FRH), which is specially designed to heat military MRE pouches.

Shelf Life

The shelf life of retort foods is approximately equal to that of canned foods. Like other foods, they last longest when stored at lower temperatures. If stored in a cool place, their emergency shelf life can be up to five years.

Cost and Vulnerability

Prices for retort foods are substantially higher than for canned or frozen foods and in the same range as freeze-dried foods, and the pouch is susceptible to puncture and rodent damage, but they are a good alternative and certainly have a place wherever refrigeration is an uncertainty. Stored in their cartons, they take up as much space as canned foods, but they do not require a can opener.

Aseptically Packaged Foods

Aseptic sterilization, or UHT (ultra-high temperature), is a unique method of processing that rapidly sterilizes foods at high temperatures and creates and maintains a sterile package without the need for refrigeration. Packages keep for up to one year. Although resembling a cardboard carton, the packaging is quite sophisticated, made up of several layers of aluminum, polypropylene, and paperboard. Milk is the most common food processed with UHT, keeps for about eight months unrefrigerated, and is often found in Europe and other parts of the world. Fruit juices keep about six months.

Dried, Dehydrated, and Freeze-Dried Foods

Dried Foods

As the term is used here, "dried foods" refers to foods whose moisture content has only been reduced to the 20 to 25 percent level. These foods are mainly fruits and include raisins, prunes, figs, dates, apricots, and apple, peach, and pear slices. They feel moist and soft to the touch, are available in supermarkets, and are usually packaged in a plastic bag or in a box. Because of their high moisture content, dried foods mold easily. Oxidization causes fruits to darken, and they are not well suited for long-term storage. However, if placed in an airtight, pest-proof container and kept in a cool, dry, dark location, they may keep for one to two years. If they become overly hard, they can be soaked overnight in warm water or stewed.

Dehydrated Foods

Commercially dehydrated foods are air-dried at temperatures between 140° and 400° F in large ovens or drums. These foods shrink and become hard and brittle when fully dehydrated and need time to be soaked in water to reconstitute and may even need to be cooked in water for ten to twenty-five minutes before they can be eaten.

A wide variety of foods are available in dehydrated form, and most people eat them every day in the form of common convenience foods, such as instant potatoes, dry soups, seasoning mixes, and dry baking mixes.

The most popular dehydrated foods are fruits and vegetables, milk products, and eggs. Except for jerky, meats are not air-dried. Dehydrated foods are often combined to make entrées, side dishes, desserts, and beverages.

PERSONALLY SPEAKING

The dehydrated food I use most often is diced onions. They are inexpensive and convenient — I use them just the way they are in cooked recipes and rehydrate them for a short time before adding them to sautéed dishes. One of my favorite ways to use them is to puree them in a small food processor to make onion powder.

The second dehydrated food I love to use is apples! You can buy them economically in #10 cans, but I highly recommend dehydrating your own if you have access to fresh apples in the fall. All you need is an apple peeler and a dehydrator. Our family likes them with a little cinnamon sugar sprinkled on them. They are also a nice addition to cooked cereals and breads.

Freeze-Dried Foods

Freeze-dried foods are flash frozen at temperatures as low as -50° F. Radiant heat is used to turn the ice crystals directly into water vapor, which is then drawn off by a vacuum. The cellular structure of the food remains unchanged and results in a spongelike, porous food that keeps its original size and shape both when dry and when reconstituted with water. Freeze-dried foods save both time and fuel. Many are precooked, and their porous nature allows them to be table-ready in just minutes.

Familiar fruits and vegetables, meats, and "space treats" are freeze-dried individually. The ready-to-eat entrée is a good choice for short emergencies when the power is out and you need a nourishing, warm meal. The best quality ready-to-eat freeze-dried entrée meals are cooked first and then freeze-dried.

The biggest drawback to freeze-dried foods is cost — a freeze-dried entrée for two costs a little less than ten dollars.

Home Freeze-Dryer

Harvest Right has developed a home freeze dryer that works as claimed and produces quality freeze-dried food, but it's pricey and takes up a lot of space.

Comparing Dehydrated and Freeze-Dried Foods

Dehydrated and freeze-dried foods have had the moisture level reduced to only a small percentage. See table 14.1 for a side-by-side comparison of dehydrated and freeze-dried products. Companies that specialize in dehydrated and freeze-dried foods are listed on pages 425–426 in the resource section.

Convenience

They are easy to use and convenient to store, conceal, and transport and are usually packaged in Mylar bags or in #2.5 and #10 cans for home storage. Oxygen absorbers

For a reasonable balance between taste and cost, choose dehydrated for most fruits, vegetables, and dairy products and freeze-dried for meats, eggs, and entrées.

Crisis Preparedness Handbook

are commonly used for long-term storage. Some dehydrated and freeze-dried foods can be eaten right from the pouch or can. Freeze-dried entrées, especially, can provide a meal in a hurry; just add boiling water to the pouch. Dehydrated foods require more rehydrating, but in either case, be sure you have enough water on hand to reconstitute dehydrated foods when water may be in short supply. Fruits and vegetables need about a gallon per pound of dehydrated product, and eggs and meats average one-third gallon per pound.

Table 14.1 Comparing Dehydrated and Freeze-Dried Foods	
Characteristics of Dehydrated	Characteristics of Freeze-Dried
Shriveled shape and smaller size than original food	Maintains shape and texture of original food
Compact and intense flavor, like cooked food when reconstituted	Good flavor, plumper, closer to fresh taste
Lightweight, takes less space than freeze-dried	Lightweight for portability but bulky
More economical than freeze-dried	Expensive but highly rated
Milk products usually come in dehydrated form.	Meat available in only freeze-dried form (except jerky)
Some foods may be dried at home in a dehydrator or oven.	Home freeze dryer makes it possible to freeze-dry at home.
Rehydration requires soaking or cooking for 20 minutes.	Short rehydration time of 6 to 9 minutes, saving time and fuel
Shelf life prolonged once opened if container is kept closed	Begins absorbing moisture after opening; must be eaten soon after opening
Shelf life varies from few years to 30 years.	Up to 30-year "best if used by"

©Patricia Spigarelli-Aston

Dehydrated fruits weigh about $\frac{1}{7}$ and vegetables about $\frac{1}{10}$ of their original weight. Freeze-dried meats weigh about $\frac{1}{3}$ of their fresh weight.

Dehydrated foods take up only $\frac{1}{3}$ as much space as fresh foods, but manufacturers often pack the cans lightly, using up some of that space savings. Freeze-dried foods take up the same amount of space as fresh foods since they do not change shape or size.

Nutritional Value

As mentioned, all foods lose nutritional value in processing and during storage, and dehydrated and freeze-dried foods are no exception. Although the water has been removed, dehydrated foods still contain some fats and/or sugars, which are susceptible to oxidation. Over time, oxidation turns sugars brown, causes oils to go rancid, and creates undesirable colors and flavors. A lower moisture content also contributes to the loss of protein.

Freeze-dried food has higher initial nutritional values because of lower processing temperatures. However, air-dried food tends to lose nutrients more slowly because its shriveled form protects it better than the porous structure of freeze-dried foods.

Despite nutritional losses, both dehydrated and freeze-dried foods are good sources of nutrition and can be an important part of your food storage.

Taste and Appearance

The flavors and appearance of dehydrated and freeze-dried foods change during processing and storage. Dehydrated foods may look a lot like cooked foods when they are reconstituted, but over time, they lose their ability to rehydrate and become tough when cooked.

Freeze-dried foods look much like they did before they were processed but with a frothy, airy texture until reconstituted.

Taste is subjective. Most reviews for freeze-dried foods and entrées speak highly of their taste. Dehydrated foods have a strong, concentrated flavor, and opinions about the taste range from delicious and fresh to cardboard-like and artificial. You really must try them yourself to decide what you think. Try a couple of freeze-dried entrées and purchase some small sizes of dehydrated foods. Food storage companies often provide samples.

Cost

While some dehydrated foods, like dehydrated onions, are inexpensive, freeze-dried foods are among the most expensive processed foods. If you want to compare the cost of dehydrated and freeze-dried foods with other kinds of foods, you'll need to put them in equivalent measures so you are comparing them fairly. For example, one-third cup of dried onions is equivalent to one onion, and one pound of dried onions is equivalent to eight pounds of fresh, chopped onions.

Shelf Life and Vulnerability

In recent years, the shelf life for dehydrated foods and freeze-dried foods has been revised upward. To achieve the best shelf life, foods should be stored in opaque, airtight containers in an oxygen-free environment and at temperatures consistently below 70° F (21° C). Foods will last increasingly longer the lower the temperature. Table 14.2 lists estimated shelf life for common storage foods.

Table 14.2 Shelf Life of Common Dry, Dehydrated, and Freeze-Dried Foods*		
Food Product	Best-If-Used By	Life-Sustaining
Grains	10–12 years	30+ years
Beans	8–10 years	25+ years
Pasta	8–10 years	20+ years
Rolled oats	8–10 years	30+ years
Dairy products	3–5 years	15 years
Dried whole egg powder	1–2 years	5–10 years
Dehydrated peanut butter	5 years	10 years
Dehydrated apple slices	5 years	30 years
Dehydrated fruits	5 years	20 years
Dehydrated carrot slices	5 years	10 years
Dehydrated vegetables	5 years	10 years
Potato flakes	5 years	30+ years
Freeze-dried vegetables	5 years	30+ years
Freeze-dried fruits	5 years	30+ years
Freeze-dried entrées	5 years	30+ years
Freeze-dried ground beef	5 years	30+ years

*Estimates only. Based on optimum storage conditions at temperatures lower than 70° F in oxygen-free #10 cans.

©Patricia Spigarelli-Aston

The Nutrition, Dietetics & Food Science Department at Brigham Young University has established some of the best research on the longevity of stored foods. Based on nutritional analysis and taste testing, they determined that dehydrated fruits and vegetables and grains have a reasonable level of nutrition and acceptable taste after thirty years. This is called their life-sustaining or emergency shelf life.

Dairy products have about half this storage life. Dehydrated egg products did not fare well in the testing and have a relatively short shelf life and should be rotated within a year or two.

Despite their longer shelf life, dehydrated foods are best if rotated and used within a four to five-year period. Check the quality of sensitive items, such as dairy products and eggs, more frequently.

Shelf Life of Freeze-Dried Foods

The shelf life of freeze-dried food depends on the quality of the food and the integrity of the packaging. Mountain House, a pioneer in freeze-dried foods, has been making freeze-dried food since 1969, and their product controls meet the demands of NASA and the military. In 2016, they revised the shelf life of their freeze-dried foods upward. The "best if used by" shelf life printed on their packaging is now thirty years from when it was manufactured.

Other companies with state-of-the-art freeze-drying equipment also claim a twenty-five- to thirty-year shelf life. Although these companies may not have existed for thirty years, they are able to make their assertions based on laboratory testing and extrapolation. If you intend to store freeze-dried food for thirty years, carefully investigate the products and the company selling those products.

Open Shelf Life

Since dehydrated and freeze-dried foods begin to oxidize and absorb moisture once opened, cover open cans with airtight lids when you're not removing ingredients. Dipping into a can rather than pouring will reduce exposure to air. Use fresh desiccants and oxygen absorbers as needed, and store open cans in a cool, dry area.

Covered properly, opened cans of air-dried foods will keep for three months to a year. High-fat foods have the shortest shelf life, fruits and vegetables the longest. Freeze-dried foods attract moisture more quickly and will only store for two to three months. Repackage unused amounts in vacuum packaging with oxygen absorbers to keep them longer.

Freeze-Dried and Dehydrated Entrées

Freeze-Dried Entrées

One way to purchase freeze-dried food is in entrée form. A freeze-dried entrée is first prepared as a complete dish, such as beef stroganoff or chicken teriyaki, and then freeze-dried. The freeze-dried food is packaged in pouches containing two to four servings or in #10 cans containing about ten servings. Freeze-dried entrées can be ready to eat in minutes by adding water. They are a good option for emergency-evacuation or shelter-in-place situations. They also provide variety to a survival diet.

PERSONALLY SPEAKING

 As I was writing this section, I decided we needed to try some up-to-date freeze-dried entrées. We sampled two Mountain House entrées—Lasagna with Meat Sauce and Mexican Style Rice & Chicken. We bought them at a local sporting goods store, and the best-by-date was thirty years out! The dry product resembled the entrée in a non-freeze-dried state but was very light and porous. I added two cups of water as directed, stirred, closed the zipper, and waited the recommended eight minutes. Nothing could be easier! They both passed the taste test. The container said it contained three servings, but it was enough for three average servings or two generous servings for big appetites.

The Lasagna with Meat Sauce had real cheese, and there were bits of ground beef throughout the tomato sauce. Although it was not layered like lasagna, the overall texture and taste were authentic. The Mexican Style Rice & Chicken had small chunks of chicken along with kidney beans, rice, sliced olives, morsels of green peppers and onions, all in a spicy tomato sauce. One of the things I liked about both entrées was that they had relatively low sodium levels—19 percent DV (daily value) and 24 percent DV, respectively. I have seen other entrées with over 30 percent.

Combo Entrées

The other kind of entrée contains dehydrated ingredients and often a meat substitute. Some may contain actual freeze-dried meats or vegetables in addition to other dehydrated ingredients. The dry components are mixed and then packaged in pouches or cans. Note that dehydrated components take longer to reconstitute than freeze-dried. Like most prepared foods, they often have a long list of additives and may contain excessive amounts of sodium to make them palatable.

Do Freeze-Dried and Combo Entrées Have a Place in Your Plan?

To decide if they're right for you, consider under what circumstances you would need them. One entrée costs about as much a fast-food meal, but they can be useful when you need something quick. Take time to sample different entrées and decide which ones your family favors. Be sure to store water for reconstitution as well as have a way to heat the water.

You do not want to count on a food storage that doesn't deliver what it promises.

"All-Inclusive" Menu Plans

Dehydrated foods are often sold in all-inclusive menu packages and promoted as the solution to storing an emergency food supply. Food storage companies often market these packages as worry-free, "gourmet" meal plans. They make statements about the convenience and nutritional superiority of their foods and claim they are cheaper and taste better than competing brands. Multilevel-marketing sales techniques and the use of celebrities to make their case are both common practices. They often use advertising come-ons like "free bonus items" or "hurry, offer ends soon."

Advantages of Long-Term Food Storage

Despite the hype and aggressive marketing, there are some advantages to long-term food storage preparedness packages. First, if stored properly, they have a legitimately long shelf life—twenty-five to thirty years. And it is true that you can store them away and basically forget about them. Second, they are especially useful for short-term emergencies, when there is a lot of uncertainty and you need hassle-free food. Third, they offer variety. And fourth, they are light and relatively portable.

Are Long-Term Food Storage Packages Worth the Cost?

You may be wondering about whether to purchase these "complete" food storage packages, buckets, or meal plans. It may be tempting to purchase a three-month, six-month, or even a year's supply of food so you can "be done with it." Unfortunately, it's not that simple if your goal is to be *thoughtfully* prepared. These products are frequently made from highly processed foods with preservatives and other additives and are low on quality components. They are very expensive for what you are getting. But most importantly, you do not want to be in a crisis and counting on something that will not deliver the taste, nutrients, and calories you need.

Take the time and effort to critically examine the products you are considering for your food storage.

It's important to spend some time and effort critically examining the food you are considering purchasing so you can be sure you are getting quality.

Take Time to Analyze Ingredient Lists and Nutrition Labels

In adherence to FDA requirements, food manufacturers publish nutritional information about their products. Most food storage companies post ingredient lists and nutritional labels on their websites, allowing the consumer to scrutinize product content and nutrition. Take the time to analyze these labels so you can make an informed decision about which products to purchase. You will find some surprising information—the ingredients may not be what you thought they were.

Table 14.3 Making Sense of Daily Value on Food Labels			
Nutrient	Daily Value	%Daily Value	Goal
Total Fat	65 g	=100%DV	Less than
Sat Fat	20 g	=100%DV	Less than
Cholesterol	300mg	=100%DV	Less than
Sodium	2,300mg	=100%DV	Less than
Total Carbohydrate	300g	=100%DV	At least
Dietary Fiber	25 g	=100%DV	At least
Total Sugars			
Added Sugars	50 g	=100%DV	Less than
Protein*	50 g	=100%DV	At least

Given Daily Value amounts are based on a 2000-calorie diet. This chart is based on information from "How to Understand and Use the Nutrition Facts Label" www.fda.gov website
*Protein amount is about what a person needs and is included for the reader's benefit, but does not have a Daily Value.

©Patricia Spigarelli-Aston

Table 14.3 on page 101 will help you determine the nutritional value of a food as you study its label. It shows the daily value of several important nutrients in a typical two-thousand-calorie diet.

Note that for some nutrients, your aim is to be lower than the daily value; for others, your aim is to be higher. For example, the daily value for total sugar is 50 grams, but the goal is to have less than that amount daily. Also, it's helpful to know that per serving of food, a nutrient listed at 5 percent is a low amount; 20 percent is a high amount.

What Exactly Are You Getting in That "Gourmet" Entrée?

A food might be called lasagna or stroganoff—two foods traditionally made with meat—but check the label! The ingredients are listed in descending order by weight. In the list of ingredients, where is the meat? At the top or way down the list? Does it even have meat in it?

Deceptive Labeling

More than one dehydrated food storage company offers chicken a la king, and there is not a single morsel of chicken in the product. Be cautious of main-course foods labeled to give the illusion that they are made with meat or labels that say "beef-flavored" or "chicken-flavored"—there is likely no meat in them at all.

Besides not listing any known meat as an ingredient, there will often be an ingredient called "textured vegetable protein," or TVP, which is a protein made from soybeans that is used to simulate meat. It is not necessarily a bad product, but the implication that you are getting and paying for meat is deceptive.

Does the label list real dairy products or dairy components, like whey or nondairy creamers? A high-quality food will have real meat, real dairy, real cheese, and fruits and vegetables near the top of the list.

> **High-quality entrées will have foods like real meat, real dairy, real cheese, and fruits and vegetables listed near the top of the ingredient list.**

How Much Is a Bowl of Oatmeal Worth to You?

Once you have scrutinized dinner-entrée ingredients, take a good look at breakfast entrées, sides, and beverages. You will realize that breakfast entrées are often nothing more than grain cereal packed with sugar and artificial creamer, and they cost about two dollars per half-cup serving. Sides are often inexpensive carbohydrates like rice and instant potatoes, and a good share of the beverages are sugar-sweetened flavored drinks or artificial milk drinks. They are often sold along with more costly-to-manufacture entrées as part of a "complete meal-plan" bucket. But there are much less expensive ways to purchase them! You must ask yourself if the convenience is worth the price you are paying. Use worksheet 14.1 on page 104 to help you evaluate the dehydrated and freeze-dried foods you are considering and to help you make an informed decision about purchasing them.

What Happened to Lunch?

One last thing to consider. How many meals are in the "complete meal plan?" More often than not, there will only be two daily meals in the meal bucket. Always check the number of meals provided as well as the number of calories provided.

Buying Hints

You can buy from local dealers or national online companies, but after buying, open all cartons and make sure you've received what you were expecting.

Local supermarkets and wholesale warehouse clubs also carry some dehydrated items. They usually cost less than items from a food storage-dealer, and the sizes may be more manageable for your family's needs, although they may need to be repackaged for long-term storage.

Look for sales and specials and don't hesitate to ask for discounts. Commercial and government contract overruns are sometimes available.

A Worksheet to Help You

Use worksheet 14.1 on page 104 to help you evaluate dehydrated and freeze-dried food plans. For a downloadable PDF file, go to CrisisPreparedness.com.

		Worksheet 14.1 **Evaluating Dehydrated and Freeze-Dried Food Plans**
		Product: Number of servings per container: Serving size: Cost per ounce:
✓Yes	✓No	Evaluate dehydrated and freeze-dried plans by checking yes or no. A yes means the plan is acceptable in the category.
		Does the food taste good when reconstituted?
		Is the food aesthetically appealing? (texture and appearance)
		Is the serving size easy to visualize? (cups as opposed to grams or 1/4 pouch)
		Is the serving size realistic? (at least 1 cup for entrées)
		Are there enough calories in a daily menu to equal at least 2,000?
		Do entrées contain quality proteins? (real meat, real cheese)
		Is there adequate protein? (about 46 g for women and 56 g for men per day)
		Are dairy products "real" milk, cheese, and cream?
		Do complex carbohydrates and fiber make up a high proportion of the carbohydrates?
		Are added processed sugars used sparingly? (less than 50 g daily)
		Are entrées low in sodium? (less than 2,300 mg daily–20 percent DV for sodium in a single food item is high.)
		Are vitamins and minerals listed in the ingredients?
		Is there a variety of breakfasts meals, not just inexpensive cereals, such as flavored oatmeal?
		Does the "complete plan" include three meals per day rather than just breakfast and dinner?
		Is there a minimum amount of food additives and preservatives? (i.e., high-fructose corn syrup, monosodium glutamate, yeast extracts, texture proteins, hydrolyzed protein, sodium or calcium caseinate, etc.?
		Are you able to sample all products you are purchasing?
		Have you compared and sampled foods from several different companies?
		Can the company's nutritional and shelf-life claims be substantiated?

©*Patricia Spigarelli-Aston. For a downloadable PDF file go to CrisisPreparedness.com.*

Short-Term Food Storage Plan

It's likely we will all face an unexpected short-term emergency—those situations that are intense but over quickly. These may include natural disasters or man-caused disruptions. Utility services are often interrupted, and you may be without power, water, or sewer. A crisis of this type will disrupt your normal food preparation. Also, if you have to evacuate, you may not have access to your normal food-preparation supplies and equipment.

PERSONALLY SPEAKING

My daughter lives in a state along the mid-Atlantic coast where during some winters they experience an unending series of arctic storms—the kind where schools close, the power is off, and everything comes to a standstill for a few days. When blizzards or hurricanes are predicted, people scramble to get prepared at the last minute. Before each storm, the grocery stores are stripped of convenience foods, and there are long lines at the gas pumps. A couple of winters ago, one of my daughter's friends tweeted, "Hey, Baltimore, it's a snowstorm, not the Apocalypse." She continued, "Do you only live on bread, orange juice, and frozen pizza?" My daughter says that milk, bananas, and ice also go fast. And batteries and generators are among the first things to disappear too.

More recently, in the COVID-19 pandemic, for a time, it was nearly impossible to find pasta, dry beans, flour, yeast, and, of course, toilet paper. And when workers at meat-processing plants were hit with the disease, meat production fell and there were shortages of beef, pork, and chicken.

People want something convenient to eat. It makes sense to have a pantry filled with easy-to-fix meals so you do not have to rush to the grocery store in a panic!

Stockpiling for a Short-Term Crisis

Along with shelter and water, a short-term food storage plan is a key part of preparing for a crisis. Most short-term food storage plans are based on the foods you normally eat. A reasonable goal is to begin by stockpiling enough food for two weeks and work up to one month. If desired, you can increase that to two or three months.

FIVE THINGS YOU CAN DO NOW

1 Clean out and organize your pantry. Throw away old food.

2 Buy two cans instead of one and store the extra can.

3 Keep track of the meals your family eats for one week.

4 Adapt one or more favorite recipes to use foods you can store.

5 Plan three menus that you could build out of foods that can be stored in your pantry.

Ready-to-Eat Meals

When the disruption and chaos of a crisis confronts you, a supply of quick and easy food that requires minimum preparation is crucial.

Plans that Duplicate Current Eating Habits

The Copy Canning and the Slice of Time plans attempt to duplicate your regular eating habits. Both are appealing because they are familiar and based on your present routine and nutritional needs. Both will help you identify the foods your family likes and uses. A rotation system will track what is used and what needs to be replaced.

Storing the foods you normally eat can be an expensive option if you try to duplicate fresh meat and produce with their freeze-dried counterparts. But if you substitute alternatives such as canned meat or canned fruits and vegetables, the expense is just a matter of purchasing more of your regular foods, which you'll eventually eat. Preparing these foods in a crisis may also be difficult if you are without normal energy sources. However, any canned food can be eaten unheated. And if you are like a typical American and eat as much as 30 percent of your meals away from home, you'll need to compensate for that. Table 15.1 summarizes the pros and cons of these plans.

Copy Canning Plan

Copy canning is simply buying two of everything you regularly use, putting one into storage and consuming the other. You will gradually stock up a food supply typical of your normal diet. Obviously, not all the food you purchase comes in a can or other storage-friendly packaging, and much of it is perishable. This approach is also a bit random and unsystematic. It assumes you buy everything on a daily or weekly basis, while that may not be your actual practice.

To get started and "do something," you may want to try this approach, and then, when you have a good stockpile, take inventory and figure out where you need to fill in the gaps.

Slice of Time Plan

This second plan is more accurate. Begin by keeping a detailed record of all the food consumed during a period, usually two weeks to a month. Multiply the total amount of food used during that period to find the amount necessary for a longer time. Make a record of both a summer and a winter period to account for seasonal variations. Use worksheet 15.1, "Slice of Time Daily Menu Log," on page 110 to help you record food consumed. It is also found as a downloadable PDF on our website, CrisisPreparedness.com.

Like the Copy Canning Plan, one challenge with this method is that normal menus

Table 15.1 Advantages and Disadvantages of Plans Based on What You Normally Eat	
Advantages	Disadvantages
Foods that are your normal preferences	Requires inventorying and planning to create
Familiar foods during stressful times	Difficult to store many foods in normal diets
You already have the needed food-preparation skills	May not be nutritionally balanced
Well suited for a two- to three-month storage rotation	Expensive alternatives may be needed to replace fresh meats and vegetables

contain fresh meat, dairy, eggs, and produce as well as perishable items, like bread, crackers, and chips. To make these plans work, you'll need to make adaptations and substitute alternatives that store well.

Rotating Menus Plan

This plan solves the "What shall I fix for dinner?" question and is built around the foods your family likes. You begin by planning meals that lend themselves to storage foods. Then you incorporate them in a rotating schedule. Select as many different meals as you desire to rotate through. First, decide how often each meal will be used. Next, calculate the total individual ingredients needed to last the length of time you determine. Use worksheet 15.2, "Rotating Menu Plan," on page 111 or on our website to help organize menus and shopping lists.

Choose at least seven main courses and three to five breakfast meals to get started. It may be easier to begin by purchasing enough food to cycle through the meals for two weeks.

Gradually build up to one month and then three months. Also add foods for lunches, your favorite comfort foods, and desserts.

This method works well in combination with long-term food storage. You can rely on the familiar foods for a crisis of short duration, and then, if the crisis is prolonged, you'll have the basic food storage to back it up. Table 15.2 summarizes the advantages and disadvantages of using rotating menus.

Table 15.2 Advantages and Disadvantages of Rotating Menus	
Advantages	Disadvantages
Uses familiar family recipes	Requires careful calculation and effort
Familiar food during stressful times	Rigid adherence to menus
Easy to rotate food	May not be nutritionally balanced
Ideal for a two to three-month storage rotation	Menu fatigue with limited meal choices

PERSONALLY SPEAKING

You may be wondering what kinds of recipes to use for the Rotating Menus Plan. I begin by thinking about what recipes I already prepare for my family. Then I think how they can be made using foods I can keep in my pantry. Recipes for soups, stews, and casseroles can usually be adapted. When I am creating a menu, I think "cans and boxes." I am always on the lookout for a recipe I can adapt to foods I can store. You will find examples of the Rotating Menu Plan, including sample menus on our website.

Stocking Your Pantry and Food Storage Room

Stocking a pantry is a time-honored, organized way to have the food you regularly use on hand so you can prepare a meal without a trip to the grocery store. It is accessible and ready-to-use. It helps you use and rotate the storage food you already have.

Typically, the pantry is a small room or cupboard in or near the kitchen. It contains cans and cartons of the nonperishable dry goods you regularly use. It is where you keep small, open containers of bulk foods, like sugar, flour, beans, rice, wheat, and oatmeal. Your refrigerator is also part of your pantry.

A food storage room is like a mini grocery store in your home and is located in a basement or may be an extension of a large pantry or closet near your kitchen. (A garage is not a good place to store food because temperatures and humidity fluctuate.) This is where you keep food packaged in buckets and #10 cans for long-term storage. It may contain cases of canned food, bottles of home-canned foods, and extra cartons, cans, and bottles of food purchased on sale. When food in the pantry is used, it is replaced with food from the storage room. A freezer is also part of the food storage no matter where it is located.

PERSONALLY SPEAKING

My pantry is my go-to cupboard when I am preparing meals. It contains the foods I regularly use. I keep flour and sugar in storage buckets with gamma seals. My beans, oatmeal, and grains are stored in airtight OXO POP containers, and I also have a big container of our favorite homemade granola. I keep smaller items in small storage containers for easy access. My pantry also contains canned goods as well as crackers, chips, cereals, snacks, and cooking ingredients.

My food storage room is where I keep a good backup supply of things like pasta, canned beans, vegetables, tomato products, tuna fish, canned chicken, soup base, brown sugar, peanut butter, chocolate chips, and condiments. I replenish my pantry from the storage room, and I replenish the storage room when items go on sale at my local grocery store. I keep boxes of home-canned foods and canning supplies on the lower shelves. This is also where my long-term food storage is stored in #10 cans and poly buckets. The meats, fruits, vegetables, and convenience foods in my freezer are also part of this storage.

Commercial Food Storage Packages

Some people like the convenience of purchasing prepackaged commercial food storage. They are offered in one-month, three-month, six-month and twelve-month modules. It might seem like a good idea to save time and effort to go with something already put together. For more detailed information about the usefulness of commercial food storage packages, see the section on dehydrated and freeze-dried foods in chapter 14.

As a review, there are three main objections to commercial packages. First, they are expensive. Second, they often do not live up to their claims of furnish-

QUICK CHECK

Evaluating a Commercial Package

✓ Does a daily menu consist of at least 2,600 calories?
✓ Is it balanced nutritionally?
✓ Does it include fruits and vegetables?
✓ Are there at least 40-60 grams of protein per day?
✓ Are there sufficient oils?
✓ Does it contain REAL meat, cheese, vegetables, and fruit?
✓ Does it have enough variety?
✓ Does it include seasonings and leavenings?
✓ Are you able to personalize the plan?
✓ Are you able to sample all items in the unit?
✓ Does everyone in your family like the food?
✓ Is the higher cost worth the convenience?

ing a delicious and nutritious diet; their goal is to provide calories and protein as profitably and, therefore, as cheaply as possible. Third, these meals all have a "sameness" about them that becomes monotonous.

If you are considering purchasing a commercial unit, use the Quick Check to help you evaluate whether the plan will work for you.

PERSONALLY SPEAKING

 You might be able to tell that I am not very impressed with commercial food storage packages. They are expensive for what they offer and, frankly, I think you are much better off taking the time to plan your own short-term food storage. If you want quick convenience foods for an emergency, then I recommend you start with Mountain House freeze-dried entrées. There are more than ten different entrées to choose from They just need hot water to reconstitute within the pouch. One of my sons is an avid backpacker and he also likes entrées from Backpacker's Pantry and Alpine Air for more variety.

I really like the Rotating Menu Plan! It helps me plan for all the ingredients to make a quick meal. And these menus are perfect for those days when you need a quick meal, emergency or not, because you have everything you need! Just remember to replace the food the next time you shop.

Worksheets to Help You

Use worksheets 15.1 on page 110 and 15.2 on page 111 to help you plan for "Slice of Time Daily Menu" or the "Rotating Menus Plan." For a downloadable PDF file, go to CrisisPreparedness.com.

| | Worksheet 15.1 | | |
	Slice of Time Daily Menu Log Week # _____		
	Breakfast	Lunch	Dinner
Day 1			
Day 2			
Day 3			
Day 4			
Day 5			
Day 6			
Day 7			

©*Patricia Spigarelli-Aston. For a downloadable PDF file, go to CrisisPreparedness.com.*

	Worksheet 15.2 Rotating Menus Plan from _____ to _____	
	Menu	Food Items to Complete Menu
Menu 1		
Menu 2		
Menu 3		
Menu 4		
Menu 5		
Menu 6		
Menu 7		

NOTES

Basic Long-Term Food Storage Plan

This chapter will give you options for basic long-term food storage. In the next chapter, we will look at planning a customized long-term food storage. The goal is to give you the tools to create the best possible long-term food storage plan so you are confident and prepared in a time of crisis. Once you learn about the different methods of planning, you can decide which one or which combination of plans works best for you.

The information in this chapter will be useful if you are just beginning to prepare or if you want simplicity. If you already have a plan, this will help you identify possible shortcomings and those areas that might need some refining. You will also be able to evaluate commercial packages.

The emphasis will be on stockpiling food for one year, but you can adapt it for shorter or longer periods. By following a few simple steps, you can determine how much of the various foods groups you need to store.

The Basic Four Plan

The Basic Four Plan is the simplest, most trouble-free of all plans. In the past, it was often referred to as the Mormon Basic Four Plan, due to its origin, and consists of the four basic food items listed in table 16.1.

The plan supplies adequate protein (ninety-four grams per day) and is absolutely the cheapest and most compact way of storing a year's supply of food. Except for the milk, it has a nearly unlimited shelf life. If stored in square containers, it takes up less than twelve cubic feet for one person.

Unfortunately, it also has some major drawbacks: Most importantly, it is lacking in fat. It is also low in vitamins A, C, and D. It offers an extremely limited variety of foods and

Table 16.1 Basic Four Plan for One Year	
Food Item	Quantity per Person*
Wheat	300 lb.
Powdered milk	100 lb.
Sugar or honey	100 lb.
Salt	5 lb.
*Based on 2600 calories per day.	

FIVE THINGS YOU CAN DO NOW

1 Decide if storing basic long-term food is a goal for your family.

2 Choose how long you want food storage for. Three months? Six months? One year?

3 Determine the number of population equivalents for your family.

4 Decide on a food storage location and how much room you can dedicate to storing food.

5 Make a list of the equipment you will need to use the food storage.

requires considerable skill and effort to prepare appetizing meals. It provides a very austere, subsistence-level diet that is drastically different from what people in developed countries are accustomed to. And finally, it is somewhat low in calories, containing less than the American average of 2,600 per day.

The 7-Plus Basic Plan

For those who want an improved but uncomplicated basic food storage plan, Jack and I developed the 7-Plus Basic Plan. It overcomes some of the problems with the Basic Four Plan by adding just a few items. It has similar advantages—low cost, minimal need for rotation, compactness, and low weight. It contains 20 percent more calories, provides better nutritional balance, includes fats, and offers a slightly improved variety of foods. Most items have a very long shelf life, but you'll need to rotate the oil, vitamins, leavening agents, and yeast because their shelf life is more limited. See table 16.2 for quantities. You will also find worksheet 16.1, "Determining Your Population Equivalent, on page 118 and worksheet 16.2, "Calculating Amounts for the 7-Plus Plan," on page 119 to help you calculate the amount of food you need for your family. The worksheets are also available as downloadable PDF files on our website, CrisisPreparedness.com.

> The 7-Plus Basic Plan is a good place to begin if you want an uncomplicated food storage plan.

The amounts given in table 16.2 provide about 2,600 calories, one hundred grams of protein, and thirty-five grams of fat per person per day for one year.

	Table 16.2 7-Plus Basic Plan for One Person for One Year		
	Food	Quantity per Person*	Shelf Life
1	Grains (wheat, rice, corn)	375 lb.	30+ years
2	Legumes (beans, peas, lentils)	60 lb.	30+ years
3	Sugar	65 lb.	30+ years
4	Milk, nonfat dry	60 lb.	20 years
5	Oil (1 gallon of cooking oil = 7.5 lb.)	21 lb.	2–3 years
6	Salt (table, pickling & canning)	10 lb.	30+ years
7	Multi-vitamins with minerals	365	4–5 years
+	Leavening agents, yeast, spices and herbs, and flavorings	1 lb. baking powder 1 lb. baking soda	5+ years
		¾ lb. yeast	2–4 years
*Based on 2,600 calories per day.			

©Patricia Spigarelli-Aston

Specialized Cooking Equipment

You will need a grain mill and specialized recipes to prepare appetizing menus. You may also want to invest in a heavy-duty mixer for making bread and a pressure cooker for cooking beans and legumes. Refer to chapter 19 for information about grain mills and other specialized kitchen equipment.

Combining Plans

Use the 7-Plus Basic Plan to complement one of the short-term food storage plans described in chapter 15, or upgrade and personalize this basic plan by substituting any of the wide variety of foods discussed under the Custom Advanced Plan in chapter 17.

How Much Food Do You Need?

Using Averages to Determine How Much Food You Need

The easiest way to estimate how much you need of each item is to simply multiply the amount for one person by the number of people you are storing food for. For example, if you have four persons in your family, the amount of wheat you would store is 1,500 pounds (4 x 375). You can increase or decrease the amounts for your situation, adding 5 percent for waste and unforeseen needs.

Using Population Equivalents to Determine How Much Food You Need

To be more exact in calculating your family's needs, you'll need to determine the population equivalents for your family.

What Is a Population Equivalent?

A population equivalent is the percentage of the average number of calories a specific population segment needs. The typical number of calories consumed by an average American is 2,600 per day, but different segments of the population require different amounts of calories, some more and some less than 2,600 calories. Table 16.3 shows you the population equivalents for different segments.

Table 16.3 Population Equivalents		
Category	Age of Person	Population Equivalent % of 2600 Calories
Infants	0-1	0.35
Children	1-3	0.52
	4-8	0.81
Males	9-13	1.04
	14-18	1.25
	19-30	1.33
	31-50	1.2
	51+	1.15
Females	9-13	0.98
	14-18	0.94
	19-30	0.88
	31-50	0.85
	51+	0.83

©Patricia Spigarelli-Aston

PERSONALLY SPEAKING

In the original Crisis Preparedness Handbook, we wanted to give you a way to be as accurate as possible in determining how much food you would need for long-term emergencies. We thought it was important to consider the unique nutritional needs of different age groups and genders, so Jack created the "population equivalent" factor. I knew this method could help you be more exact, but I wondered if it might be too complicated and if it would really be used.

So I talked it over with my daughter who has three little girls. She told me, "Mom, it was totally easy. I just followed the steps the way Dad explained it. For sure, include it." So I was convinced that many of you'll want to use population equivalents as you calculate food storage quantities. It was ideal for my daughter since her girls have tiny appetites—not anywhere near as much as a teenage boy or grown man.

I decided to provide both methods of calculating—using averages and population equivalents. That way you can choose which one works best for you.

The total quantities of food will still be an estimate but closer than just using averages. Remember to add 5 percent to your final quantities for waste or unforeseen needs.

Follow the Example Family

To help you follow this and later calculations, I will illustrate by using an "example family." You will see exactly what they would do at each step of designing their plan for two years out. The specific food selections for the example family are not recommendations but are representative of practical, economical choices. You should decide what selections are best for you.

The example family consists of two middle-aged adults, one male and one female, and two teenagers—a seventeen-year-old boy and a fourteen-year-old girl.

Population Equivalents for the Example Family

The example family begins by determining the population equivalent for each person in the family. Their calculation is based on two years. Next, they calculate the total. See example 16.1.

Calculating How Much Food the Example Family Needs

The example family determines their total population equivalent is 4.35. They multiply 4.35 times the quantities for one person to determine the quantities they will need of each food item. They also add 5 percent for potential waste and unforeseen needs. Next, based on their preferences, they allocate the total for each basic

Example 16.1 Determining Population Equivalents for Your Family				
Category	Age of Person	Number of Persons in the Category	Population Equivalent % of 2,600 Calories Required	Population Equivalent Totals
Infants	0–1	_____	x 0.35 =	
Children	1–3	_____	x 0.52 =	
	4–8	_____	x 0.81=	
Males	9–13	_____	x 1.04 =	
	14–18	_____	x 1.25 =	
	19–30	1	x 1.33 =	1.33
	31–50	1	x 1.23 =	1.23
	51+	_____	x 1.15 =	
Females	9–13	_____	x 0.98 =	
	14–18	1	x 0.94 =	0.94
	19–30	_____	x 0.88 =	
	31–50	1	x 0.85 =	0.85
	51+	_____	x 0.83 =	
		Total Population Equivalent		4.35

food among specific foods. Example 16.2 shows their choices. (The choices are representative and not recommendations.)

	Example 16.2 Example Family Storage List		
Basic Food	Calculation	Approximate Amount	Amounts to Store
Grains	4.35 x 375 = 1,632.5 0.05 x 1,630 ≈ 81 Total = 1,713.5 lb.	1,700 lb. grain	1,250 lb. wheat, 25 lb. barley, 200 lb. rice, 50 lb. corn, 100 lb. oats, 25 lb. rye, 50 lb.
Legumes	4.35 x 60 = 261 0.05 x 261 ≈ 13 Total = 274 lb.	275 lb. legumes	100 lb. pinto, 30 lb. black, 30 lb. red, 15 lb. lentils, 30 lb. navy, 15 lb. peas, 25 lb. soy, 30 lb. Great Northern
Sugar	4.35 x 65 = 282.75 0.05 x 283 ≈ 14 Total = 296.75 lb.	300 lb. sugar	250 lb. granulated sugar, 30 lb. brown sugar, 20 pounds powdered sugar
Milk, nonfat dry	4.35 x 60 = 261 0.05 x 261 ≈ 13 Total = 274 lb.	275 lb. milk, nonfat dry	275 lb. milk, nonfat dry
Oil	4.35 x 21 = 91.35 0.05 x 91 ≈ 4.5 Total = 95.85 lb.	100 lb. oil	8 gallons liquid oil (7.5 lb./gal) 40 lb. shortening)
Salt	4.35 x 10 = 43.5 0.05 x 44 = 2.2 Total 45.7 lb.	45 lb. salt	25 lb. table salt, iodized, 20 lb. pickling and canning
Multivitamins with minerals	4 x 365 = 1,460		1,460 multi-vitamins with minerals
Yeast	4.35 x .75 ≈ 3.25 lb.		3.25 lb. yeast
Baking soda	4.35 x 1 = 4.35 lb.		4.5 lb. baking soda
Baking powder	4.35 x 1 = 4.35 lb.		4.5 lb. baking powder

7-Plus Basic Plan Calculator

To make it even easier, we created a food storage calculator based on the 7-Plus Basic Plan that uses population equivalents to determine quantities. You can access it on our website at CrisisPreparedness.com.

Worksheets to Help You

Use worksheets 16.1 on page 118 and 16.2 on page 119 to help you determine the population equivalents for your family and to help you calculate how much food you need. For a downloadable PDF file, go to CrisisPreparedness.com.

Worksheet 16.1 Determining Your Family Population Equivalent				
Category	Age of Person	Number of Persons in the Category	Population Equivalent % of 2,600 Calories Required	Population Equivalent Totals
Infants	0–1	_____	x 0.35	=
Children	1–3	_____	x 0.52	=
	4–8	_____	x 0.81	=
Males	9–13	_____	x 1.04	=
	14–18	_____	x 1.25	=
	19–30	_____	x 1.33	=
	31–50	_____	x 1.23	=
	51+	_____	x 1.15	=
Females	9–13	_____	x 0.98	=
	14–18	_____	x 0.94	=
	19–30	_____	x 0.88	=
	31–50	_____	x 0.85	=
	51+	_____	x 0.83	=
		Total Population Equivalent		=

©*Patricia Spigarelli-Aston. For a downloadable PDF file go to CrisisPreparedness.com*

Worksheet 16.2
Calculating Amounts for the 7-Plus Basic Plan

Total population equivalent for your family: _____ (Use calculation from worksheet 16.1 on page 118)
Use your family-population-equivalent number in the calculations below.

Basic Food	Calculation	Amount	Amounts to Store
Grains	_____ x 300 =	_____ lb. grain	*List selected grains and amounts to store.*
Legumes	_____ x 60 =	_____ lb. legumes	*List selected legumes and amounts to store.*
Sugar	_____ x 65 =	_____ lb. sugar	*List selected sugars and amounts to store.*
Milk, nonfat dry	_____ x 60 =	_____ lb. milk	*List the total pounds of milk to store.*
Oil	_____ x 21 =	_____ lb. oil	*List selected oils and amounts to store.*
Salt	_____ x 10 =	_____ lb. salt	*List selected salts and amounts to store.*
Multivitamins with minerals	Multiply number of people by 365. _____ x 365 = _____ vitamins		
Yeast	Multiply number of people by 0.75 lb. _____ x 0.75 = _____ lb. of yeast		
Baking soda and Baking powder	Multiply number of people by 1.0 lb. _____ x 0.75 = _____ lb. of baking soda and baking powder		

©Patricia Spigarelli-Aston. For a downloadable PDF file go to CrisisPreparedness.com

Custom Advanced Long-Term Food Storage Plan

The great advantage of the Custom Advanced Food Storage Plan is that it closely matches a normal diet, will work with any storable food, and is nutritionally balanced. It is also very flexible and will allow as much variety as you want. But it will cost substantially more than the 7-Plus Basic Plan because it includes fruits, vegetables, and meats. It will also require more rotation, weigh more, and take up more space. Unlike one-size-fits-all plans, it is personalized and will require some calculating as well as a commitment in time and effort.

PERSONALLY SPEAKING

If you are looking for a thorough, comprehensive food plan tailored to the uniqueness of your family, you'll like the Custom Advanced Plan. But I should warn you that it is quite complicated, and the math might be intimidating. I have done my best to give you a step-by-step process you can manage with a simple calculator. In my family, some of my adult children say it's not hard to figure out, while others say they are sticking with the simplicity of the 7-Plus Basic Plan.

Using the Custom Advanced Food Storage Plan

As in the previous chapter, each calculation is illustrated using a hypothetical family. Where possible, I have given you options for making it simpler.

Throughout the chapter you'll find references to worksheets to help you plan and calculate your food storage. These are located at the end of the chapter and are also available as downloadable PDF files on our website, CrisisPreparedness.com.

Modify to Meet Your Needs

The Custom Advanced Plan is designed to provide a year's worth of food storage. If the sheer amount of food is overwhelming or does not fit your needs, modify it. Consider planning for a shorter period, perhaps three or six months, rather than a year. Or just choose the parts most useful for you. For example, you may only want to add fruits and

---

FIVE THINGS YOU CAN DO NOW

1 Calculate how many calories your family needs for one day.

2 Figure out how much salt your family needs and get it.

3 Give powdered milk a try. Check out several brands.

4 Take inventory of how many servings of fruits and vegetables you currently have in your pantry.

5 Experiment with grains or beans you haven't used before.

---

vegetables to the 7-Plus Basic Plan, or you may decide not to add meat at all, or to add a greater variety of grains and legumes.

The foods discussed in the Custom Advanced Food Storage Plan are described individually and in detail in chapter 18.

> **The Custom Advanced Plan lets you eat as close to a normal diet as possible.**

PERSONALLY SPEAKING

The first food storage plan Jack created was the 7-Plus Basic Plan. It was an improvement over earlier plans and was appealing because of its simplicity, but he was not satisfied with it. He wanted a way for people to have food storage that was closer to the way they lived and ate as opposed to "just surviving." He wanted a plan that was comprehensive, practical, adaptable, and appetizing, and so he labored over this section of the book, researching, studying, and calculating, often discussing his ideas with me. At last he came up with his most powerful food storage creation: the Custom Advanced Food Storage Plan. This chapter is the heart and soul of Jack's original contribution to food storage for preparedness.

He wanted to help people to be thoroughly prepared for any crisis and knew many people felt it a personal responsibility to be as self-reliant as possible. He wanted the Custom Advanced Food Storage Plan to help people reach their preparedness goals. I hope it will make preparing easier and give you peace of mind.

STEP 1 – Determining Calorie Requirements

The first task in the Custom Advanced Plan is to calculate how many calories each person in your family requires daily. It will only be an estimate since varying physical activities and individual metabolisms affect caloric needs. There are three ways to ways to calculate your caloric requirements.

Using Averages to Determine Caloric Needs

The first is to use averages. Table 17.1 shows the average number of calories used by different genders and ages based on the U.S. government's *Dietary Guidelines for Americans, 2015*.

Totaling the caloric needs for your family is simple. Assume a high level of physical activity, as that will be likely during a crisis. Also, be generous and use the high number of the range. In addition, consider adding 15 percent extra as a cushion. Example 17.1 illustrates how this is calculated for the example family.

Example 17.1 Example Family Estimated Daily Caloric Needs			
Category	Age	Activity Level	Calories Needed
Male	31-50	High	3,000
Female	31-50	High	2,200
Male	14-18	High	3,200
Female	14-18	High	2,400
		Subtotal Calories	10,800
		Additional 15%	1,620
		Total Daily Calories	12,420

		Table 17.1 Estimated Daily Caloric Needs* by Age, Gender, and Physical Activity		
Category	Age	Sedentary Activity Level	Moderate Activity Level	High Activity Level
Child	2–3	1,000–1,200	1,000–1,400	1,000–1,400
Male	4–8	1,200–1,400	1,400–1,600	1,600–2,000
	9–13	1,600–2,000	1,800–2,200	2,000–2,600
	14–18	2,000–2,400	2,400–2,800	2,800–3,200
	19–30	2,400–2,600	2,600–2,800	3,000
	31–50	2,200–2,400	2,400–2,600	2,800–3,000
	51+	2,000–2,200	2,200–2,400	2,400–2,800
Female	4–8	1,200–1,400	1,400–1,600	1,400–1,800
	9–13	1,400–1,600	1,600–2,000	1,800–2,200
	14–18	1,800	2,000	2,400
	19–30	1,800–2,000	2,000–2,200	2,400
	31–50	1,800	2,000	2,200
	51+	1,600	1,800	2,000–2,200

*Based on "Estimated Caloric Needs per Day by Age, Gender and Physical Activity Level," *Dietary Guidelines for Americans, 2015*, published by the USDA and U.S. Department of Health and Human Services.
Average Male: 5 feet 10 inches tall, weighing 154 pounds. Average Female: 5 feet 4 inches, weighing 124 pounds.

©*Patricia Spigarelli-Aston*

Using Body Weight to Determine Caloric Needs

The second way to determine the number of calories is more precise and based on body weight. Table 17.2 on page 124 gives you the number of calories used per pound of body weight for three activity levels.

To determine daily caloric needs, multiply body weight by the multiplier for the specific age, gender, and activity level. It requires a little more calculating, but it will give you more accurate caloric requirements if your body weight and height are significantly different from those for the average heights and weights used in table 17.1.

Of course, this method is still based on average metabolism, and yours may differ. If you anticipate even greater exertion and caloric demands, you can increase the calories. For reference, a large man doing hard labor or hunting on a cold day with snow on the ground can require more than six thousand calories!

Example 17.2 shows how the example family calculates their caloric needs based on body weight. Use worksheet 17.1 on page 151 to help you calculate your family's caloric needs. The worksheet can also be found on our website, CrisisPreparedness.com.

Table 17.2 Estimated Daily Caloric Needs per Pound of Body Weight* by Age, Gender, and Physical Activity				
Category	Age	Sedentary Activity Level	Moderate Activity Level	High Activity Level
Child	2–3	54	54	54
Male	4–8	49	49	49
	9–13	47	47	47
	14–18	28	30	32
	19–30	22	24	26
	31–50	18	20	22
	51+	16	18	20
Female	4–8	16	18	20
	9–13	14	16	18
	14–18	17	19	21
	19–30	16	18	20
	31–50	15	17	19
	51+	14	16	18

The calories per pound of body weight are only estimates and should be adjusted to fit your personal needs. Caloric data is based on the estimated caloric needs tables published in the 2015 *Dietary Guidelines for Americans*.

©*Patricia Spigarelli-Aston*

Example 17.2 Example-Family Estimated Caloric Needs Based on Body Weight					
Category	Age of Person	Activity Level	Weight	Calories per Pound of Body Weight	Caloric Needs
Male	31–50	High	175 x	22 =	3,850
Female	31–50	High	136 x	19 =	2,584
Male	14–18	High	165 x	32 =	5,280
Female	14–18	High	116 x	21 =	2,436
				Subtotal Calories	14,150
				Additional 15%	2,123
				Total Daily Calories	16,273

If you compare the daily caloric needs for the example family based on averages with those based on body weights, there is almost a four-thousand-calorie difference in estimated daily caloric needs.

PERSONALLY SPEAKING

 I prefer to figure the number of calories we need based on body weight. This would be good for you, too, if your family members, like mine, are taller and bigger than average. For example, my six-foot, four-inch sons who weigh close to 200 pounds need approximately 4,000 calories a day at a high activity level as opposed to 2,800 to 3,000 for the average-sized man their age. That is almost a third more!

Using an Internet Calculator to Determine Caloric Needs

A quick internet search will give you several calculators with which to determine your caloric needs. Most require that you input your gender, age, height, weight, and activity level. Any of them will do, but the downside is that most focus on weight loss, and so the emphasis is on keeping calories low.

STEP 2 – Determining How Much Salt

Salt is one of the easiest things to acquire and store. It does not take up much space and stores indefinitely. It is very inexpensive, so there is no excuse not to have it.

According to the Centers for Disease Control and Prevention (CDC), Americans should consume between 1,500 and 2,300 mg of sodium per day, which is between ¾ and 1 teaspoon of salt. For that amount, one person would need between three to five pounds of salt for a year.

In comparison, the actual average amount of sodium consumed by Americans is between 3,500 and 4,000 mg per day. For one year, that's between seven and eight pounds per person.

These amounts do not consider that a small amount of sodium is found naturally in many foods and that processed foods often have high levels of added salt. Also, sodium needs may increase with extra exertion, higher temperatures, or illness. In a long-term crisis, you may want extra salt for brining and canning and possibly for bartering.

> **QUICK FACTS**
>
> ### Uses for Salt
> - Maintaining bodily functions
> - Enhancing flavors
> - Preserving food
> - Controlling fermentation in baking
> - Making cheeses and butter
> - Smoking meats
> - Tanning leather
> - Gargling and brushing teeth
> - Bartering

> **To keep it simple, store five pounds of table salt and another five pounds for other uses per person.**

With these considerations in mind, it's wise to store a minimum of ten pounds of salt per person. Store half of it as table salt and half as pickling salt. See example 17.3 on page 126.

STEP 3 – Determining How Much Milk

The Custom Advanced Plan uses the recommended daily allowance (RDA) for calcium to determine how much powdered milk you should store. Powdered milk is easy to store and is a good way to get calcium in a survival diet. Milk products also supply potassium and protein and are usually fortified with vitamins A and D. If you prefer not to store milk, you'll need to plan on other sources for these nutrients.

Example 17.3 Example Family Salt Needs		
Table salt	5 lb./ person	20 lb./ family
Pickling salt	5 lb./ person	20 lb./ family
	Total	40 lb.

The amount of milk in the Custom Advanced Plan is based on the calcium recommendations by the 2015 *Dietary Guidelines for Americans*. The quantities suggested assume that 85 percent of the total calcium RDA come from milk and milk products. Table 17.3 shows you the RDA for calcium based on age and gender and the amount of powdered milk that supplies 85 percent of that RDA.

Determining How Much Milk You Need

Using an Estimation

You can determine how much milk you need in two ways. The first lets you skip any calculations. Simply use an estimate of between 50 and 70 pounds of milk per person. The high amount is generous and gives you plenty of extra milk for making yogurt or cheese.

Using Calculations

Table 17.3 Calcium RDA by Age and Gender and Pounds of Powdered Milk or Milk Equivalents to Meet 85 Percent of the Calcium RDA for One Year			
Category	Age of Person	RDA for Calcium	Pounds of Powdered Milk for One Year*
Children	1–3 yr.	700 mg	36.75 lb.
Children	4–8 yr.	1,000 mg	52.50 lb.
Male	9–13 yr.	1,300 mg	68.25 lb.
	14–18 yr.	1,300 mg	68.25 lb.
	19–30 yr.	1,000 mg	52.50 lb.
	31–50 yr.	1,000 mg	52.50 lb.
	51–70 yr.	1,000 mg	52.50 lb.
	71+ yr.	1,200 mg	63.00 lb.
Female	9–13 yr.	1,300 mg	68.25 lb.
	14–18 yr.	1,300 mg	68.25 lb.
	19–30 yr.	1,000 mg	52.50 lb.
	31–50 yr.	1,000 mg	52.50 lb.
	51–70 yr.	1,200 mg	63.00 lb.
	71+ yr.	1,200 mg	63.00 lb.

* The recommended amounts fulfill 85% of the calcium RDA. Amounts are based on 5.25 pounds of powdered milk supplying 100 mg of calcium per day for one year.

©Patricia Spigarelli-Aston

The most accurate method calculates the amount of milk you need based on the calcium requirements for different ages and genders. Table 17.3 outlines this for you. Use it to calculate how many pounds of milk your family needs. Remember that this calculation is for 85 percent of the total calcium needs and that other foods will supply the rest. This calculation is demonstrated with the example family in example 17.4. To help you calculate your family's milk needs, use worksheet 17.2 on page 152 and at CrisisPreparedness.com.

Using Milk Equivalents

You can purchase all your milk in powdered form, but for more variety, substitute 10 percent to 20 percent with some of the milk sources listed in table 17.4. Notice that milk products listed in the table have different amounts of calcium per serving. A multiplier is listed for each product to help you convert it to the calcium equivalent of powdered milk. See chapter 18 for a complete description of individual milk sources.

Example 17.4 Example Family Calcium Needs for One Year			
Category	Age of Person	RDA in mg of Calcium	Pounds of Powdered Milk or Equivalent Milk Products
Male	31–50	1,000	52.50 lb.
Female	31–50	1,000	52.50 lb.
Male	19–30	1,000	52.50 lb.
Female	14–18	1,300	68.25 lb.
Total Pounds of Powdered Milk or Equivalent Milk Products			225.75 lb.

Table 17.4 Calcium Equivalents for Milk Products				
Additional Milk Sources	Serving Size	Mg of Calcium* in a Serving	Calcium Equivalent (Equivalent to 1 pound of powdered milk)	Calcium Equivalent Multiplier
Powdered milk, regular	3 tbsp.	300 mg	1.0 lb.	1
Powdered milk, instant	4 tbsp.	300 mg	1.0 lb.	1
Powdered buttermilk	4 tbsp.	300 mg	1.0 lb.	1
Canned evaporated milk	½ cup	300 mg	6 12-oz. cans	6
Fresh cheese	1 oz. (28 g)	200 mg	1.5 lb.	1.5
Parmesan cheese, shredded	1 oz. (28 g)	250 mg	1.2 lb.	1.2
Processed cheese product	1 oz. (28 g)	200 mg	1.5 lb.	1.5
Freeze-dried shredded cheeses	0.50 oz. (14 g)	200 mg	.75 lb.	0.75
Dehydrated cheese powder	0.25 oz. (7 g)	200mg	0.375 lb.	0.375

*The quantities are approximate. One pound of powdered milk contains about 4,800 mg of calcium.

©Patricia Spigarelli-Aston

Buttermilk

Useful in cooking and baking, you may want one or two pounds per person. It can be substituted pound for pound for powdered milk.

Evaporated Milk

It takes six twelve-fluid ounce cans to equal the calcium in one pound of powdered milk. I suggest six to twelve cans per person for one year.

Cheese

The average American consumes about twenty-two pounds of cheese a year. Cheese is a more expensive source of calcium than milk, but the variety and added enjoyment is worth it. Equivalents vary with the type of cheese, so use table 17.4 on page 127 to determine specific equivalents.

The Example Family Calculates Their Milk Needs

The example family calculates the amount of milk they need to fulfill 85 percent of their calcium needs. It adds up to 225.75 pounds. See example 17.4 on page 127.

They decide to purchase two hundred pounds of powdered milk and the rest as powdered-milk equivalents. This equals about 226 pounds. Example 17.5 shows their choices and how they've calculated the quantities they plan to store. Use worksheet 17.3 on page 153 to calculate the amount of each milk product you desire. The worksheet can also be found on our website, CrisisPreparedness.com.

Example 17.5 Example Family Milk Products Selection for One Year				
Milk Product	Calcium Equivalent Multiplier	Calcium Equivalents	Conversion Calculation	Quantity to Store
Powdered milk	1	200	1 x 200 =	200 lb.
Powdered buttermilk	1	5	1 x 5 =	5 lb.
Evaporated milk	6	8	6 x 8 =	48 cans
Parmesan cheese	1.2	5	1.2 x 5 =	6 lb.
Processed cheese product	1.5	4	1.5 x 4 =	6 lb.
Freeze-dried cheese	0.75	4	0.75 x 4 =	3.75 lb.
Calcium Equivalents Total		226		

STEP 4 – Determining How Many Fats and Oils

Oil is one of the most important items in your food storage. Fats and oils are necessary for baking, frying, and cooking. They are also important for getting adequate calories and for making food palatable. Liquid oils and shortening should comprise about 8 percent of the total calories.

Oil is one of the most overlooked but most valuable items in your food storage plan.

The 2015 *Dietary Guidelines for Americans* recommends that 20 percent to 35 percent of the calories in an adult diet should come from fat. Growing children require even more. In a survival diet it may be difficult to store enough fats and oils. Aim for 10 to 15 percent of the total calories in your diet coming from the fats and oils group. The rest of the needed fats will come from other foods.

Choosing Fats and Oils

Vegetable oils, shortening, and lard have about 4,000 calories per pound. Bacon, mayonnaise, peanut butter, and nuts are common high-fat foods, but they are not as high in fat as oil or shortening. For example, it takes about 1.5 pounds of peanut butter to equal the fat in one pound of oil or shortening. In table 17.5, the calories for common high-fat foods are compared with the 4,000-calorie baseline for vegetable oils and shortening. Descriptions of individual fats are found in chapter 18.

Figuring Out How Many Fats and Oils You Need

Follow the steps in calculation 17.1 to determine how much fats and oils you need. You can calculate the amount of fats and oils your family will need for one year or use it to calculate the amount for a shorter period.

Oil has a short shelf life. The best-if-used-by date for oil is only two years. With that in mind, choose to store what you can easily rotate in two years. It may be less than a year's supply. In step 4 multiply by the number of days you would like to have oil for in your food storage. Use worksheet 17.4 on page 153 to help you determine the amount of fats and oils you need. The worksheet is also found on our website, CrisisPreparedness.com.

Table 17.5 Calories for Common High Fat Foods		
Fat Source	Approximate Calories per Pound	Equivalent to One Pound of Oil or Shortening
Canola oil	4,000	1.0 lb.
Corn oil	4,000	1.0 lb.
Olive oil	4,000	1.0 lb.
Soybean oil	4,000	1.0 lb.
Lard	4,000	1.0 lb.
Shortening	4,000	1.0 lb.
Shortening, powdered	3,400	1.2 lb.
Butter or margarine	3,300	1.2 lb.
Butter, powdered	3,000	1.3 lb.
Mayonnaise	3,000	1.3 lb.
Bacon	2,500	1.6 lb.
Peanut butter	2,700	1.5 lb.
Peanut butter, powdered	1,900	2.1 lb.
Nuts and seeds	2,500	1.6 lb.
Unshelled nuts	1,300	3.1 lb.

©Patricia Spigarelli-Aston

Calculation 17.1 Calculating How Many Fats and Oils You Need	
Step 1	Determine your family's total daily calories. (Refer to table 17.1 or table 17.2 for caloric requirements.)
Step 2	Decide what percentage of calories you want to come from fats and oils. The recommendation is for 10%–15%.
Step 3	Calculate the daily calories that will come from fats and oils by multiplying your total daily caloric needs by this percentage in decimal form (e.g., 15% =0.15).
Step 4	Multiply the daily calories needed from fats and oils by the number of days you want the oil for. For a year, multiply by 365; for six months, by 180; for three months by 90, etc.
Step 5	Divide this total by 4,000. (There are approximately 4,000 calories in a pound of fat or oil.) This number will be the total number of pounds of fat or oil you need for your selected timeframe.

The Example Family Calculates Their Fats and Oils Needs

The example family decides they will get 10 percent of their calories in fats and oils, their calculations based on 12,400 total daily calories.

In example 17.6, they determine that their daily caloric need from fats and oils is 1,240. They multiply that by 365 and find they need 452,600 calories from fats and oils for a year. They divide that by 4,000 (the number of calories in a pound of oil or shortening) and find out they need 113 pounds of fats and oils or their equivalents, for a year.

Next, they decide which fats and oils to store. Example 17.7 shows that they begin with 40 pounds of vegetable oil and 40 pounds of vegetable shortening. They need 33 more pounds to complete their 113 total. They spread that among several different foods and use table 17.5 on page 131 to find equivalent amounts of fat to calculate conversions.

Use worksheet 17.5 on page 154 to help you select the quantity of different fats and oils to store. It can also be found on our website CrisisPreparedness.com.

Example 17.6 Example Family Fats and Oils Calculation for One Year		
Total daily calories		12,400
Percentage (decimal) from fats and oils	0.10 x 12,400 =	1,240
Total yearly calories from fats and oils	365 x 1,240 =	452,600
Total pounds of fats and oils	452,600 ÷ 4,000 =	113

Example 17.7 Example Family Fats and Oils Selection for One Year				
Oil or Fat	Fats and Oils Equivalent Multiplier	Fats and Oils Equivalent Amount	Conversion Calculation	Quantity to Store
Vegetable oil	1	40	1 x 40 =	40 lb.
Vegetable shortening	1	40	1 x 40 =	40 lb.
Peanut butter	1.5	12	1.5 x 12 =	18 lb.
Bacon	1.6	8	1.6 x 8 =	12.8 lb.
Mayonnaise	1.4	5	1.4 x 5 =	7 lb.*
Dehydrated butter	1.33	6	1.33 x 6 =	8 lb.
Nuts	1.5	2	1.5 x 2 =	3 lb.
Total Oil Equivalents		113		
*About 3 quarts. One quart of mayonnaise contains about 2.25 pounds.				

STEP 5 – Determining How Many Fruits and Vegetables

Fruits and vegetables are the primary sources of vitamins A and C and other important vitamins and minerals in the regular diet. Although multivitamins are a good supplement, real fruits and vegetables are far superior. Besides the nutritional benefit, most people prefer the variety of flavors, textures, and colors fruits and vegetables bring to a diet.

How Many Servings?

Though the USDA nutritional guidelines now suggest as many as thirteen total servings a day, the typical American still eats fewer than five servings, and a crisis would make getting additional servings even more difficult.

If storing a year's supply of fruits and vegetables is daunting, you can use the information to help you plan for fewer servings and shorter periods if you prefer. If your goal is a year's supply, begin with three months and gradually increase the amount. The total servings for three months would simply be one-fourth the amount for a year. Also, remember that a vegetable garden and fruit trees will supplement your stored food.

Fruits

The malnutrition in war-ravaged Europe after World War II was due more to the lack of fruit than anything else. For preparedness purposes, store a minimum of two half-cup servings of fruits per day. This is what the average American eats and is both realistic and practical. As you select fruits and vegetables, be sure to choose at least some that are good sources of vitamin C. Fruit-drink powders fortified with vitamin C can help supply this nutrient.

Vegetables

I recommended three-and-a-half, half-cup servings per person per day. Since most of your vitamin A will come from vegetables, be sure you include items such as carrots, spinach, tomatoes, or sweet potatoes daily.

Potatoes are also important to store. You may want to double the serving size if they are one of your staples. Remaining servings should be divided up among a good selection of vegetables of your choice. You can exchange ten to twelve pounds of sprouting or microgreen seeds to replace one serving of vegetables for a year. They will add fresh variety and crispness to salads, sandwiches, etc. See chapters 24 and 25 for more information about sprouts and microgreens.

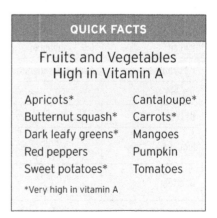

QUICK FACTS

Fruits and Vegetables High in Vitamin A

Apricots*	Cantaloupe*
Butternut squash*	Carrots*
Dark leafy greens*	Mangoes
Red peppers	Pumpkin
Sweet potatoes*	Tomatoes

*Very high in vitamin A

Figuring Out How Many Fruits and Vegetables You Need

Follow the steps in calculation 17.2 on page 132 to determine how many fruits and vegetables to store. The calculation is for one year, but you can use it to figure out servings for a shorter period.

Calculation 17.2	
Calculating How Many Fruits and Vegetables You Need	
Step 1	Begin with fruit. The recommended number of servings is two. Determine the number of servings your family desires per day, then multiply this by the number of days you want it for. A serving size is generally a half-cup serving. Repeat for vegetables. The recommended number of servings for vegetables is three and a half.
Step 2	Decide which fruits and vegetables you would like to store. Consider family preferences, cost, and storage space.
Step 3	For many common fruits and vegetables, refer to tables 17.6, 17.7, and 17.8 for the amounts required to provide one serving per day for one year.* Multiply the amounts by the total number of servings desired or divide the number if you are planning for less than a year.
Step 4	For an item not listed in the tables, you can determine the amount you need by dividing the number of days by the number of servings listed on the container. This will give you the number of containers of that item needed for one serving per day.
Step 5	Continue selecting fruits and vegetables until you have reached the total number of servings your family needs.
* The total servings for three months would be one-fourth the amount for a year; the total for six months would be one-half.	

Canned Fruits and Vegetables

Table 17.6 shows about how many cases of common can sizes you need for one serving per day for one year. You can also use this information for shorter periods by dividing the number of cases.

Table 17.6				
Amount of Canned Fruits or Vegetables for One Serving per Day for One Year				
Can or Jar Size	Typical Contents	Number of Cups per Container	Number of ½-Cup Servings per Case*	Number of Cases* for 1 Year
#303 can	Fruits and vegetables	2 cups	48	8 cases
#2 can	Pineapple	2 ½ cups	60	6 cases
#2½ can	Pumpkin, sauerkraut	3 ½ cups	84	4 ½ cases
# 3 can, 46 fl. oz.	Fruit juices	6 cups	144	2 ½ cases
#10 can	Institutional size	12–13 cups	288–312	1 case
Pint jars	Home canning	2 cups	48	8 cases
Quart jars	Home canning	4 cups	96	4 cases
*For this chart a case is twelve cans or jars. The number of cases is rounded.				

©Patricia Spigarelli-Aston

Fresh Fruits and Vegetables

Some fresh fruits and vegetables will store from six months to a year and can be part of your food storage allotment if you store them in cool conditions. Table 17.7 helps you calculate your fresh fruit and vegetable servings. Approximate quantities are given for both thirty days and one year.

Table 17.7 Amount of Storable Fresh Fruits or Vegetables for One Serving per Day for Thirty Days and for One Year				
Fruit or Vegetable	Serving Size	Serving Weight	Fresh Weight for Thirty Days	Fresh Weight for One Year
Apples	1 apple	5.5 oz.	10.5 lb.	126 lb.
Dried prunes	3 prunes	1.0 oz.	2 lb.	24 lb.
Raisins	¼ cup	1.4 oz.	2.5 lb.	30 lb.
Cabbage	1 cup	3.0 oz.	6 lb.	72 lb.
Carrots	½ cup	2.5 oz.	5 lb.	60 lb.
Sweet potatoes	½ cup	3.0 oz.	6 lb.	72 lb.
Potatoes	1 potato	5.3 oz.	10 lb.	120 lb.

©Patricia Spigarelli-Aston

Dehydrated and Freeze-Dried Foods

Dehydrated and freeze-dried vegetables are available in many varieties. They may come packaged in Mylar pouches or in #2½ and #10 cans.

Serving Size

There is not a standard serving size for dehydrated and freeze-dried foods as they vary significantly among manufacturers. Vegetables especially have a wide range of calories and volume. Also, some dried vegetables, such as onions and celery, are more suitable as recipe ingredients, so the serving size is small.

The calories in a full serving should be 75 to 100 calories for fruits and starchy vegetables and 25 to 50 calories for other vegetables.

The best way to guarantee serving size is by weight. Table 17.8 on page 134 shows the weight you should expect from a serving. It is important to check the weight because the serving size offered by some food storage companies is very small. Calories can also help you check serving size. Fruits and starchy vegetables should be from 75 to 100 calories and vegetables should be from 25 to 50 calories.

Table 17.8 on page 134 shows the amount of dehydrated or freeze-dried foods that make one serving per day for one year. You can adjust amounts for shorter periods. The serving amount is about ½ cup reconstituted.

Table 17.8 Amount of Dehydrated or Freeze-Dried Fruits and Vegetables for One Serving per Day for One Year				
Fruits or Vegetables	Dehydrated		Freeze-Dried	
	Serving Size	Amount Per Year	Serving Size	Amount Per Year
Fruits: apples, bananas, mangoes, peaches, pineapple, berries	1 oz.	22.8 lb.	10 g ~0.4 oz.	8–9 lb.
Vegetables: asparagus, broccoli, carrots, green beans, tomatoes, zucchini	0.5 oz.	11.4 lb.	10 g ~0.4 oz.	8–9 lb.
Vegetables: corn, peas	1 oz.	22.8 lb.	21 g ~0.75 oz.	17 lb.
Onions, celery, peppers, spinach, mushrooms	0.25 oz.	5.7 lb.	2.5 g ~0.1 oz.	3.65 lb.
Dehydrated potatoes	1 oz.	22.8 lb.	NA	NA
Tomato powder, FD tomato chunks	0.5 oz.	11.4 lb.	10 g ~0.4 oz.	8–9 lb.
Fruit-flavored drink mix	0.78 oz./ 22 g	~18 lb.	NA	NA

©Patricia Spigarelli-Aston

The Example Family Calculates Their Fruits and Vegetable Needs

In example 17.8, the example family begins by figuring out how many servings of fruits and vegetables they'll need for one year. (Use worksheet 17.6 on page 155 to help you calculate your family's fruit and vegetable needs.)

Example 17.8 Example Family Fruits and Vegetable Needs			
Category	Age of Person	Recommended Servings of Fruit per Day	Recommended Servings of Vegetables per Day
Male	31–50	2 servings	3 ½ servings
Female	31–50	2 servings	3 ½ servings
Male	19–30	2 servings	3 ½ servings
Female	14–18	2 servings	3 ½ servings
Family Total Servings per Day		8 servings	14 servings
Family Total Servings for 360* Days		8 x 360 = 2,880	14 x 360 = 5,040
*360 days for simpler calculations			

Next, they decide which foods they would like. They look at fruits and vegetables tables 17.6, 17.7, and 17.8 to select quantities for one year. Examples 17.9 and 17.10 show their choices. The 365 days in a year are rounded to 360 for easier computations. Use worksheets 17.7 on page 156 and 17.8 on page 157 to select fruits and vegetables for your storage plan. The worksheets can also be found on our website, CrisisPreparedness.com.

Example 17.9
Selecting Fruits (Refer to Tables 17.6, 17.7, 17.8)

Number of Days: 360* – Number of Servings per Day for Family: 8 – Total Number of Servings for Family: 8 x 360 = 2,880

Fruits	Servings per Day of Listed Fruit	~Total Servings of Listed Fruit	Quantity to Store
Vitamin C fortified drink mix	1	360	18 lb
Fresh apples	0.50	180	63 lb
Dried raisins	0.50	180	15 lb
Dehydrated apple slices	0.50	180	12 lb
Freeze-dried fruit	0.50	180	5 lb
Assorted #303 cans of fruit (4 half-cup servings)	1.50	540	11~12 cases (12 cans per case)
Assorted #2 cans of fruit (5 half-cup servings)	0.50	180	3~4 cases (12 cans per case)
46 fl. oz. cans of fruit juice (11.5 half-cup servings)	1	360	2~3 cases (12 cans per case)
Quarts of home-canned fruit (8 half-cup servings)	2	720	7~8 cases (12 jars per case)
Total Daily Servings of Fruit	8	2,880	

*360 days for simpler calculations

Example 17.10
Selecting Vegetables (Refer to Tables 17.6, 17.7, 17.8)

Number of Days: 360 – Number of Servings per Day for Family: 14
Approximate Total Number of Servings for Family: 14 x 360* = 5,040

Vegetables	Servings per Day of Vegetable	~Total Servings of Vegetables	Quantity to Store
Assorted dehydrated potatoes	4	1,440	91.2 lb.
Dehydrated tomato powder	1	360	11.4 lb.
Assorted dehydrated vegetables	1	360	11.4 lb.
#303 cans of vegetables (4 half-cup servings)	4	1,440	30 cases (12 cans per case)
#303 cans of tomatoes (4 half-cup servings)	1	360	7-8 cases (12 cans per case)
8 oz. cans of tomato sauce (2 half-cup servings)	1	360	7-8 cases (12 cans per case)
46 fl oz cans of tomato juice (11.5 half-cup servings)	1	360	2-3 cases (12 cans per case)
Fresh carrots	0.5	180	30 lb.
Fresh potatoes	0.5	180	60 lb.
Total Daily Servings of Vegetables	14	5,040	

*360 days for simpler calculations

STEP 6 – Determining How Much Sugar

Sugars are an important energy source and increase the palatability of food. The average American consumes about 150 pounds of refined sugar a year, or about 30 percent of total daily calories. The USDA guidelines recommend that this be lowered to 10 percent of total calories, or about 50 pounds of sugar a year. Use this 10 percent recommendation as a base for the refined sugar in your plan. You may want to add 1 to 2 percent for children to help them meet their relatively high-energy needs and an additional ten pounds per person for canning fruit, if desired.

> On average, store about fifty pounds of sugar per person for a year.

Use calculation 17.3 to help you calculate how much sugar you need for one year. You can adjust this amount for shorter periods.

	Calculation 17.3 Calculating How Much Sugar You Need
Step 1	Decide what percentage of your daily calories you want to come from sugars. The recommendation is 10 percent.
Step 2	Multiply this percentage times your total daily caloric needs to find the number of calories that will come from sugar. (See tables 17.1 or 17.2 for caloric requirements.)
Step 3	Multiply the daily calories that will come from sugars by 365 to find the amount for one year. You can adjust this for shorter periods.
Step 4	Divide this total by 1,775 (there are approximately 1,775 calories in a pound of granulated sugar). This number will be the approximate total pounds of sugar you need for a year.

Choosing Sugar and Sugar Equivalents

Regular sugar has about 1,775 calories per pound. Honey, molasses, maple syrup, and other sugar products can be substituted for sugar. Table 17.9 shows the equivalent amounts of alternative sugar products that equal the same as one pound of white, granulated sugar.

The Example Family Calculates Their Sugar Needs

The example family uses their daily calories to total up how many calories should come from sugar. Next, they calculate how many pounds of sugar their family needs for one year. See example 17.11. They select which sugar equivalents they want from table 17.9. Example 17.12 show their choices. Use worksheet 17.9 on page 158

Example 17.11 Example Family Sugar Needs		
Total daily calories		12,420
Percentage of calories from sugar	0.11* x 12,420 =	1,366
Total yearly calories from sugar	365 x 1,366 =	498,590
Total pounds of sugar (rounded)	498,590 ÷ 1,775 =	280

*Since the family has children, they increase their sugar percentage from the recommended 10% to 11%.

to help you calculate how much sugar your family needs to store. Use worksheet 17.10 on page 159 to help you select the sugar products for your food storage. Both worksheets can be found on our website CrisisPreparedness.com.

Example 17.12
Example Family Sugar Selection for One Year

Sugar	Sugar Equivalent Multiplier	Sugar Equivalent Amount	Conversion Calculation	Quantity to Store
Sugar	1 lb.	100	1 x 100 =	100 lb.
Brown sugar	1 lb.	30	1 x 30 =	30 lb.
Powdered sugar	1 lb.	30	1 x 30 =	30 lb.
Honey	1.25 lb.	60	1.25 x 60 =	120 lb.
Maple syrup	1 pt.	5	1 x 5 =	5 pt.
Fruit-flavored gelatin	6 pkg.	10	6 x 10 =	60 pkg.
Pudding mixes	6 pkg.	5	6 x 5 =	30 pkg.
Jams and jellies*	1 pt.	20	1 x 20 =	20 pt.
Vitamin C fortified flavored drink mix	1 lb.	20	1 x 20 =	20 lb.
Total Sugar Equivalents		280		

Table 17.9
Sugar Equivalents for One Pound of White Granulated Sugar

Type of Sugar	Calories per Measurement	Approximate Equivalent of One Pound of Sugar	Type of Sugar	Calories per Measurement	Approximate Equivalent of One Pound of Sugar
White sugar	1,775 / lb.	1 lb.	Agave syrup	1,920 / pt.	1 pt.
Brown sugar	1,676 / lb.	1 lb.	Pancake syrup	1,470 / pt.	1 pt.
Powdered sugar	1,758 / lb.	1 lb.	Jams and jellies	1,600 / pt.	1 pt.
Honey (lb.)	1,382 / lb.	1.25 lb.	Flavored drink mix	1,775	1 lb.
Honey (pt.)	2,048 / pt.	1.15 pt.	Pudding mix	1,705 / six 3.4-oz. pkg.	six 3.4-oz. pkg.
Molasses	1,954 / pt.	1 pt.	Fruit-flavored gelatin	1,786 / six 3-oz. pkg.	six 3-oz. pkg.
Maple syrup	1,664 / pt.	1 pt.	Hard candy	1,775/lb.	1 lb.

©Patricia Spigarelli-Aston

STEP 7 – Determining How Much Animal Protein

Animal products, including meat, eggs, and dairy, provide a complete, balanced protein. Approximately 60 percent of the total protein in a normal American diet comes from meat and eggs. However, because of its expense and difficulty in storing, the recommendation given here provides 10 to 30 percent from meat and eggs. With milk products added, there is sufficient animal protein for a high-quality diet.

Vegetarian Alternative

It is not necessary to add animal protein to your plan. If you prefer to avoid animal protein, skip to step 8. Legumes and grains can supply adequate and complete protein.

Eggs

Eggs are an excellent protein source. You may want them for scrambling, omelets, French toast, and for use in baking. If eggs are an important part of your diet, store enough for three eggs per person per week. Eggs can be stored in dehydrated, freeze-dried, and crystallized form. Follow the steps in calculation 17.4 to determine how much egg product to store. See chapter 18 for a discussion of egg products.

	Calculation 17.4 Calculating How Much Egg Product You Need
Step 1	Decide on the number of servings you want per person for a year. One egg per week is about fifty servings.
Step 2	Add up the total number of servings you want for your family.
Step 3	Decide what kind of egg products you want to store.
Step 4	Distribute the number of servings among the different egg products.

Serving Sizes

On average, a fresh, large, whole egg has 72 calories. Dehydrated egg products vary in calories, ranging from as low as 30 to as much as 84 per serving! Double-check to make sure you are getting at least 72 calories per serving and increase quantities if necessary.

The equivalent egg product for fifty servings should be at least 3,600 calories with 300 grams of protein.

Table 17.10 shows you how much egg product provides fifty servings. This amount is about one egg serving per week for a year.

Table 17.10 Amount of Egg Product for Fifty Servings	
Dehydrated whole eggs	1.50 lb.
Dehydrated scrambled egg mix	1.60 lb.
Crystallized whole eggs	1.25 lb.
Freeze-dried scrambled eggs	4.00 lb.

©Patricia Spigarelli-Aston

The Example Family Calculates Their Egg Needs

The example family decides on two servings of eggs for each person per week for a total of four hundred servings of eggs for a year (see example 17.13). Use worksheet 17.11 on page 160 to help you determine the amount of egg product to store.

Meat, Poultry, and Fish

The average American eats two to three servings of meat daily, but a survival diet may consist of substantially less. For example, even adding just 5 percent of meat to a bean dish gives you approximately the same quality of protein as an all-meat dish.

Store a minimum of one hundred servings and up to 350 servings of meat or fish per person for one year. The suggested serving size is three ounces, or about the size of a deck of cards. Canned meats, either commercially or home-canned, are the most economical and will probably be the bulk of your storage. Freeze-dried meat is an excellent but expensive product. You may also want to consider dehydrated (jerky), smoked, and retort products.

Nutrition

Most lean meat and fish will have between twenty-one and twenty-eight grams of protein per three-ounce serving. Fifty three-ounce servings of meat should have between 1,000 and 1,400 grams of protein. Low-quality meat, meat trimmings, or meats with more fat or other added ingredients have less protein.

Example 17.13			
Example Family's Egg Selection for One Year			
Egg Product	Quantity for Fifty Servings	Servings	Amount of Egg Product
Dehydrated whole eggs	1.50 lb.	200	4.50 lb.
Dehydrated scrambled eggs	1.60 lb.	100	3.20 lb.
Crystalized whole eggs	1.25 lb.	50	1.25 lb.
Freeze-dried scrambled eggs	5.00 lb.	50	5.00 lb.
Total Number of Servings		400	

Table 17.11			
Average Grams of Protein, Grams of Fat, and Calories in a Three-Ounce Serving of Fresh Meat			
Meat	Grams of Protein	Grams of Fat	Calories
Round roast	27	4	153
Sirloin	26	6	165
Lean ground beef	25	7	162
White chicken meat	26	4	127
Turkey breast	26	3	133
Pork shoulder	22	13	207
Ham	21	5	133
Tuna (water-packed)	25	1	111
Salmon, fresh	23	9	183

©Patricia Spigarelli-Aston

Comparing Cost

The cost for animal protein varies greatly. If you are interested in comparing protein value, you can divide the cost by the total grams of protein to find cost per gram of protein.

Serving Size

Although three ounces is generally considered a serving, be aware that stated serving sizes might vary and are often only two ounces for both canned and freeze-dried meat products. Also, serving sizes are often lower in protein than you would expect based on the protein content of their fresh counterparts.

Table 17.11 on page 139 gives you a basis for determining the quality of canned or freeze-dried meats. It shows you the protein, fat, and calories in a three-ounce serving of common fresh meats. Use it to gauge the quality of meat products you're considering.

Canned Meat

Canned meats, both commercial and home-canned, are popular sources for meat protein for storage. Besides cans, meat can be processed in retort pouches to make shelf-stable products. Common in Asia and Europe, examples of these products include tuna fish, ham, and chicken breast.

> **For fifty three-ounce servings of meat, you'll need 150 ounces of canned meat.**

Calculating How Much Meat to Store

You have two options. First, decide how many pounds of meat you want to store. For example, let's say you want two three-ounce servings a week. For that amount you would multiply 52 x 6 = 312 ounces, which is equivalent to 19.5 pounds of meat per person (312 ÷ 16 = 19.5).

The second option is to approximate the number of servings based on the number of servings in common can sizes. Table 17.12 shows the number of cans you'll need to equal fifty servings. Fifty servings will give you about one serving a week for a year.

Use calculation 17.5 to help you distribute the total number of servings of canned meat you desire among the various meats you want to store.

Table 17.12 Amount of Canned Meat that Provides Fifty Three-Ounce Servings of Meat		
Can or Jar Size	Approximate Number of Three-Ounce Servings	Number of Cans or Jars Equal to Fifty Servings
6–7 oz.	2	22–25 cans
12–13 oz.	4	12–13 cans
14.5 oz.	4.5	11 cans
16 oz.	5	10 cans
28 oz.	9	5–6 cans
Pint jars (16 oz.)	5	10 jars
Quart jars (32 oz.)	10	5 jars

For can sizes not listed in the tables you can determine the number of cans you need by dividing 150 (the number of ounces in 50 three-ounce servings of meat) by the number of ounces in the can. You will also need at least 1,000 grams of protein to equal 50 servings.

©Patricia Spigarelli-Aston

Calculation 17.5 Calculating How Much Canned-Meat Product You Need	
Step 1	Decide on the number of servings of canned meat you want per person per week, then calculate the total for the family. The recommendation is one hundred total servings of meat per person.
Step 2	Multiply the total number of servings by three to determine the total number of ounces of canned meat.
Step 3	Decide which kinds of meat and how much will come from each kind. Consider which sizes of cans or jars you prefer.
Option 1	
Step 4	Distribute the total ounces among the various kinds of canned meat until you have reached the total number of ounces of canned meat you desire to store.
Option 2	
Step 4	For many common can sizes, refer to table 17.15 for the amounts required to provide fifty three-ounce servings. Multiply the amounts by the total number of servings desired.
Note: When you have completed your calculations for either option 1 or option 2, check to be sure the total amount of protein for 50 servings is between 1,000 and 1,400 grams of protein. You may want to increase the amount of meat if it's less.	

PERSONALLY SPEAKING

When it comes to long-term emergency meat storage, my first choice is home-canned. That way I know the source and exactly what else is in it—usually just a little salt. Canning meat requires an investment in a pressure canner and jars. Your cooperative extension service and the *Ball Blue Book Guide to Preserving* are good sources for step-by-step instructions on how to safely can meats. Through experience, I've learned that canned meats have at least a five-year shelf life. We have canned farm-grown chicken, turkey, beef chunks, salmon, chopped razor clams, and Dungeness crab meat.

If you decide to go with commercially canned meat, I highly recommend you examine the labels and try out the brands before you buy large quantities for storage. As you evaluate products, you will find a wide range of claims for protein content. Look for products that are all meat with a little salt added. Compare grams of protein per ounce and look for a protein content that is seven to eight grams per ounce to be sure you are getting the value you want. I personally avoid products with starches, flavor enhancers, sugar, and MSG or MSG substitutes. I also consider the country of origin.

Freeze-Dried Meats

Freeze-dried meats are another popular option for meat protein. Many people claim that freeze-dried meats have much the same look and taste as the original, natural fresh foods. You should try several varieties so you can decide for yourself.

Freeze-dried meats are packaged in pouches, #2½ cans, and #10 cans, but serving size, weight per serving, and protein content vary by company, making it extremely hard to identify a consistent weight and serving size. So for consistency, it's best to calculate servings by grams of protein or freeze-dried weight. Use Calculation 17.6 on page 142 to calculate the amount of freeze-dried meat you need.

Table 17.13
Calories and Protein in a Half-Cup Serving of Freeze-Dried Meat and Amount Needed for Fifty Servings

Meat	Serving Size	Calories	Expected Grams of Protein per Serving	Pounds to Equal Fifty Servings
Diced beef	½ cup / 24 g	110	19 g	2.6 lb.
Ground beef	½ cup / 38 g	230	19 g	4.2 lb.
Diced chicken	½ cup / 24 g	170	21 g	2.6 lb.
Turkey	½ cup / 24 g	100	21 g	2.6 lb.
Ham	½ cup / 30 g	140	21 g	3.3 lb.
Sausage	½ cup / 40 g	280	12 g	4.4 lb.

©Patricia Spigarelli-Aston

Calculation 17.6	
Calculating How Much Freeze-Dried Meat Product You Need	
Step 1	Decide how many total servings of freeze-dried meat you want and decide which kinds of meat you want to store.
Step 2	Refer to table 17.13 for the pounds of freeze-dried meat in fifty servings.
Step 3	Distribute the number of servings among different freeze-dried meats. The total amount of protein should be between 1,000 and 1,400 grams.

Jerky

Brining and dehydrating is a time-honored method of preserving meat and has a place in the preparedness diet. Because the protein content is more concentrated than regular meat, high-quality jerky will usually have a protein content between twelve and fifteen grams per ounce.

Five pounds of high-quality beef jerky is equivalent to about fifty servings of animal protein.

Jerky can be made directly from meat or extruded from a mixture of meat, meat products, and additives. Depending on the quality, the fat content, and the way it's processed, it may have as little as six grams and as much as sixteen grams of protein per ounce. To keep protein servings standardized, again, it is best to compare protein amounts per ounce. Use table 17.14 to help you determine how much jerky you will need to equal fifty servings of meat protein.

Table 17.14			
Amount of Beef Jerky Equivalent to Fifty Three-Ounce Servings of Meat Protein			
Meat	Approximate Grams of Protein per Ounce	Grams of Protein in Fifty One-Ounce Servings	Pounds to Equal Fifty Servings
Jerky	12–16 g	600–800 g	5 lb.–7.5 lb.
Jerky product	6–8 g	300–400 g	10 lb.–15 lb.

©Patricia Spigarelli-Aston

PERSONALLY SPEAKING

 Commercially prepared meats are expensive. Part of becoming prepared is developing skills and using those skills to improve your quality of life. It's quite easy to make jerky at home. Jerky is ideal for tougher and less-flavorful cuts of meat, like rump roast or top round. It is also a good way to use wild game. We make jerky by first slicing the meat very thinly—it's easier if partially frozen. We marinate it for a day and then preserve it by processing it in our dehydrator or smoker. Then we package it in a vacuum pack. It should be refrigerated or frozen.

I recommend you invest not only in the supplies and tools you need, but especially in the time to learn the skills. In chapters 31 and 32, you will find more details about dehydrating and preserving meat.

Combination Foods and Entrées

Combination foods, such as canned beef stew, chili con carne, hash, etc., or dehydrated and freeze-dried entrées can also be used as meat-protein sources. If meat is listed among the first few ingredients on the label and appears to make up a substantial portion of the product, count half the total protein listed as animal protein. You will have to use your judgment.

Besides sampling the product yourself, the most important thing you can do is inspect the nutritional label and ingredient list. Check protein content, calories per serving, sodium level, number of additives, and actual ingredients. Look for serving sizes of at least 200 calories and protein content of at least ten grams per serving. Fifty servings should have between 1,000 and 1,400 grams of protein.

Chapters 14 and 18 provide more detailed information about canned and freeze-dried main-course entrées.

The Example Family Calculates Their Meat-Protein Needs

The example family decides they want at least one serving of meat every other day for 180 servings per person (4 x 180 = 720). They increase that to 750 total servings for the family of four. They use tables 17.12, 17.13, and 17.14 to help them select meat-protein products. Use worksheet 17.12 on page 161 to help you select meats for your storage. It can also be found on our website, CrisisPreparedness.com.

Example 17.14 Example Family's Meat Selection for One Year			
Meat Product	Can Size	Servings	Amount
Cans of beef chunks	13-ounce	100	26 cans
Cans of corned beef	13-ounce	50	13 cans
Cans of chicken	13-ounce	100	26 cans
Cans of turkey	13-ounce	50	13 cans
Freeze-dried ground beef	NA	50	4.2 lb.
Freeze-dried chicken	NA	50	2.6 lb.
Cans of tuna fish	7-ounce	150	75 cans
Cans of salmon	16-ounce	50	10 cans
Beef jerky	NA	50	5 lb.
Spam*	12-ounce	50	24 cans
Cans of stew**	38-ounce	50	24 cans
	Total Servings	750	

*Spam has roughly two-thirds to half the protein per ounce as the leaner meats listed in the chart. The recommended number of cans compensates for this.
**Typical purchased stew has 10 grams of protein per 8.5 ounce serving and 24 cans have about 1,100 grams of protein.

STEP 8 – Calculating Remaining Calories Needed

This section will help you determine how many of your remaining calories need to be provided by grains and legumes, the last big portion of your food storage. To determine how much you need, it is necessary to calculate the total calories included in the five major groups in your plan up to this point. The approximate number of calories of each food group is found in table 17.15. Follow the steps in calculation 17.7. Use worksheet 17.13 on page 162 to help you calculate your total calories. It can also be found on our website, CrisisPreparedness.com.

Calculate with Confidence

This is where math experts can have some fun. However, if the math seems challenging, just take it one step at a time. It may seem like a big task to calculate the calories in your plan, but if you've made all the calculations

Table 17.15 Approximate Number of Calories for Each Food Group		
Storage Food Group	Typical Unit Measure	Average Calories for Unit Measure
Milk products	Pounds	1,600
Fats and oils	Pounds	4,000
Vegetables	½-cup serving	50
Fruits	½-cup serving	75
Sugar	Pounds	1,800
Eggs	1-egg serving	70
Meats	3-oz. serving	150
Entrées	1-cup serving	200
Legumes	½-cup serving	125
Grains	10 servings per pound	150

©Patricia Spigarelli-Aston

described so far in this chapter, it will be simple. It's basic arithmetic a calculator can help you with.

It's important to do this so you can see for yourself that you have enough calories. This is also a good time to fine-tune the amounts you are planning to store. Use common sense to determine if there is too much or not enough of an item. It's especially important to assess how many calories you need if you are relying on a prepackaged food storage plan.

	Calculation 17.7 Calculating How Many Remaining Calories Come from Grains and Legumes	
Step 1	Make a list of the categories and the total quantity for each category you have selected so far in your plan. The categories include pounds of milk products, pounds of oils, servings of fruits and vegetables, pounds of sugar, and servings of eggs, meats, and entrées.	
Step 2	Next, calculate the calories in each category and add them up to determine the total number of calories so far for a year.* Use table 17.18 for calorie listings.	
Step 3	Divide the total calories so far by 365 to determine calories per day.	
Step 4	At the beginning of the chapter you calculated the total daily number of calories your family needs per day using tables 17.1 and 17.2. Use this number in step 5.	
Step 5	Subtract the total daily calories so far (step 3) from the total calories needed (step 4) to find out how many remaining calories you need each day. These calories will come from the last storage group, grains and legumes.	

*If you have chosen another length of time for your food storage, you will need to adjust the steps of this table accordingly.

The Example Family Calculates Their Total Calories and How Many Calories Will Come from Grains and Legumes

The example family is ready to calculate their total calories from other food groups and how many calories will come from grains and legumes. There calculations are shown in example 17.15 on page 146. They learn that they will get 5,284 calories from the foods selected thus far. They calculate that they will need to provide 7,116 calories from grains and legumes.

Why We Calculate Calories by the Day

Wondering about the value of calculating the number of daily calories rather than just using a year's total? Knowing how many calories you need per day lets you figure out how many you would need for thirty, or sixty, or any other specified number of days.

You can also calculate the daily calories in any commercial food storage package and know if you are getting an adequate number of calories and how much more storage food you need so that you have the necessary calories.

Storage Food Group	Typical Unit	Average Calories for Unit	Multiply by Total Pounds or Servings	Total Calories for Category
Milk products	Pounds	1,600	1,600 x 226 =	361,600
Fats and oils	Pounds	4,000	4,000 x 113 =	452,000
Vegetables	½-cup serving	50	50 x 5,040 =	252,000
Fruits	½-cup serving	75	75 x 2,880 =	216,000
Sugar	Pounds	1,800	1,800 x 280 =	504,000
Eggs	1-egg serving	70	70 x 400 =	28,000
Meats	3-ounce serving	150	150 x 700 =	105,000
Entrées	1-cup serving	200	200 x 50 =	10,000

Example 17.15
The Example Family Calculates Total Calories for Each Food Category and How Many Calories Come from Grains and Legumes

Total Calories So Far =	1,928,600
Divide total calories so far by 365* to get daily calories needed: 1,928,600 ÷ 365 =	5,284
List total daily calories needed: (See tables 17.1 and 17.2)	12,400
Subtract daily calories so far from total daily calories to get remaining calories to come from grains and legumes: (Total Daily Calories: 12,400) −(Daily Calories So Far 5,284) =	7,116

*If you have chosen another length of time for your food storage, you will need to adjust the steps of this worksheet accordingly.

STEP 9 – Determining How Many Grains and Legumes

Grains and legumes make up the balance of your plan's calories. They supply vegetable protein, complex carbohydrates, and other important nutrients. High-quality proteins can be made from the right combination of complementary vegetable proteins or from a small amount of animal protein combined with vegetable protein. See table 17.16 for excellent and good combinations.

Complementary Proteins

As discussed, the animal products milk, eggs, and meat provide complete protein with all the amino acids in balance. In the right ratio,

Table 17.16
Complete Protein Combinations

Protein Combinations	Quality of Protein
Dairy and grains	Excellent
Grains and legumes	Excellent
Legumes and seeds	Excellent
Dairy and seeds	Good
Dairy and legumes	Good
Grains and seeds	Good

©Patricia Spigarelli-Aston

vegetable proteins will also provide a complete, balanced protein. A good ratio is four pounds of grain to one pound of legumes. With added animal protein, a ratio of eight pounds of grain to one pound of legumes is sufficient.

Grains

Current annual grain consumption is about 150 pounds per person, but in 1910, it was double that. Grains include whole grains, flours, pastas, breakfast cereals, crackers, and mixes. Regardless of what else you have in your plan, you should have at least 250 pounds of grains per person for one year. They are relatively inexpensive, store well, and are the core of most diets. The majority should be whole grains with the rest coming from flour and other processed grain products.

Unless rice or corn are your basic grains, you will probably want 60 percent to 80 percent of your total in wheat and wheat products. Although flour is convenient, you shouldn't store more than fifty pounds per person because its shelf life is limited. That's also about the amount you can use and rotate before it becomes stale.

> Regardless of what else you have in your food storage, store at least 250 pounds of grains per person, the majority being whole grains.

Increasing Grains and Legumes in Your Diet

Thanks in part to the eat-clean, whole-foods, and plant-based movements and an emphasis on healthy eating, there are more wholesome grains and legumes to choose from than ever before! The example family's storage reflects some of those possibilities. If you are serious about storing wheat and other grains, this is a good time to consider how you can increase the whole grains, legumes, and seeds in your normal diet. As you use a wide variety of grains and legumes, your family will become accustomed to the foods you store. You will rotate through them and have the bonus of a healthier diet.

Deciding What to Store – Where to Begin

To decide what to store, begin by considering which grains and legumes your family already eats. So, for example, if your family consumes one pound of

QUICK FACTS
How Much Wheat?

- A bushel of wheat weighs 60 pounds.
- There is a pound of wheat in a one-and-a-half-pound loaf of bread.
- A 50-lb. bag of wheat yields fifty loaves of bread.
- Fifty pounds will give you five loaves of bread per week for ten weeks.
- 260 pounds of wheat will give you five loaves of bread per week for a year.

pasta a week, you might want to start with twelve pounds—enough for three months. Then build up to six months and then a year. Or if your family eats rice twice a week and you use about two cups of rice a week, that equals about one pound. That means a twenty-five-pound bag will be about right for six months. You can calculate other foods your family uses this way.

In a crisis where you must rely on stored food, you will likely need more than the amounts you are presently using, but this

> A one-and-a-half pounds loaf of bread contains about a pound of wheat.

will give you a comparison for the recommended amounts and the numbers you get in calculation 17.8.

	Calculation 17.8 Calculating How Many Grains and Legumes
Step 1	Divide the total number of remaining calories (Calculation 17.6) by 1500 to determine the total pounds of grain and legumes needed per day. (There are between 1400 and 1600 calories in a pound of grain or legumes.)
Step 2	Decide on a ratio for pounds of grains to one pound of legumes. In other words, how many pounds of grain will you select for each pound of legumes? (See the discussion on page 148 to help make your decision.)
Step 3	Add the two sides of your chosen ratio together (e.g. a ratio of 4:1 will be 4 + 1 = 5, and 8:1 will be 8 + 1 = 9). Divide the total from Step 1 by this number to find the pounds of legumes for one day.
Step 4	Subtract the pounds of legumes in Step 3 from the total number of pounds of vegetable proteins for one day to find the total pounds of grains for one day.
Step 5	Multiply the pounds of legumes and the pounds of grains needed per day by 365* to determine how many pounds of each are needed for a year.
Step 6	Distribute these amounts among various types of grains and legumes.

*Or you could find the amount of grains and legumes you would need for thirty days, ninety days, etc.

Legumes

Select a variety of legumes for multiple uses. Include pinto, navy, black, red, and garbanzo beans, as well as split peas, lentils, etc. The choice is based on your family's preferences.

Because the versatile, nutritional soybean is the only legume that's a complete protein by itself, it's worth learning how to use soybeans in your recipes.

Seeds

Seeds are another nutritious option for making complete proteins. Some seeds to consider are sunflower, sesame, chia, flax, pumpkin, and hemp. Consult table 17.16 on page 146 for ideas of complementary protein combinations.

The Example Family Calculates Their Grains and Legumes Needs

The example family uses their previous calculations to figure out that 7,116 calories should come from grains and legumes each day (example 17.15 on page 146). They decide on an 8:1 ratio of grain to legumes. That means $\frac{1}{9}$ of the total will be legumes, $\frac{8}{9}$ grains.

Example 17.16 Example Family Calculates How Many Pounds of Grains and Legumes They Need per Day	
Remaining Daily Calories:	7,116
Approximate daily amount of grains and legumes:	7,116 ÷ 1,500 ≈ 4.75 lb.
Chosen grain-to-legumes ratio:	8 : 1
Add the two parts of the ratio:	8 + 1 = 9
Approximate amount of legumes:	4.75 ÷ 9 ≈ 0.53 lb. (⅑)
Approximate amount of grain:	4.75 − 0.53 ≈ 4.25 lb. (⅑)
Approximate amount of legumes needed for a year:	0.53 x 365 ≈ 200 lb.
Approximate amount of grain:	4.25 x 365 ≈ 1,500 lb.

Next, they calculate how many pounds of legumes and grains they need per day. Follow their calculations in example 17.16. The numbers are rounded to make calculations easier. Use worksheet 17.14 on page 163 to calculate how many pounds of grains and legumes you need per day. It can also be found on our website, CrisisPreparedness.com.

Now the example family selects grains and legumes to fill these amounts. Examples 17.17 and 17.18 show you a representative selection and the number of servings these would provide for each person per week. Use worksheets 17.15 on page 164 and 17.16 on page 165 to help select grains and legumes. They can also be found on our website, CrisisPreparedness.com.

It seems like a staggering amount of food, but it is intended to be generous. Remember, a normal diet includes calories from many outside sources, and these complex carbohydrates make up for that. Also, a standard serving size is often less than what most people consume as a serving. The quantity of wheat is intentionally high because it is economical and most of it would be used for bread and other baked goods.

Example 17.17 Pounds of Legumes			
Total legumes: 200 pounds			
Legumes	Amount	Servings	Servings per Person per Week
Pinto	50 lb.	600	3
Black beans	20 lb.	240	1.2
Kidney beans	20 lb.	240	1.2
White beans	20 lb.	240	1.2
Navy beans	20 lb.	240	1.2
Garbanzo beans	20 lb.	240	1.2
Soy beans	20 lb.	240	1.2
Lima beans	10 lb.	120	0.6
Split peas	10 lb.	80	0.4
Lentils	10 lb.	100	0.4
Total Servings		2,340	10.6

Example 17.18 Pounds of Grains			
Total grains: 1,500 pounds			
Grains	Amount	Servings	Servings per Person per Week
Pasta	100 lb.	800 (2-oz dry)	4
Brownie and cake mixes	20 lb.	240	1.2
Macaroni-and-cheese mix	40 lb. (80 pkg)	240	1.2
Wheat	700 lb.	7,000	35
Rice	100 lb.	1,200	6
Assorted grains	200 lb.	2,000	10
Oatmeal	100 lb.	800	4
9-grain cereal	50 lb.	500	2.5
Farina	50 lb.	500	2.5
Popcorn	40 lb.	800	4
Flour	100 lb.	1,000	5
Total Servings		15,080	75.4

Servings per Pound of Grains and Legumes

It is useful to know how many servings are in a pound of legumes or a pound of grains. If you convert the pounds you select to serving sizes, it will make it easier to imagine the amounts. It will also help you tweak the quantities if they do not seem quite right to you. (See table 17.17.)

Table 17.17 Servings per Pound of Grains and Legumes							
Common Grains and Legumes	Dry Amount per Pound	Serving Size	Servings per Pound	Common Grains and Legumes	Dry Amount per Pound	Serving Size	Servings per Pound
Wheat	2 ½ cups	2–3 oz.	6-8	Corn	2 ¾ cups	1 cup	11
Wheat flour	3 ½ cups	½ cup	7	Oatmeal	5 cups	½ cup	10
Grains	2 cups	2 oz.	8	Pasta	4 cups	1 cup	6-8
Quinoa	2 cups	½ cup	8	Beans	2 cups	½ cup	12
Barley	3 cups	2 oz.	8	Split peas	2 cups	½ cup	12
Rice	2 cups	1 cup	6	Lentils	2 cups	½ cup	12

STEP 10 – Choosing Add-ons

Add-ons enhance and complete your plan. Consult the list of add-ons in chapter 18 to help you make your choices. These extras are relatively inexpensive, and a little goes a long way toward enlivening simple foods. Variety is important. Table 17.18 will help you determine how many servings are in a quantity of cooking and baking essentials.

Cooking and Baking Essentials

You will also need baking and cooking essentials, like yeast, baking soda, baking powder, and cornstarch.

Herbs and Spices

Do not skimp on herbs, spices, and other seasonings. Flavor-enhancing herbs and spices, black pepper, dehydrated garlic, onions, celery, and green bell peppers will jazz up many recipes.

Flavor Add-ons

Bouillon, dehydrated sweet and sour cream, and assorted vinegars will also add to your cooking possibilities.

PERSONALLY SPEAKING

 It is hard to come up with an all-inclusive list when it comes to herbs and spices. We all have our personal can't-do-without items, but I try to have the basics for sweet, savory, and ethnic foods. Black pepper, garlic powder, and onion powder are essential. After that, my top ten herbs and spices are bay leaves, chili powder, cinnamon, crushed red pepper, cumin, ginger powder, Italian blend, nutmeg, paprika, and sage. The best way to decide on your personal choices is to take stock of what you already use and procure a backup supply. I have used prepackaged seasoning mixes in the past, but I now prefer to make my own blends for things like chili, tacos, and spaghetti sauce. But prepackaged seasonings have their place, so if they make life easier for you, use them.

Worksheets to Help You

This section contains the worksheets that accompany the explanations in chapter 17 for how to plan a long-term food storage. You will need to refer to the tables, calculations, and examples throughout chapter 17 as you complete the worksheets.

The worksheets are also available as downloadable PDF files at CrisisPreparedness.com.

Worksheet 17.1 Family Estimated Total Daily Calories					
Family Members	Age	Activity Level	Weight	Calories per Pound of Body Weight	Calories Needed
1			X _____	=	
2			X _____	=	
3			X _____	=	
4			X _____	=	
5			X _____	=	
6			X _____	=	
				Total Family Calories	

©Patricia Spigarelli-Aston
Refer to table 17.2 on page 124 for information to complete this table.

Worksheet 17.2 Calcium Needs and Pounds of Powdered Milk for One Year			
Family Members	Age	RDA in mg Calcium	Pounds of Powdered Milk Equivalencies
1			
2			
3			
4			
5			
6			
Total Pounds of Powdered Milk and Powdered-Milk Equivalents			

©Patricia Spigarelli-Aston
Refer to table 17.3 on page 126 for information to complete this table.

Worksheet 17.3 Calculating Amounts of Milk Products				
Milk Product	Calcium Equivalent Multiplier	Calcium Equivalent Amount	Conversion Calculation Multiplier x Calcium Equivalent Amount	Quantity to Store
Powdered milk, regular	1.0 lb.		1 x _____ =	
Powdered milk, instant	1.0 lb.		1 x _____ =	
Powdered buttermilk	1.0 lb.		1 x _____ =	
Canned evaporated milk	6 12-oz. cans		6 x _____ =	
Fresh cheese	1.5 lb.		1.5 x _____ =	
Parmesan cheese, shredded	1.2 lb.		1.2 x _____ =	
Processed cheese product (Velveeta)	1.5 lb.		1.5 x _____ =	
Freeze-dried shredded cheeses	0.75 lb.		0.75 x _____ =	
Dehydrated cheese powder	0.375 lb.		0.375 x _____ =	
Total Milk and Powdered-Milk Equivalents This amount should equal the total in worksheet 17.2.				

©Patricia Spigarelli-Aston
Refer to table 17.4 on page 127 for information to complete this table.

Worksheet 17.4 Determining the Amount of Fats and Oils for Your Food Storage		
Step 1	List the total daily calories for family.	_____
Step 2	List the percentage you want from fats or oils. The recommendation is between 10% and 15%. Write it as a decimal (e.g., 15% = 0.15).	_____
Step 3	Multiply the total daily calories in step 1 times the percent as a decimal in step 2. This will give you the total daily number of calories you need from fats and oils.	_____
Step 4	Multiply the number of calories in step 3 by the number of days you want the oil to last. For a year, multiply by 365; for three months, by 90; for six months, by 180, etc.	_____
Step 5	Divide the total calories from fats and oils in step 4 by 4,000. This is the total number of pounds of fats and oils you need for the period you have chosen. Use it in your calculations in worksheet 17.5.	_____

©Patricia Spigarelli-Aston
Refer to tables 17.1 and 17.2 on pages 123 and 124 for help for calculating total daily calories.

Worksheet 17.5 Selecting Your Fats and Oils				
Oil or Fat	Fats or Oils Equivalent Multiplier	Fats or Oils Equivalent Amount	Conversion Calculation Multiplier Equivalent Amount	Quantity to Store
Vegetable oils	1 lb.		1 x _____ =	
Lard	1 lb.		1 x _____ =	
Vegetable shortening	1 lb.		1 x _____ =	
Vegetable shortening, powdered	1.2 lb.		1.2 x _____ =	
Butter or Margarine	1.2 lb.		1.2 x _____ =	
Butter, powdered	1.3 lb.		1.3 x _____ =	
Mayonnaise	1.4 lb.		1.4 x _____ =	
Miracle Whip	3.1 lb.		3.1 x _____ =	
Bacon	1.6 lb.		1.6 x _____ =	
Peanut butter	1.5 lb.		1.5 x _____ =	
Peanut butter, powdered	3.0 lb.		3.0 x _____ =	
Nuts and Seeds	1.6 lb.		1.6 x _____ =	
Unshelled nuts	3.1 lb.		3.1 x _____ =	
Total Number of Pounds of Fats and Oils to Store This amount should equal the total in worksheet 17.4				

©Patricia Spigarelli-Aston
Refer to previous 17.4 and example 17.7 on pages 129 and 130 for information to complete this table.

Worksheet 17.6 Fruit and Vegetable Needs			
Family Members	Age	Servings of Fruit per Day*	Servings of Vegetables per Day*
1			
2			
3			
4			
5			
6			
Family Total Servings Per Day			

*The recommended number of servings is 2 servings of fruit and 3 ½ servings of vegetables.

©Patricia Spigarelli-Aston
Refer to calculation 17.2 on page 132 for information to complete this table.

	Worksheet 17.7 Selecting Fruits			

Number of Servings per Day for Family:
Number of Days*:
Total Number of Servings for Family: (number of servings x number of days) =
*It is recommended you select days in increments of 30 for easier calculation, i.e., 30 days, 90 days, 360 days).

Fruits	Servings per Day of Listed Fruit	Total Servings of Listed Fruit	For Reference: Amount for One Serving per Day per Year/360 Days	Quantity to Store
Vitamin C-fortified fruit drink mix			18 lb.	
Canned fruits in # 303 cans			8 cases	
Canned fruits in # 2 cans			6 cases	
Canned fruits in # 2 ½ cans			4 ½ cases	
Canned fruit juice			2 ½ cases	
Canned fruit in #10 cans			1 case	
Home canned fruit in pints			8 cases	
Home canned fruit in quarts			4 cases	
Fresh apples			125 lb.	
Dried prunes			25 lb.	
Raisins			30 lb.	
Dehydrated apples			22.8 lb.	
Dehydrated bananas			22.8 lb.	
Assorted dehydrated fruits			22.8 lb.	
Assorted freeze-dried fruit			8–9 lb.	
Other fruit				
Other fruit				
Other fruit				

©Patricia Spigarelli-Aston. Refer to calculation 17.2; tables 17.6, 17.7, and 17.8; and examples 17.8 and 17.9 on pages 132 to 135 for information to complete this table.

Worksheet 17.8
Selecting Vegetables

Number of Servings per Day for Family:
Number of Days*:
Total Number of Servings for Family: (number of servings x number of days) =
*It is recommended you select days in increments of 30 for easier calculation, i.e. 30 days, 90 days, 360 days).

Vegetables	Servings per Day of Listed Vegetable	~Total Servings of Listed Vegetables	For Reference: Amount for One Serving per Day per Year/360 Days	Quantity to Store
Canned vegetables in # 303 cans			8 cases	
Canned vegetables in # 2 ½ cans			4 ½ cases	
Canned vegetable juice			2 ½ cases	
Fresh cabbage			75 lb.	
Fresh carrots			60 lb.	
Sweet potatoes			70 lb.	
Fresh potatoes			120 lb.	
Assorted dehydrated vegetables			28.8 lb.	
Dehydrated onions			15 lb.	
Dehydrated potatoes			22.8 lb.	
Dehydrated tomato powder			22.8 lb.	
Assorted freeze-dried vegetables			8–9 lb.	
Freeze-dried corn, green beans, peas			22.8 lb.	
Freeze-dried onion			3.65 lb.	
Other vegetables				
Other vegetables				
Other vegetables				

©*Patricia Spigarelli-Aston. Refer to calculation 17.2; tables 17.6, 17.7, and 17.8; and examples 17.8 and 17.10 on pages 132 to 135 for information to complete this table.*

		Worksheet 17.9 Determining the Amount of Sugar You Need	
Step 1	List the total daily calories for family.		_____
Step 2	List the percentage you want from sugar. The recommendation 10%–12%. Write it as a decimal (e.g., 12% = 0.12).		_____
Step 3	Multiply the total daily calories in step 1 times the percent as a decimal in step 2. This will give you the total daily number of calories from sugar.		_____
Step 4	Multiply the number of calories in step 3 by the number of days you want the sugar to last. For a year, multiply by 365; for three months, by 90; for six months, by 180, etc.		_____
Step 5	Divide the total calories from sugar in step 4 by 1,775. This is the total number of pounds of sugar you need. Use it in your calculations in worksheet 17.12.		_____

©*Patricia Spigarelli-Aston*
Refer or tables 17.1 or 17.2 on pages 123 and 124 to complete this table.

Sugar	Sugar Equivalent Multiplier	Sugar Equivalent Amount	Conversion Calculation Multiplier x Sugar Equivalent Amount	Quantity to Store
Worksheet 17.10 Selecting Your Refined Sugars				
Sugar	1 lb.		1 x ____ =	
Brown sugar	1 lb.		1 x ____ =	
Powder sugar	1 lb.		1 x ____ =	
Honey	0.8 lb.		1 x ____ =	
Molasses	1 pt.		1 pt. x ____ =	
Maple syrup	1 pt.		1 pt. x ____ =	
Agave syrup	1 pt.		1 pt. x ____ =	
Pancake syrup	1 pt.		1 pt. x ____ =	
Jams and jellies	1 pt.		1 pt. x ____ =	
Fruit-flavored drink mix	1 lb.		1 lb. x ____ =	
Pudding mix	6 3.4-oz. pkg.		6 pkg. x ____ =	
Flavored gelatin mix	6 3.0-oz. pkg.		6 pkg. x ____ =	
Hard candy	1 lb.		1 lb. x ____ =	
Total Number of Pounds of Sugars to Store This amount should equal the total in worksheet 17.9.				

©Patricia Spigarelli-Aston
Refer to calculation 17.3 on page 136 to complete this table.

Worksheet 17.11
Selecting Egg Products

Number of servings* per person:
Total number of servings for family:
*For easy calculations, group in fifty-serving units. One serving per person per week for a year is about fifty servings.

Egg Product	Quantity for 50 Servings	Servings	Total Amount of Egg Product
Dehydrated whole eggs	1.50 lb.		
Dehydrated scrambled eggs	1.60 lb.		
Crystalized whole eggs	1.25 lb.		
Freeze-dried scrambled eggs	5.00 lb.		
Total Number of Servings			

©*Patricia Spigarelli-Aston*
Refer to calculation 17.4, table 17.10, and example 17.13 on pages 138 and 138 for information to complete this table.

Worksheet 17.12 Selecting Meat			

Number of servings* per person:
Number of servings for family:
*The recommendation is from one serving per week to one hundred servings per year.

Canned Meats	Can Size	Amount	Servings
1			
2			
3			
4			
5			
6			
Freeze-Dried Meats	xxx	Amount	Servings
1	–		
2	–		
3	–		
4	–		
Other Meat Products	xxx	Amount	Servings
1	–		
2	–		
3	–		
4	–		
		Total Servings of Meat	

©Patricia Spigarelli-Aston
Refer to calculations 17.5 and 17.6 and tables 17.11, 17.12, 17.13, and 17.14 on pages 139 to 144 to complete this table.

Worksheet 17.13
Determining Total Daily Calories Used So Far and Calculating How Many Calories Come from Grains and Legumes

Storage Food Group	Typical Unit	Average Calories for Unit	Multiply by Total Pounds or Servings	Total Calories for Category
Milk products	Pounds	1,600	1,600 x _____ (total pounds) =	_____
Fats and oils	Pounds	4,000	4,000 x _____ (total pounds) =	_____
Vegetables	½-cup serving	50	50 x _____ (total servings) =	_____
Fruits	½-cup serving	75	75 x _____ (total servings) =	_____
Sugar	Pounds	1,800	1,800 x _____ (total pounds) =	_____
Eggs	1-egg serving	70	70 x _____ (total servings) =	_____
Meats	3-oz. serving	150	150 x _____ (total servings) =	_____
Combination foods and entrées	1-cup serving	200	200 x _____ (total servings) =	_____
			Total Calories So Far =	_____
			List total daily calories needed: (See table 17.1 and 17.2)	_____
			Divide total calories so far by 365 to get daily calories needed: _____ ÷365=	_____
			Subtract daily calories so far from total daily calories to get remaining calories to come from grains and legumes: Total Daily Calories _____ - Daily Calories So Far _____ =	_____

©Patricia Spigarelli-Aston
Refer to calculation 17.7 and table 17.15 on pages 145 and 146 to complete this table. Also decide how many servings of combination foods and entrées you desire.

	Worksheet 17.14		
	Calculating How Many Pounds of Grains and Legumes You Need per Day		
Step 1	Divide the result in worksheet 17.13 by 1,500 (this is the average number of calories in a pound of grain or legumes). This will give you the number of pounds of grains and legumes needed daily for your family.	_____ ÷ 1,500 =	_____
Step 2	Choose a ratio of pounds of grains to one pound of legumes: (4:1 or 8:1 is recommended)		_____
Step 3	Add the numbers of the ratio together:	____ + ____ =	_____
Step 4	Divide the result in step 1 by the result in step 3: This will equal the pounds of legumes needed daily for your family.	____ ÷ ____ =	_____
Step 5	Subtract the result in step 4 from the result in step 1. This will equal the number of pounds of grains needed daily for your family.	____ - ____ =	_____
Step 6	Multiply the pounds of legumes needed per day by 365 to determine how many pounds you need for one year	____ x 365 =	_____
Step 7	Multiply the pounds of grains needed per day by 365 to determine how many pounds you need for one year	____ x 365 =	_____

©Patricia Spigarelli-Aston
Refer to calculation 17.7 on page 145 to complete this table.

Worksheet 17.15 Selecting Legumes			

Number of pounds of legumes for family for one year: _____

Beans			
Bean Variety	Pounds	Servings per Pound	Number of Servings
1			
2			
3			
4			
5			
6			
7			
8			
Total		Total	

Split Peas and Lentils			
Split Peas or Lentil	Pounds	Servings per Pound	Number of Servings
1			
2			
3			
Total		Total	

©Patricia Spigarelli-Aston
Refer to calculation 17.8 and examples 17.16 and 17.17 on pages 148 and 149 to complete this table.

Worksheet 17.16
Selecting Grains and Grain Products

Number of pounds of grains and grain products for family for one year: _____

Grains

Grain Variety	Pounds	Servings per Pound	Number of Servings
1			
2			
3			
4			
5			
6			
7			
Total Pounds		Total Servings	

Grain Products

Grain Product	Pounds	Servings per Pound	Number of Servings
1			
2			
3			
4			
5			
Total Pounds		Total Servings	

©Patricia Spigarelli-Aston
Refer to calculations 17. 6 and examples 17.16 and 17.18 on pages 148 and 149 to complete this table.

NOTES

Individual Storage Foods

This chapter presents detailed information on individual storage foods and is organized in the same order as the previous chapter. Air-dried and freeze-dried dehydrated foods are abbreviated as AD and FD. Frozen foods are not included because of their vulnerability.

> Whenever storage life is given, it is assumed the item will be stored in ideal conditions. That means in an oxygen-free container in a dark, dry place at a temperature lower than 70° F (21°C).

Storing Salt

Salt is not affected by air, light, or heat, and noniodized salt will store indefinitely if protected from moisture in a tightly covered container. In low-humidity areas, salt can be left in the original cardboard or paper container. In high-humidity areas, it should be transferred to a plastic or glass container as it corrodes most metals.

Table Salt

Table salt is used for normal cooking and eating. It may contain an anticaking agent called calcium silicate to improve its pouring qualities. Salt is iodized to ensure that enough iodine is in the diet, especially for areas with iodine-deficient soil. Aging may cause the iodine to separate and produce a yellow discoloration, reducing shelf life to about five years, but the salt is still useable despite the discoloration. Kosher, gourmet, and specialty salts are also suitable for storing if desired.

Salt for Food Preservation

Pickling and canning salt is the purest of the salts and is used for preservation. It does not contain an anticaking agent or iodine and will not cloud water as table salt can when used in pickling and canning.

Rock Salt

Rock salt comes in blocks, large pieces or coarsely-crushed, and is used to feed domestic animals, attract wild animals, preserve meats, tan hides, and make ice cream. It is generally not for human consumption although specialty edible rock salt is used in salt mills.

Storing Milk

Powdered Milk

Powdered milk is most commonly available for home storage as instant nonfat dry milk. In addition to powdered whole milk and buttermilk, it's also available from some food storage suppliers as

> A pound of powdered milk makes five quarts (1¼ gallons) when reconstituted.

regular (noninstant) powdered milk. Powdered milk is made in a spray-drying process. Instant milk is processed further to make it easier to dissolve.

When purchasing, check that the product is 100 percent milk and fortified with vitamins A and D. Look for extra-grade because it has a lower moisture content, fewer bacteria, and higher excellence than the standard grade, which is intended for food manufacturing.

Storing Powdered Milk

Keeping powdered milk cool and dry is essential for a long storage life. Moisture causes caking, darkening, stale flavor, and rapid deterioration of nutritional value. Temperatures and ultraviolet and fluorescent lights turn the fat content rancid and destroy vitamins. Powdered milk will also absorb strong odors if not protected.

If you're purchasing ready-to-store powdered milk, an ideal packaging is a sealed Mylar liner with oxygen absorbers, which is then placed inside a bucket. Powdered milk stored in sealed #10 cans with oxygen absorbers is also an excellent choice.

It's less expensive to buy it in twenty-five or fifty-pound bags, but you'll need to repackage it for longer storage soon after purchasing. Use Mylar bags in plastic buckets, or metal or glass containers with oxygen absorbers. Powdered milk pours and "splashes" like a liquid and creates dust, so I suggest repackaging it outside on a windless day. Fill containers as full as possible and vibrate to compact. Clean the container lip of any milk powder before placing the lid on to ensure a good seal.

Shelf Life

The open shelf life for powdered milk is about six months. After opening a large storage container, transfer a two- to four-week supply to a smaller container to minimize exposure to moisture. Keep containers tightly covered except when quickly removing milk. Consider using smaller long-term storage containers so that open shelf life will not be exceeded once they are open.

It's a challenge to determine the actual long-term shelf life of powdered milk. Although the acknowledged shelf life for powdered milk is twenty to twenty-five years under good conditions, independent studies at Brigham Young University found that although twenty-year-old powdered milk was "life-sustaining," its taste was not equal to that of fresh powdered milk. For optimum quality, powdered milk should be used and rotated every three to five years. The good news is that milk that has developed off-flavors can often still be used in cooking.

Using Powdered Milk

Getting "Used to" Powdered Milk

In recent years, the drying process for making powdered milk has been improved with lower temperatures and shorter drying periods, making a better product that's more palatable, especially when stored properly. Because the quality and taste of brands vary, try several brands before purchasing large quantities.

Powdered nonfat milk that has been mixed and then allowed to blend and chill for a few hours—overnight is best—is close to the taste of regular skim milk. If your family objects to the taste of skim milk, you can try to improve the taste by adding cream, sugar, powdered cream, vanilla, evaporated milk, UHT whole milk, or chocolate syrup, but frankly, trying to jazz it up is more bothersome than just getting used to it.

PERSONALLY SPEAKING

 You may not be very enthusiastic about nonfat powdered milk. And I know it can be hard to convince kids, or even yourself, to drink it. So how can you make it work for your family? It took a while for our family to switch to skim milk. We were already drinking 2 percent, so we began by buying a gallon of fresh whole milk (4 percent butterfat) and mixing it with a gallon of nonfat powdered milk. We gradually reduced the amount of whole milk until we were mixing a gallon of fresh whole milk with three gallons of nonfat powdered milk. That turns out to be about 1 percent butterfat. The biggest challenge was getting in a routine of mixing up the milk every couple of days.

Containers for Mixing Milk

If you plan to use and rotate your powdered milk storage, you'll need containers for mixing it. Select containers that can be easily washed so that no milk residue builds up. To avoid bacteria growth, periodically wash them with detergent, rinse with a light chlorine-bleach solution, and let them air dry.

Cost of Powdered Milk

Regular nonfat, powdered milk can cost as little as half the price of regular liquid milk. Mixing it with whole milk will save you one-fourth to one-third your normal cost of milk. This will also allow you to rotate your powdered-milk storage in one and a half to two years.

> My preferred form of powdered milk for storage is regular nonfat powdered milk. It is the best buy and takes up the least amount of space.

In large quantities and depending on your source, the cost of nonfat dry milk commercially packed in plastic buckets or #10 cans should cost about the same as liquid milk.

But be cautious about the price you pay for powdered milk. Look past the hype made by some food storage companies and scrutinize their claims. And watch out for companies who advertise that their milk is freeze-dried when it is just plain powdered milk. The product may be packaged in neat little Mylar pouches, but if 160 servings cost $140.00—that makes

Table 18.1	
Calculating How Much a Gallon of Milk Costs	
Step 1	Multiply the number of 8-ounce servings per commercial package by 8 for the total number of ounces.
Step 2	Divide the total number of ounces by 128 for the number of gallons.
Step 3	Divide the total cost of milk by the number of gallons for cost per gallon.
If a 25-pound bucket of powdered milk makes 493 8-oz. servings of milk, that equals 3,944 ounces of milk. Dividing 3,944 by 128 equals about 30 gallons of milk. If the bag costs $120, the cost for a gallon of milk would be $4.00.	

it a very expensive $14.00 a gallon for powdered milk! Take time to calculate the costs so you are not misled! It is best to compare the number of gallons the products make rather than container size. (See Table 18.1 "Calculating How Much a Gallon of Powdered Milk Costs.")

Regular Nonfat Powdered Milk

Regular nonfat powdered milk is denser, takes up less storage space, and costs less than instant nonfat powdered milk. It is primarily used in food manufacturing and is more difficult to reconstitute than instant powdered milk. In addition, it's not as easy to find as instant powdered milk, but it is available from Rainy Day Foods, USA Emergency Supply, and The Church of Jesus Christ of Latter-day Saints home storage centers.

Using Regular Nonfat Powdered Milk

To use regular nonfat powdered milk at normal strength, mix one-half to two-thirds cup of powder per quart—you may want to experiment to find out what your family prefers. Mixing is easiest with a blender, but an eggbeater, wire whisk, or gravy shaker can be used if electricity is unavailable. Mix the milk powder with about a third of the required water, and when it dissolves, whisk in the remaining water. Refrigerate overnight for best results. Any small lumps that remain should dissolve overnight.

Using Powdered Milk in Baking

Baking is a particularly good way to use powdered milk and to rotate the product. You even can skip reconstituting the milk and add the powder and water separately. Try blending the dry powdered milk with flour and other dry ingredients to keep it from clumping. It's especially convenient for making bread or pancake and waffle batter. Replace regular milk in recipes with three tablespoons of powdered milk and one cup of water for every cup of milk required.

Instant Nonfat Powdered Milk

There are two varieties of instant nonfat powdered milk. The first is the traditional instant nonfat powdered milk in small boxes commonly available at grocery stores. These boxes are not suitable for storage because they offer little protection from oxygen and moisture. The second kind is a more finely crystallized nonfat powdered milk and is the variety recommended for storage. It is easy to find at food storage retailers, wholesale outlet stores, and on internet sources.

Using Instant Nonfat Powdered Milk

Today's processes for making nonfat powdered milk are much improved, and instant nonfat powdered milk is the product of choice in many parts of the world. However, some powdered milk is still made with older processes and suffers from "powdered-milk taste." Because of the inconsistency among various instant nonfat powdered milk products, be sure to sample before purchasing large quantities.

The amount of dry powder needed to make a gallon of milk also differs among brands. You will need to take this into account when you calculate how much milk a specific brand will make, how much storage space it requires, and price per gallon.

Milk Alternatives

Some food storage companies promote imitation milk as a viable alternative to real milk. The main ingredient in these imitation milks is usually whey, one of the two main milk proteins, but they also include nonfat milk solids, hydrogenated coconut oil, corn syrup, sugar, and other additives. And they are often fortified with an assortment of vitamins and calcium. The manufacturers claim they taste "just like natural milk," but you will have to decide if they're what you want.

Compare their protein and calcium content, price, and taste with regular milk. Scrutinize labels. Alternative milks may have less protein and less calcium than regular powdered milk. Check serving sizes. Test any claims that are made. And, of course, compare prices. Consider that a gallon of fresh whole milk costs about four dollars, or about twenty-five cents for an eight-ounce serving.

> **QUICK CHECK**
>
> ### Evaluate Alternative Milk Products
>
> ✓ Are the ingredients wholesome?
> ✓ Does a serving have 30% calcium?
> ✓ Does a serving have 8-9 grams of protein?
> ✓ Is the cost of a serving about the same as for powdered milk?
> ✓ Is the serving size reasonable?
> ✓ Does it taste good?
> ✓ Is it healthy?

The storage life for imitation milk products is ten to fifteen years in #10 cans, with an open shelf life of six to twelve months.

Almond, Coconut, and Soy Milk

Plant-based, lactose-free milks made from almonds, coconut, and soy are popular substitutes for dairy milk. They come in shelf-stable forms that have a storage life of six to eight months.

Buttermilk Powder

Buttermilk powder is made from real sweet-cream cultured buttermilk and can be reconstituted to make buttermilk or used directly in baking. A buttermilk blend made from sweet buttermilk and whey solids is recommended only for baking. It's used in batters for biscuits, cakes, muffins, and pancakes.

If packaged in metal cans and stored in ideal conditions, buttermilk powder has a shelf life of three to five years. In cardboard cans, its storage life is about a year, longer if refrigerated. It takes approximately two-thirds cup per quart to reconstitute. Prices can vary considerably, so be sure to compare.

Canned Milks

Canned liquid milk is another option for your storage and includes canned whole milk, evaporated milk, and sweetened condensed milk. Canned milk is sterilized in a heat process and may be in cans or in heat-resistant cardboard-like containers. As you might expect, the milk will have a cooked taste. The storage life for all canned milks is from one to two years, depending on storage conditions.

UHT Canned Whole Milk

Canned whole milk, or UHT (ultra-high temperature) milk, was first developed by the Parmalat Company, in Italy. It's used throughout Europe, is the standard for milk there, and comes in a variety of butter-fat concentrations. UHT products do not taste like the milk most Americans are used to, so some manufacturers in America add a variety of flavors to camouflage the cooked taste. Cream is also available in UHT form.

The shelf life is one year but may be longer if kept in cool conditions. Once open, UHT milk must be refrigerated. It is available in America from a variety of manufacturers and is usually more expensive than whole or powdered milk.

Evaporated Milk

Available in skim or whole, evaporated milk can be mixed with equal amounts of water to replace regular milk. It is useful in soups and chowders, as a cream substitute in sauces, or as a whipped topping. In an emergency, it can be used as part of baby formula.

Over time, the fats and other solids in canned milk separate and settle out. Turning or agitating the cans every few months can prevent this. Shaking the can vigorously before opening will also help, although if lumps do form, the milk is still usable. Also, the darkening and strong stale taste it may develop with age is not harmful and can be used in an emergency if necessary.

Condensed Milk.

Sweetened condensed milk is 40 to 45 percent sugar and is used almost exclusively for making candy, cookies, and desserts.

Storing Cheese

Cheese can be stored as fresh, dehydrated, or freeze-dried. Each type has advantages, and all of them contribute to a varied menu.

Hard and Semi-Hard Fresh Cheese

A general guideline is that the harder the cheese, the longer it will keep. There are some things you can do to extend the life of hard and semi-hard cheeses. Most initial spoilage to these cheeses comes from mold, which is a fungus that grows in dark, damp, warm conditions. So if you can minimize introducing mold spores or limit the growing conditions that foster mold growth, you can prolong the life of the cheese. Since mold spores are all around us, keeping cheese in its original packaging and keeping it cool and dry is key in prolonging its shelf life.

Hard brick cheese will keep several years if tightly wrapped in heavy plastic or foil to exclude air. It will also keep six to eight months by wrapping it in a cloth that is first soaked in vinegar and then allowed to dry. Repeat and rewrap as needed. Placing the wrapped cheese in a closed plastic bag can further extend the storage life. A paraffin coating will also keep the cheese from drying out. It is perfectly safe to trim the mold from cheese and use what remains.

Canned Cheese Products

Pasteurized cheese products are available in spreads in both cans and jars. Bega Brand canned cheese is a high-quality firm cheese from Australia that is a recent product in the American market. It is reasonably priced and has a minimum shelf life of two years and up to six or seven years if stored in a cool, dry environment. Cheese products, including cheese spreads and nacho sauces, are sold in glass jars and will keep from one to three years if kept cool. Their open shelf life is about two months.

Dehydrated Cheeses

This processed cheese product is made with whey, starches, cheese flavorings, and assorted food additives. It is versatile and can be reconstituted with water or oil or used directly in casseroles, soups, sauces, dips, and baked goods. Its shelf life is three to five years, and it keeps about four months after opening.

The most common varieties are Parmesan, cheddar, Swiss, or a blended cheese. You can also find more uncommon cheese powders, like asiago, blue, feta, Romano, and provolone on specialty food websites.

Freeze-Dried Cheese

Freeze-dried cheese is widely available in cheddar, Colby, Monterey Jack, and mozzarella. It differs from dehydrated cheese in that it's actual cheese. It works best as an ingredient in recipes and can be reconstituted or used as is. Serving sizes vary by seller, but a common serving size is one tablespoon, or about fifty calories, which is equivalent to a little less than a half-ounce of regular cheese. It makes about four times its original weight when reconstituted.

The shelf life is five to ten years in #10 cans with oxygen absorbers if kept cool and dry. Freeze-dried cheese is about two to three times more expensive than fresh cheese.

Storing Fats and Oils

Fats and oils should be kept as cool as possible, preferably below 60° F (16° C), and containers must be airtight and lightproof.

Liquid Vegetable Oils

Oil contains a high ratio of calories to weight. Common vegetable oils include safflower, canola, sesame, peanut, soybean, each having unique smoking points and fat compositions.

Oil does not have a long shelf life and should be rotated regularly. Unopened opaque plastic containers will keep two to three-plus years at 70° F (21° C) or lower. Open shelf life is four to six months. If stored in metal containers and kept cooler, oil may keep longer, but its quality will two to three-plus years at 70° F (21° C) or lower. The open shelf life of oil is four to six months.

Olive Oil

This unique vegetable oil has a high monounsaturated fat content that gives it a longer shelf life than other vegetable oils. In ideal conditions, high-quality olive oil can have a shelf life of three to four years, though its antioxidant benefits diminish over time. If this is your oil of choice, you'll want to use and rotate it regularly.

Shortening

Shortening has a slightly longer shelf life than oil and comes in baking sticks as well as one, three, and six-pound cans. It has a two-year shelf life, one-year when opened, and you can use it after its best-if-used-by date. You can tell when shortening is no longer good by its color and taste. It should be bland tasting, with no off-putting odor or color.

Powdered Shortening

Powdered shortening is most commonly used in commercial baking mixes, but it is available for home storage. It has 70 percent of the fat of regular shortening and is made from hydrogenated oil, corn syrup solids, and sodium caseinate. It can be used in place of shortening, measure for measure, and added directly to dry ingredients. It is used in baking breads, cakes, and cookies but is not suitable for frying because it leaves a residue that curdles. It has a longer shelf life than regular shortening—up to ten years.

Lard

Made from the rendered fat of hogs, lard was a staple in past generations and is still commonly used in many countries of the world. It's regaining favor as people transition back to basics and because making your own is a simple process. It can be used for frying and in baked goods and is especially prized in biscuits and pie crusts. It stores well in the refrigerator, although traditionally it was kept cool and replenished each year when the hogs were slaughtered.

Butter and Margarine

Butter and margarine normally make up 25 percent of added dietary fat. Regular butter needs to be refrigerated or frozen for optimal storage. Perhaps the best way to store butter is in the freezer. It does not take up much room, lasts at least a year, and is available whenever you need it. But for circumstances when refrigeration is not possible and for long-term storage, butter can be stored in both dehydrated powder and canned form.

Powdered Butter

This spray-dried powder has added milk solids and preservatives. For general use, it is reconstituted by adding water with a little vegetable oil until desired consistency, about a 1:1 ratio of powder to water. When reconstituted it will spread on bread and add flavor to vegetables. It is ideal for adding to dry homemade baking mixes, though it does not melt like regular butter so it is not useful for frying.

The shelf life of powdered butter is three to five years with an open shelf life of six to twelve months. It can be purchased in #2 ½ and #10 cans and in bulk packaging. Depending on the manufacturer, a #10 can is equivalent to between five and eight pounds of butter and is comparable to or a little more than the cost of regular butter.

Canned Butter

Canned butter is an option if you are looking for a way to store real butter. It is a unique product manufactured by Ballantyne of New Zealand under the Red Feather Brand and has recently become available in the United States through internet sources. It contains only pasteurized cream and salt, requires no refrigeration until opened, and has a long shelf life. It comes in twelve-ounce cans and costs about twice as much as fresh butter.

Mayonnaise and Salad Dressing

Both mayonnaise and salad dressing have a high fat content and a shelf life of one-plus years. Mayonnaise has ninety calories per tablespoon, salad dressing about forty They keep two to three months after opening if refrigerated. You can make mayonnaise with egg, vegetable oil, vinegar or lemon juice, and a little salt. Just combine in a blender or with a whisk.

Bacon

Bacon is a versatile meat that goes a long way toward minimizing menu fatigue and is valuable for its high fat content. If bacon is important in your diet, you'll find several options for storing it. The first is simply to freeze fresh bacon, use it, and rotate it. It will last at least a year in your freezer and has all the benefits of fresh bacon, including the fat content. The problem, of course, is that frozen food is susceptible to power failure.

Canned Bacon

Canned bacon is precooked to be preserved for long-term storage. The bacon is cooked, laid flat on parchment paper, and rolled before it is placed in the can and processed. The bacon is very thin and must be handled carefully as it is unrolled. It can be reheated for table use or used to flavor eggs, casseroles, and soups. This bacon is not a substantial fatty food since much of the fat has been rendered, and it will keep from five to seven years.

Bacon Pieces

Bacon is also available in fully cooked pieces, crumbles, and bits sold by major meat producers. It's available in jars, pouches with oxygen absorbers, and is found in most grocery stores. This product has a one to two-year shelf life and is easy to rotate because of its multiple uses.

Nuts and Seeds

Nuts and seeds are a great source of fats and protein and other nutritional benefits, but their high fat content makes them prone to becoming rancid, so they are best kept in a short rotation. They need protection from oxygen and high temperatures as well as excess moisture, which can cause mold.

Hard-shelled nuts, including almonds, Brazil nuts, filberts, pecans, and walnuts are somewhat protected by their shells and may last two years if stored unshelled in metal or plastic containers in cool, dry conditions. Ideally, you should store them in airtight containers with oxygen absorbers or treated with dry ice.

Shelled nuts can be dried in the oven on low heat—below 110° F (43°C) for an hour. They are best kept refrigerated or in the freezer. Sunflower and sesame seeds can be stored the same way.

Peanut Butter

Peanut butter is a versatile high-energy, high fat protein food. If your family uses it regularly it is a simple matter to stock up and then rotate, constantly replenishing it. Creamy homogenized peanut butter has a three-year shelf life and keeps two to four months after opening. Crunchy peanut butter keeps about half that long, and the old-fashioned kind, containing no anti-oxidants, keeps only six to nine-plus months.

A pound of peanut butter is about two cups or thirty-two tablespoons. Ten pounds provides about a tablespoon a day. You may want more if it's a staple for your children, less if not a family favorite.

Dehydrated Peanut Butter and Peanut Butter Powder

Dehydrated peanut butter powder is available in both the mass market as a fitness food and in the preparedness food storage market, usually in #10 storage cans. It has about a third the calories, a quarter the fat, and half the protein as regular peanut butter. It has a shelf life of five to ten years, has a variety of uses, and is comparable in price to regular peanut butter.

Peanut butter powder is made from peanut butter but has added sugar and salt and is found in regular grocery stores, big warehouse stores, and health-food stores. It has about one-tenth the fat of regular peanut butter but costs about twice as much.

Both are reconstituted with water and oil and are useful for making smoothies and in baking.

Storing Fruits

Canned Fruit

Canned fruits and fruit juices come in a great variety and keep best at cool temperatures, ideally just above freezing, 35° to 40° F (2° to 5° C). Their shelf life is two to four years, with fruit juices and highly colored fruits such as berries, cherries, and plums on the lower end. Regularly invert the cans to redistribute their contents.

Dehydrated Fruit

Many dehydrated fruits can be eaten as they are or reconstituted with cold or hot water for use as cooked fruit in side dishes, salads, or baking. Freeze-dried fruits are easy to overhydrate and will become mushy.

Dehydrated fruits will keep three to five years in ideal storage conditions. The open shelf life for air-dried fruit is about six months to a year, but freeze-dried fruit quickly absorbs air moisture and keeps only two to four weeks after opening. The shelf life for home-dehydrated fruit varies, but ideally, it should be used within a year.

Apples

Apples are one of the most economical and popular fruits and can be used in a variety of ways to increase menu options. They are good for snacks, salads, cakes, muffins, breads, apple pies or crisps, and to enhance meat dishes. Fresh apples will keep six or more months if you are able to store them just above freezing. Canned slices or applesauce stores for three years. Freeze-dried apple slices or applesauce will store up to twenty-five years if protected from heat, moisture, and oxygen.

PERSONALLY SPEAKING

 Apples are one of the easiest foods to dehydrate at home. To simplify the process, I use an apple peeler that both peels and slices in one operation. Our favorite way to enhance our dehydrated apples is to lightly sprinkle them with cinnamon sugar before drying. Some people recommend a lemon or citric-acid wash before drying, but I have not found that necessary. Some people also dehydrate them with the peel on, but we find that the skin makes them leathery. We don't have to worry about shelf life because they are eaten quickly! We plan to make a year's supply every fall when apples are in season, but so far, our apple stash has never made it that long.

Bananas

Bananas come in both freeze-dried and dehydrated slices. Slices or chips may be plain, honey-coated, or coated with coconut oil and sugar. They are delicious both as snacks and added to yogurt, pudding, oatmeal, cereal, etc. Those coated with coconut oil and sugar or honey are difficult to rehydrate but are still good for snacking. Both can be made into a powder with a food processor and then used in breads and muffins. Dehydrated chips have a shelf life of ten to fifteen years, and freeze-dried shelf life may be up to twenty-five years under ideal storage conditions.

Bananas can also be dried at home but tend to be chewy rather than crisp and typically have an unappetizing brown color.

Raisins

Regular raisins will keep two to three years. Freeze-dried raisins, which have lower moisture content than regular raisins, have a longer shelf life.

Assorted Fruits

Freeze-dried or dehydrated fruits are available in a wide variety, including apricots, tart cherries peaches, pears, grapes, blueberries, blackberries, raspberries, strawberries, mangos, pineapples, and oranges. Dehydrated fruits have a maximum shelf-life of five years in ideal storage conditions, freeze-dried fruits thirty years.

Citrus Juices

Grapefruit, lemon, lime, and orange juices are sometimes available as dehydrated crystals but those crystals are difficult to find.

Be cautious of food storage companies that claim to sell freeze-dried fruit juices. Check the ingredients label, and most likely you'll find a sugar-filled, fruit-flavored drink that sells for about three times the cost per cup of frozen orange juice.

Fruit juices are available canned or in bottles. They have a cooked flavor but are fortified with vitamins. Be sure to check the label to see what juices are actually contained in them—they are often primarily apple. Their shelf life is about two years.

Fruit flavored powders come in either sweetened or unsweetened form and may or may not be fortified with vitamin C. Those fortified with 100 percent RDA of vitamin C may be your best inexpensive option.

Storing Vegetables

Canned Vegetables

Canned vegetables can last three to five-plus years. Inverting the cans on a regular basis can help maintain the quality. Home-canned vegetables are a good option, especially for tomatoes and green beans.

Dehydrated Vegetables

Commercially dehydrated and packaged vegetables last three to eight years. Home-dehydrated vegetables keep a year or more. Open shelf life is one-plus year for air-dried but only four to eight-plus weeks for freeze-dried.

Potatoes

A South American native, potatoes are fourth in world production behind wheat, corn, and rice, and yield more food more quickly on less land than the other major crops. Potatoes are 99.9 percent fat-free and a good source for vitamin C, many B vitamins, potassium, and fiber. The annual consumption of potatoes in the United States is about thirty-five pounds of fresh potatoes and 116 pounds of processed potatoes per person.

Store quality fresh potatoes such as russets in a well-ventilated, cool, dark place around 45° to 50° F (7° to 10° C). Cooler temperatures will force the starch in potatoes to turn to sugar. Light will cause chlorophyll to activate and give the potato skin a green tinge, but in small amounts, this is not harmful. Potatoes will store several months in ideal conditions. If potatoes sprout, you may remove the sprouts and still use the potatoes.

Potatoes are also available in a variety of dehydrated forms that are ideal for storage. If stored in cans with oxygen absorbers in a cool, dark place, dehydrated and freeze-dried potatoes have a shelf life of up to twenty years.

Sweet Potatoes

Unrelated to white potatoes, sweet potatoes are all-star performers in nutrition, especially as a source of vitamin A. They are available in canned or dehydrated forms. Canned sweet potatoes have a shelf life of four to five years, dehydrated or freeze-dried sweet potatoes ten to fifteen years, in ideal storage conditions. Fresh sweet potatoes will last several months when stored around 60° F (15°C).

Tomatoes

Technically a fruit, tomatoes are used as a vegetable and are a good source of vitamins A and C. Tomatoes are an important ingredient in many Italian and Mexican dishes and for meat dishes, stews, casseroles, and in ketchup. Tomatoes keep two to four years in enameled cans and even longer in glass jars. Dehydrated tomato powder can be mixed to almost any consistency for sauce or paste, but it does not make great tomato juice. It has a ten- to fifteen-year shelf life. Home-canned tomatoes will last several years but are easy enough to replenish each harvest.

Sun-dried tomatoes add another dimension to your stored food menu and are another way to use tomatoes grown in home gardens. They are easy to dry and may be kept in oil. In regular packaging, they have a shelf life of about a year, longer if vacuum-packed.

Storing Sugar

The primary function of these simple carbohydrates is to provide fuel for the body. Sugars occur naturally in whole, unprocessed foods, like fruits, vegetables, milk, grains, and nuts. Sucrose is common table sugar, chemically made of equal parts fructose and glucose, and is refined from sugar cane or sugar beets. Fruits, vegetables, grains, nuts, and honey contain varying percentages of glucose, fructose, and sucrose. Lactose is the sugar in milk and milk products. Glucose, also called dextrose, is the form of sugar our bodies use for fuel. Sugars, including sucrose and high-fructose corn syrup, are added to many processed foods and, in excess, contribute to obesity, heart disease, diabetes, and other health issues.

Sugar is necessary for the survival diet since a survival diet is likely to have a much lower intake of processed foods containing sugar. Children, especially, have high caloric needs for their weight and may be unable to obtain sufficient energy if restricted from concentrated calorie sources. Therefore, sugars are an important food in a preparedness plan.

White Sugar

White sugar is made from juice squeezed from cane or sugar beets that have been physically cleaned of impurities. The molasses is removed and concentrated, and although chemicals are used in the processing, white sugar is not chemically altered in any way.

White sugar is most common as granulated table sugar and as confectioners, or powdered, sugar. It is a very stable food and will keep indefinitely if protected from moisture and contamination. High temperatures may compromise the quality, but it is safe to eat. Sugar is best stored in opaque, airtight food-grade containers, such as polyester film bags, metal cans, and plastic buckets. It can be stored in its original bags in low-humidity areas, although the paper bags will deteriorate over time and are susceptible to puncture. Hard, lumpy sugar can be broken up with a hammer or placed in water and made into syrup.

Brown Sugar

Brown sugar is made from white sugar that has a small amount of molasses added to it. The more molasses, the darker and stronger the flavor is. You can make your own by adding four tablespoons of molasses per cup of white sugar. The higher moisture content of brown sugar needs to be retained or it will harden. It can be softened by placing a slice of apple or fresh bread, or a water-soaked clay disk in the container. Or for immediate softening, sprinkle water over it and place it in an oven at 250° F for a few minutes. It keeps indefinitely if protected from too much moisture.

Honey

Honey is formed by a honeybee enzyme from the diluted sucrose found in plant nectar. The average worker honeybee makes about one-tenth of a teaspoon of honey in its lifetime. To make one pound of honey, the bees in a hive must visit two million blossoms and fly fifty-five thousand miles. A thriving honeybee colony will make sixty to one hundred pounds of honey a year. The nectar source determines the flavor, aroma, and color of honey, and usually the lighter the color, the milder the flavor. Its exact composition also depends on the nectar source, but its chief sugars are fructose and glucose.

> **QUICK LOOK**
>
> ## Qualities of Premium Honey
>
> • Minimally processed, gently heated, and lightly strained
> • Produced exclusively from flora
> • No added water, corn syrup, or sugar syrup
> • Low water content of 15-18%

Qualities of Honey

Most honey is strained or heated and filtered to remove the suspended particles and fine air bubbles that cause honey to crystallize faster. It is also graded by its clarity and lack of impurities. There are over three hundred distinct varieties of honey produced in the United States.

Honey is somewhat sweeter than sugar and a tablespoon of honey has one-third more calories than a tablespoon of sugar. Like sugar, honey is easily and quickly digested and absorbed into the blood.

Honey is also more expensive, usually costing about eight to ten times more per pound. In many cases it can be substituted for sugar if desired. It gives baked goods a distinctive flavor and aroma and may improve the texture and browning. Honey also attracts water, and baked goods made from it tend to remain moist and stay fresh longer.

Honey is not recommended for infants who are under a year old because it is a possible source of infant botulism.

Buying Honey

According to the National Honey Board, raw honey is "honey as it exists in the beehive or as obtained by extraction, settling, or straining without adding heat." It has not been pasteurized or ultrafiltered—processes that remove the beneficial vitamins, enzymes, antioxidants, and other natural nutrients.

Beginning in 2014, FDA guidelines required companies to label honey diluted with other sweeteners as a "blend." Be aware that the processed, crystal-clear, golden honey sold in most mainstream grocery stores is likely ultrapasteurized. It may be a blend with added sugars and may, at least in part, be imported from foreign sources with questionable production practices.

For quality honey, purchase directly from a reputable beekeeper.

The most important thing you can do to ensure you're getting quality, pure honey, is to purchase it directly from a trusted, reputable beekeeper. The National Honey Board has a list of honey producers in each state. I recommend purchasing minimally processed local raw honey that is slightly heated and strained of impurities. Some recommend you purchase honey made near where you live because it has a positive effect on potential pollen allergies. Some appreciate the minute quantities of additional nutrients found in honey.

Do not hesitate to ask a beekeeper about their honey—how it is made, the source of the nectar, whether the bees are supplemented with sugar water during the producing season, etc. Reputable beekeepers will be proud of their products and very transparent, glad to tell you all about their honey. You may even find a beekeeper who will fill your containers.

Storing Honey

Honey will last indefinitely if stored properly. All good honey will crystallize over time and become darker and more strongly flavored with age. Small amounts of crystallized honey may be reliquefied by placing the container in a pan of water on low heat. Ideally, hold the honey as close to 105° to 115° because higher heat can damage the honey. Place larger containers near a heater, or set on a low rack in the oven or in the sun for a few days.

Honey is slightly acidic and is best stored in glass, ceramic, or food-grade plastic containers, as the acid in honey causes metal containers to corrode over time. Protect transparent containers from light because light will cause color and flavor changes.

Look for pure, undiluted honey containing 15 to 18 percent water, and avoid honey that has added water. Although honey may legally contain up to 20 percent added water to prevent crystallization, this allows fermentation and mold within a short time unless kept refrigerated. Also, honey attracts moisture and should be kept airtight.

PERSONALLY SPEAKING

 I recently ran out of honey and decided to try the honey we put in our storage over twenty-five years ago. Originally, it was pure spun honey, and it had been stored in a plastic tub with just a snap-on lid and kept in our basement food storage room. I opened it up to find a solid crystal mass floating in a dark-brown liquid. It smelled fine, so I tasted it. It tasted just like honey, maybe a little more flavorful. I put the plastic container on a small rack in a water bath at low temperature, and it gradually liquefied. Although it was darker than its normal amber color, it tasted like honey. It would have been better to use the honey in a rotation cycle, but it was still usable after years of storage.

Corn Syrup

Corn syrup is glucose syrup produced from corn. Regular corn syrup is an important ingredient used in syrups and candy to control crystal formation. It will keep two-plus years in an airtight container and is different from high-fructose corn syrup (HFCS), the pervasive, highly refined sweetener in many processed food products.

Molasses

Molasses is useful for adding flavor and variety to baking. Most molasses is a by-product of making white sugar. It is largely sucrose and keeps for four to ten years. Blackstrap molasses, the third extraction, or final by-product, offers some vitamins and minerals, but a tablespoon per day only provides 10 percent of the RDA for calcium and iron.

Maple Syrup

While a mature sugar maple tree (*Acer saccharum*) will yield approximately ten gallons of sap, it takes forty gallons of sap to make one gallon of maple syrup. Primarily sucrose, it will keep for five to ten years in its pure, undiluted form.

Agave Nectar

Agave nectar is not really nectar but a concentrated, highly processed liquid-fructose sweetener made from the agave plant in much the same way high-fructose corn syrup is made from corn. It has a different chemical composition than regular sugar and is sweeter, so less is needed. It has a mild, sweet taste, does not crystallize, and can be used for sweetening beverages and other recipes that call for liquids. Experiment to find your preferred amount for sweetening, but as a general rule, for each cup of sugar, use two-thirds cup of agave and reduce the liquid by one-fourth cup. Agave is more expensive than white sugar and has a shelf life of two to three years.

Jams and Jellies

Jams, jellies, marmalades, and preserves are high in sugar content. Keep them in a cool, dark place to preserve their flavors and colors. Their shelf life is two to three years.

Gelatin Desserts and Puddings

Often referred to by the brand name Jell-O, gelatin desserts are almost straight sugar. If kept in a cool, dark place, they will keep five to ten years before losing their color, flavor, and setting-up ability. Puddings are also high in sugar.

Presweetened Drink Mixes

Pre-sweetened drink mixes are available under brand names like Tang, Kool-Aid, Wyler's, and Hawaiian Punch. They come in many flavors and are high in sugar. Although convenient, they are more expensive than their unsweetened counterparts. Breakfast drinks like Tang usually provide 100 percent of the RDA for vitamin C in each four-ounce serving, but other drinks often have only one-tenth as much. Check the label to be sure.

Companies that sell commercially packaged food storage often offer fruit drinks as part of their whole package. If you look carefully at the ingredients label, however, you'll see that they are frequently little more than fruit-flavored sugar drinks. Though the advertising suggests you are getting "real" fruit juice, they are merely expensive forms of sugar.

Storing Animal Proteins

Animal proteins, including milk, cheese, and the other dairy products listed previously, are an expensive way to store protein but indispensable in most Americans' diets.

Eggs

Eggs are commonly regarded as the best quality protein available. They are the basis for many dishes and perform vital functions in many recipes.

Fresh Eggs

Although the stated shelf life for fresh eggs is about a month, they may store up to six months if refrigerated. Before refrigeration was widely available, fresh eggs were commonly stored without it, which practice continues in Europe today, as fresh eggs do not necessarily require refrigeration for short-term storage.

It may seem counterintuitive, but washing your eggs increases the likelihood of bacterial contamination. Washing removes the cuticle, or bloom, a naturally occurring protective covering. Refrigeration promotes condensation and has the same effect as washing.

Dehydrated Eggs

Dehydrated, or powdered, eggs are the most common way to store eggs. Air-dried powdered eggs come as whole eggs, egg whites, and egg yolks. Dry eggs are also available as a scrambled-egg mix with added milk and vegetable oil. Dehydrated eggs are reasonably priced.

Dehydrated eggs are useful in baking and can be added directly to the dry ingredients. The scrambled eggs and omelets offered in hotel breakfast buffets are made from dried powdered eggs. Although powdered eggs today are an improvement over those in the past, do not expect them to taste like fresh eggs. Whether they are suitable for scrambled eggs, in omelets, or in other fresh-egg applications is a matter of personal taste.

Dehydrated egg products are sold in #2½ and #10 cans, though the quantities in similar cans may not be the same. To determine how many eggs a can holds, assume there are thirty-two to thirty-six equivalent eggs per pound of whole-egg powder. You can also determine the egg equivalents knowing that one egg is about seventy-two calories, despite the serving size that might be stated on the product. Table 18.2 shows you how many calories are in different sizes of eggs.

QUICK LOOK
Storing Fresh Eggs
• Unwashed fresh eggs have a natural protective coating
• Purchase eggs from farmers with certified salmonella free hens
• Store unwashed fresh eggs at room temperatures for several weeks
• Store refrigerated eggs up to five months
• Low water content, 15-18%
• Coating eggs with sodium silicate solution may extend storage life
• Coating eggs with lard may extend storage life.
• Crack eggs into a separate bowl before using to check freshness

Table 18.2 Calories in Eggs	
Small	55
Medium	70
Large	75
Extra large	80
Jumbo	90

Storing Dehydrated Eggs

Be aware that shelf-life claims for dehydrated eggs may be exaggerated, especially if you want to use them in scrambled eggs or omelets. For best quality and to most resemble fresh taste, plan on a storage life of about two years and for use in baking five to ten years. Egg powder readily absorbs moisture, making it lumpy and strong-flavored. High storage temperatures will damage flavor, solubility, and thickening ability.

Shelf life after opening is three to four months. You may extend this by repackaging with oxygen absorbers. Of course, after reconstituting, treat them like fresh eggs.

Egg Crystals

Egg crystals are sold under the brand name Ova Easy and manufactured by Nutriom. They are another form of dehydrated egg that many find to be fresh tasting and an improvement over regular dehydrated egg. They have the functionality of fresh eggs; however, they are more expensive than fresh or dehydrated eggs. Their shelf life is seven years.

Freeze-Dried Eggs

Freeze-dried eggs are cooked scrambled eggs that have been freeze-dried. They require no further cooking and are reconstituted with boiling water. They come plain or with additions like bacon, ham, sausage, peppers, and onions.

They are packaged in pouches or #10 cans. The shelf life is thirty-plus years. Because the open shelf life is only about a week, you may want to repackage the large cans into smaller pouches.

Table 18.3 highlights the characteristics of different types of dehydrated eggs.

Table 18.3 Characteristics of Different Egg Forms				
Type of Egg	Best Use	Rating for Taste	Shelf Life	Relative Cost
Dehydrated Egg Powder	Baking, cooking ingredient	Low for scrambled	2 years	Low
		Moderate for baking	5 years	
Dehydrated Scrambled Egg Mix	Scrambled eggs, omelets, French toast	Moderate to good	2 years	Low
Egg crystals	Scrambled eggs, omelets, French toast, baking, cooking ingredient	Good	7 years	Moderate
Freeze-Dried Scrambled Eggs	Scrambled eggs, breakfast burrito	Good	30 years	High

Meats, Poultry, and Fish

Meat will make your menus more palatable and much more like the food most people are used to eating. Even the addition of small amounts of animal protein to vegetable proteins raises the protein quality substantially and for little cost.

Dried Fish and Meat

Dried and smoked meat and fish can be kept several years in appropriate containers. Common dried meats are chipped beef, jerky, and pemmican. These products can be purchased commercially, but you can also make your own to save money and ensure a quality product. It is a survival skill worth acquiring.

Canned Meats

Canned meats, poultry, and fish have a shelf life of three to five-plus years if kept at 70° F (21° C) or cooler. Canned tuna (oil or water-packed), canned chicken, stewed chicken, and canned beef are all good storage options. Combination meats, such as Spam are usually lower in protein but add variety to your storage. Again, these products can be purchased or home-canned. Home canning is discussed in chapter 29.

Table 18.4 shows you the advantages and disadvantages of using canned meats.

Table 18.4 Advantages and Disadvantages of Canned Meat	
Advantages	Disadvantages
· Good variety	· Heavy
· Widely available	· May contain excessive sodium
· Moderately priced	
· Durable packaging	· May contain unwanted additives
· Ready to use	
· Cooking not necessary	· Susceptible to freezing
· Familiar in regular diet	· Cans susceptible to rusting
· Moderately long shelf life	· Jars vulnerable to breaking

Retort Meats

Meats may be canned in retort pouches, which are processed under pressure in retort kettles. Although not widely popular, retort packaging is flexible, sturdy, takes up less space, and offers less waste than traditional cans or jars. The retort sterilization process requires less time than traditional canning, which results in a more delicate product. The process is especially nice for fish and seafood. Retort meats have at least a one-year shelf life.

Food storage companies often offer retort meals known as MREs (Meals Ready to Eat). These combination meals, originally developed for in-the-field military operations, may contain adequate amounts of meat protein, but they are often very high in sodium and other food additives. Be sure to read the labels and sample the products before making a large investment.

Retort foods may be eaten directly from the pouch in an emergency or heated in water or with a flameless ration heater (FRH), also developed for the military. The shelf life is five years or more.

Freeze-Dried Meats

Freeze-dried beef, chicken, pork, and turkey are already cooked and only require boiling water to reconstitute for eating.

Freeze-dried meat is more expensive than comparable fresh or canned meat. The flavor and texture depends on the quality of the meat and how it was prepared before freeze-drying. Do not expect the meats to be exactly like fresh meat, but freeze-dried meat should be handled like fresh meat once water is added. Diced or ground meats are most flexible and can be used in many recipes.

Freeze-dried meat is also found in a large selection of combination entrées that are less economical but save you from opening half a dozen cans to prepare a single dish. The taste is satisfactory, but it can be worrisome if there is a long list of additives. Some people claim there is a sameness about the entrées. The stated serving sizes are often small, and you could need two or three per person! Despite these limitations, entrées are a good option for convenience and short-term emergencies.

The shelf life claims for freeze-dried food range from fifteen to thirty years when packaged in an oxygen-free #10 can. The open shelf life is only four to eight weeks in a closed container because freeze-dried foods attract moisture.

Table 18.5 summarizes the advantages and disadvantages of freeze-dried meats.

Table 18.5 Advantages and Disadvantages of Freeze-Dried Meat	
Advantages	Disadvantages
· Good variety	· Expensive
· Durable packaging if stored in cans	· Requires water to prepare
· Ready to use, cooking not necessary	· Bulky despite being lightweight
· Reconstitutes quickly	· Foil pouches susceptible to puncture
· Retains original shape, texture, and color	· Cans susceptible to rusting
· Lightweight	
· Long shelf life	

Meat Substitutes

Other sources of animal protein may be less expensive than meat products and are easy to store. These highly processed foods are usually made from milk protein or soy protein.

Protein Concentrates

Frequently used by athletes to supplement normal protein intake, protein concentrates can be useful in a survival diet as well. Powdered protein concentrates are normally made from the milk proteins whey or casein or from soy protein or a combination with lesser amounts of other ingredients and sweeteners. They come in a variety of flavors and can be dissolved in milk or fruit juice or mixed directly with other foods.

Look for protein concentrates that are complete proteins with all nine amino acids and that dissolve readily. Price varies. Compare the cost of protein gram for gram with other potential storage foods. Since protein concentrates use one-third the space and about half the weight, they are useful if space is limited or if concealment and portability are major concerns. Shelf life is three to seven years, open shelf life about one year.

Textured Vegetable Protein (TVP)

Textured vegetable protein, known as TVP (also known as total soy protein or TSP), was developed for the food industry as a meat replacement or extender. It is made from soybean protein that has been spun into fibers and then fashioned into various forms to imitate different meats. Because of its relatively lower cost and long shelf life, it has made its way into the food storage market.

Manufacturers of these pseudomeats claim they duplicate the look, texture, and taste of the real thing at reduced prices. TVP comes plain or flavored and, except for the bacon bits, is reconstituted and then cooked as real meat would be. Shelf life is three to five years.

While cost is its major selling point, palatability is its biggest problem. Most who have tried it straight would not put it in the same class with real meat. Its protein quality, even though made from soybeans and normally fortified with extra methionine, is still inferior to nearly all animal-protein sources. There is also concern with the safety of soy products in general and the effect of the large quantities of additives in TVP. Try it before you buy it.

Storing Grains and Legumes

Whole grains and legumes provide more nutrition for the money than any other food. Processing, such as milling, rolling, or being formed into pasta, raises the price. Mixes, premade casseroles, and breakfast cereals are even more expensive. Although grains and legumes are incomplete proteins, when combined, they create a complete and nourishing high-quality protein. The quality of protein is also improved when a small amount of meat is added to them.

> **Grains and legumes bought in bulk provide more nutrition for the money than any other food.**

Many sources will tell you that grains and legumes will store for thirty-plus years. While it may be true that they will not be spoiled and may sustain life, the quality of micronutrients will be compromised. Ideally, both grains and legumes should be part of your daily menu and used on a regular basis. The Utah State University Cooperative Extension recommends you rotate your wheat so that no stored grain is older than five years. Grains with high oil content, such as brown rice or buckwheat, have a shorter shelf life. The shelf life for all grains will be longer if they are stored in with oxygen absorbers in a cool, dark location. Because beans tend to develop a hardened shell, it's wise to store an amount you'll cycle through in two to three years.

Store grains and legumes in as close to ideal storage conditions as possible to have good results. And finally, always try a product before buying it in quantity.

Grains, Cereal, and Flour

The main grains for long-term storage are corn, rice, and wheat. In addition, to add variety and to enhance nutrition, consider the specialty grains described in table 18.7 on page 191. Oatmeal and other cereals, as well as flour and products made from flour also enhance our diet. However, they have a shorter shelf life and must be rotated.

Nutrition in Cereal Grains

Cereal grains are excellent sources of protein, vitamin-B complex, vitamin E, essential fatty acids, minerals, and fiber. Table 18.6 on page 188 shows how the nutrients are distributed in a kernel of grain.

Whole grains are much more nutritive and keep much longer than milled grains. The bran and germ are normally removed during commercial milling to give flour its lighter color and to prevent it from becoming rancid, but doing so lowers its nutritional value.

Table 18.6 Parts of a Whole Grain Kernel			
Parts	Use in the Grain	Nutrients	Processing
Germ	Sprouts and contains the wheat embryo	Antioxidants, vitamins E and B, essential oils	Removed during milling
Endosperm	Energy supply for the germ	Carbohydrates, proteins	Source of white flour
Bran or hull	Surrounds, seals, and protects the kernel	Fiber, B vitamins	Removed during milling

Corn

The American Indians first grew corn, or maize. During early pioneer times, corn was eaten by nearly everyone in some form for breakfast and dinner almost every day. Dried field corn can be made into hominy and corn grits, or ground into meal and flour for flat cakes, mush, and quick breads. The flour can be mixed with other flours to create a crunchy, crumbly texture and a sweet taste. The nutritious germ is removed from commercial cornmeal during milling so it must be enriched, but you can grind your own corn for better flavor and higher nutrition. Whole dried field corn keeps five to ten-plus years, while cornmeal keeps three-plus years. Fumigation is normally not necessary. Moisture content should be 12 percent or less.

Popcorn

Popcorn is a special strain of corn with a very hard hull. It will keep five to ten-plus years, but it must be kept tightly covered to retain its moisture content so it will pop when heated. It makes for an inexpensive and nourishing treat that adds variety, lift spirits, and eases hard times. Use a handheld popcorn popper to pop it over a fire.

Oats

Native to Central Asia, oats are the most popular cooked breakfast cereal. The kernel is a long, light-brown grain with a bland, sweet taste. Since oats cannot be easily processed at home, it's best to buy them already in the form you want: quick rolled, whole rolled, or steel-cut. Whole oat groats are the whole grain after the hull is removed.

> **QUICK FACTS**
>
> ### Oatmeal Equivalents
>
> 1 lb. = 5 cups dry
> ½ cup dry oats = 1 cup cooked oatmeal
> ½ cup dry oats = 150 calories
> ½ cup dry steel-cut oats = 170 calories

Steel-cut groats, sometimes called Scotch oats, have been cut into several pieces that can then be rolled into flakes. They are steamed lightly, and the thinness of the flake determines whether they are regular or quick-cooking oats. Shelf life for oatmeal is less than for whole grains—about two to three years with good storage practices.

Rice

Rice the most widely consumed cereal grain, rice is considered by half the world to be the "staff of life." It comes as long, medium, or short kernels covered with a green-brown husk.

Brown Rice

Brown rice has the husk removed, while white rice has been polished to also remove the germ and several layers of bran. Brown rice is superior in nutrition and has a much higher fat content, which causes it to become rancid more quickly. It should contain less than 12 percent moisture and be rotated since its optimum storage period is six to twelve months, possibly two years, in an airtight container with oxygen removed.

White Rice

White rice, although considerably less nutritious, is preferred for storage because it will last for thirty years or more if stored properly. Almost all white rice is enriched to make up the nutrients lost during processing.

Long grains cook up fluffy and separate; medium grains are tender, moist and clingy; and short grains are very clingy and cook to a creamier texture. Rice is one of the first foods infants eat and is helpful for people suffering from diarrhea. It can be used in a gluten-free diet. White rice should have less than 14 percent moisture content.

Basmati and Jasmine Rice

Two popular varieties of rice you might want to consider for storage are Basmati and Jasmine. Both are pleasant and aromatic. Basmati is a long-grain rice commonly used in Indian and Middle Eastern cooking. It is best if soaked a half hour before cooking. Jasmine is a shorter-grained rice that has a soft, sticky texture when cooked. Both types will store well at cool temperatures and in sealed containers.

Wheat

Also known as the "staff of life," wheat is a nutritious, versatile grain that's excellent for storage and used primarily for breads and cereals. It is the least-expensive grain and is rich in B vitamins, vitamin E, essential fatty acids, iron, and fiber. The two most common varieties of storage wheat are hard red and hard white. Both have a high-protein and low moisture content that makes them ideal for storage.

Durum wheat is used to make semolina flour, which is used for pasta and pizza doughs. Soft wheat varieties are used for pastry flour.

Ancient Wheat Varieties

Spelt, emmer, einkorn, and Khorasan are types of ancient wheat that originated in the Fertile Crescent region. These grains have become popular because they are less hybridized than modern wheat. They have a reputation for being easier to digest and are sometimes tolerated by those with wheat sensitivities. They also have more micronutrients and a higher protein content than modern varieties. They are valued for their rich flavors and interesting textures.

Heritage Wheat Varieties

Red Fife, Turkey Red, and White Sonoma are examples of heritage varieties of wheat that predate modern industrial wheat. They have been cultivated regionally and adapted to specific soils and growing conditions and are often prized by artisan bakers for their superior baking properties and subtle and unique flavors. Like ancient varieties, they may be helpful for those with wheat sensitivities.

PERSONALLY SPEAKING

People often ask whether they should store hard red or hard white wheat. It really boils down to personal preference. Both make quality bread. If you prefer a heartier flavor, select the red wheat. If you prefer a closer-to-white-bread taste, go with the hard white. Whatever kind you choose, learn how to use your wheat. Bread making is an art that takes practice, but there are plenty of resources to help you get started. Invest in the tools and time to use your wheat. As you become an expert, you may want to explore using other varieties of wheat.

Storing Wheat

If you plan to store wheat, buy it from a reputable source. Look for a protein content of 13 to 15 percent and moisture content 12 percent or lower. Protein content that is extremely high can be undesirable. Kernels shriveled by frost or drought may have abnormally high protein, but they are also low in starch and make poor bread due to low gluten content. Many varieties grown in different areas of the country are suitable, and the actual protein content and other characteristics depend not only on the variety but on the environmental and soil conditions in a given growing season. Spring wheat tends to be larger and plumper, with a higher protein and moisture content.

Wheat can be purchased in bulk from flour mills, grain dealers, and food storage retailers. You can purchase it in bulk in paper or reinforced plastic bags, but only purchase it in bags if you intend to transfer it into more permanent storage containers because wheat stored in sacks without a moisture barrier will draw moisture from the air.

Preparing wheat for storage is easy and inexpensive, especially if you have an economical source for storage buckets. Use oxygen absorbers to delay oxidation and prohibit insect infestation. If you must purchase storage buckets, it will probably cost about the same to buy wheat already in buckets. Some companies offer wheat in buckets with a Mylar bag lining for added protection. This is a good option if you plan to store it for a long time. Purchasing wheat in #10 cans will cost about twice as much as buying it in bulk.

If you do not have a local supplier, you'll have to consider freight costs in your purchase. Some companies offer "free shipping," but the higher cost of the items often makes up for the savings in shipping. Take the time to compare the true costs. If you live where grains are unavailable, you may want to join or form a co-op and order in semi-loads to consolidate shipping costs.

Other Grain Varieties

In addition to corn, rice, and wheat, consider using and storing a few lesser known grains listed in table 18.7. They will give you a more interesting diet and offer enhanced nutritional benefits. Many of these grains can be found at Rainy Day Foods, Honeyville, Pleasant Hill Grain, and Whole Grains Council.

	Table 18.7 Grain Varieties and Their Characteristics	
Grain	Characteristics	Taste/Uses
Amaranth	Small brown seeds high in protein, gluten-free	Light, nutty, earthy flavor; use in soups, salads, side dishes; add to baked goods; pop like popcorn
Barley, pearl	Short, stubby, creamy-white, polished kernel with outer husk removed; low in gluten	Mild flavor with chewy texture; use as a hot breakfast cereal; add to risottos, soups, and casseroles; roast and steep to make hot drinks; mix with wheat flour to make sweet, cakelike breads
Buckwheat	Small, triangular brown seed high in protein, gluten-free	Toasty, nutty flavor; use as buckwheat flour in pancakes, crepes, noodles
Chia	Tiny brown or black seeds, high in fiber and omega-3 fatty acids, nutritious and gluten-free	Mild, nutty flavor that complements both savory and sweet foods; add to other foods for nutritional benefits; use to thicken sauces and puddings
Flax	Small, flat, oval brown seeds; especially nutritious; gluten-free	Mild, nutty flavor; use whole or grind into meal; add to hot cereal, granola, breads, baked goods, crackers
Millet, hulled	Tiny, round, pale-yellow seed; chewy texture; easily digested	Sweet, nutty flavor, chewy texture; use as a breakfast cereal, in soups, stews, casseroles, desserts, or as a stand-alone side dish; adds crunchy texture to wheat breads; use as a baby cereal
Quinoa	Small grain-like seed; creamy white, red, or black; gluten-free	Sweet, earthy taste; cook as cereal; serve hot or cold in salads; substitute for rice in casseroles and stir-fries
Rye	Small grain-like seed; creamy white, red, or black; gluten-free test	Hearty, sweet, nutty flavor; best mixed with wheat for flavorful, full-bodied breads
Sorghum	Roundish brown seeds with yellow and red mixed in, gluten-free	Light-colored, mild taste; can be milled into flour for baking and used with other flours; use in risotto-like dishes
Triticale (pronounced trit-ih-KAY-lee)	Hybrid of wheat and rye; large, plump kernels	Sweet, nutty taste; good baking properties; makes hearty whole-grain bread; rolled into flakes for cereal

Enriched Flour

Flour costs one-fourth to one-third more than whole grains. It is more sensitive to heat and moisture and does not store anywhere near as long as whole wheat. It is convenient, but do not store more than you'd normally use in a year's time. Moisture can drastically reduce its shelf life, so store it in airtight plastic or metal containers. Fumigation is useless because of its density.

Enriched, all-purpose white flour contains less protein, fiber, vitamin E, essential fatty acids, and calcium than the whole wheat it came from. It keeps two to three years at 70° F (21° C) with a 12 to 13 percent moisture content.

Whole-wheat flour has a high fat content and will become rancid, so it keeps for much less time than white flour. A better solution is to invest in a grain mill and grind your own fresh whole-wheat flour as needed. Breads made from whole-wheat flour tend to be heavier and crumblier, but with good recipes and practice, they need not be that way. Properly prepared, home-baked whole wheat bread will make most commercial breads lose their appeal forever.

Pasta

Pasta comes in a wide variety of shapes and sizes and is made from semolina, the starch portion of durum, the hardest wheat grown. The quality of the pasta changes with the quality of the durum wheat available each year. Low-quality pasta is often made from blends or even other varieties of wheat that are starchier, cook up gluey, and are less delicious. The best quality pastas give the best results.

Pasta does not deserve the "empty calorie" label it sometimes gets. It contains the basic wholesomeness of wheat and provides essential B vitamins, iron, and calcium. It is easy to store, quick to prepare in a wide variety of dishes, high in protein, low in fat, and easily digested. Italians eat sixty pounds yearly per person, the average American about twenty.

> **QUICK FACTS**
>
> ### Pasta Equivalents
>
> 2 oz. dry pasta = 1 cup cooked
> 2-oz. serving = 210 calories
> 1 lb. pasta = 8 servings

Pasta is inexpensive and can be purchased in a variety of package sizes. It will store three to five-plus years if kept sealed in a dry place at 70° F or less. Like grain, it can be fumigated and will keep longest in quality containers. Pasta sloughs layers and loses quality with age.

Prepared Mixes

Prepared mixes, such as those for cakes, brownies, and muffins, are both convenient and less costly than the individual ingredients. Most can be kept in the original containers in a dry area for a short period of time but should be repacked in better containers for long-term storage, especially in humid areas. Shelf life is one to two years.

Cereals for Babies

Simple cereals for babies can be made from home-milled whole grains, particularly rice and rolled oats, or ready-made baby cereals. They come in half and one-pound boxes and normally keep about two years before becoming rancid. Canned baby food will also keep one to two-plus years. Powdered formula keeps longer than ready-mixed.

Beans, Peas, and Lentils

Dried legumes are the richest source of vegetable protein and are an important complement to grains. They are high in vitamin B1, iron, calcium, phosphorus, potassium, fiber, and, except for soybeans, low in fat.

The proper storage of beans begins with moisture control. Moisture content is best in the 8 to 10 percent range. Too much moisture rots the beans, and too little causes "hard shell," a condition where the beans become hard with a papery skin. They are nearly impossible to rehydrate and will not absorb water during soaking or cooking. Though it's better to prevent the problem with proper storage, this condition can be overcome somewhat by blanching with steam or boiling for one to two minutes prior to soaking.

According to the Utah State University Cooperative Extension, legumes will store for about one year in a plastic bag. However, if sealed in an airtight, moisture-proof container such as a #10 can or a Mylar bag with oxygen absorbers, they will keep five to ten-plus years.

Beans

The quality of beans depends on their selection from that year's crop, good or bad. Look for beans with as few defects as possible, a uniform size for cooking, and a bright, even color. A dull or low-colored bean is usually old. And even the best seed cleaner occasionally passes bean-sized rocks and dirt clods, so you should always hand sort before using. Do not buy beans treated for planting.

Although dried beans are a far better value, canned beans are convenient enough to have a small supply on hand.

There are many varieties of beans depending on local preferences. See table 18.8 on page 194 for a review of the most widely available beans.

Peas

Dried peas may be green or yellow and are field peas, with split peas being the most common form. They are easier to cook but usually cost more, are slightly less nutritious due to oxidation, and will not keep as long. Yellow peas have a mild flavor, while green have a more distinctive, earthy taste. Whole peas can be sprouted, used whole, or ground up for soup and other dishes. Peas are rich in vitamins A and B, iron, calcium, potassium, and phosphorus.

Lentils

Lentils are small, nutritious, disc-shaped legumes with a mild flavor that blends well with other foods. Like all legumes, they are high in protein and fiber and low in fat. Only soybeans are more nutritious. They do not need to be soaked, and they cook in fifteen to thirty minutes. They come in three varieties: brown, red, and green. Brown lentils, the most common and least expensive, are frequently used in soups. Red lentils have a mild, sweet taste and are typically used in purees and Indian cuisine. Green lentils have a nuttier flavor, hold their shape when cooked, and are used in salads.

Table 18.8 Bean Varieties and Their Characteristics		
Grain	Characteristics	Taste/Uses
Black, turtle beans	Small, dark ovals, dark cream to gray	Sweet tasting, soft texture, mild flavor, vegetable side, main dish, salads, Brazilian feijoada, Latin American
Blackeye, cowpeas, black-eyed peas	Small, kidney shape, tan with a black spot	Creamy texture, Southern cuisine with chicken and pork
Cranberry, Roman	Small, rounded, oval, mottled tan-and-red skin	Creamy texture, chestnut-like flavor, red specks disappear during cooking; Italian, Mediterranean
Garbanzo, chickpea, ceci beans	Round, nut shape, light tan	Nutty flavor, great in salads, soups, roasted, main ingredient in hummus, Middle Eastern
Great northern	Medium, oval, creamy	Delicate flavor, used in bean soups, baked beans, salads, French cassoulet
Kidney	Large, kidney-shaped, dark or light red	Firm texture, hold up well in soups, salads, chilies, Louisiana red beans and rice
Lima	Large, flat, white	Rich, buttery taste
Navy	Small, oval, creamy white skin	Delicate flavor, soups, famous in Boston baked beans
Pink	Small, oval, pale pink	Used in chili and barbecue style with spicy seasonings, Caribbean cuisine
Pinto	Speckled brown and white, become brown when cooked	Mexican cuisine, refried beans, chili. baked beans, bean paste
Red	Small, oval, dark red, delicate, softer texture than kidney beans	Featured in Creole red beans and rice, also in chili, salad, and refried
Triticale (pronounced trit-ih-KAY-lee)	Hybrid of wheat and rye, large, plump kernels	Sweet, nutty taste; good baking properties, makes hearty whole-grain bread; rolled into flakes for cereal

Storing Specialty Foods and Add-ons

Beverages

The most important beverage in your emergency preparation is water. Chapters 6 through 9 thoroughly discuss how to store, purify, and filter it. Milk is also a significant beverage and is also discussed in previous chapters. Next important are those beverages your family regularly uses.

Sodas and juices do not have a long shelf life, so store and rotate just the amount your family uses within a month or two. Unsweetened beverage mixes and sugar-based beverage mixes both have a longer shelf life and add calories and flavor to your diet.

Electrolyte-balanced drinks, such as Gatorade, are beneficial for emergency situations, especially for persons suffering from burns and diarrhea. However, they only have a shelf life of about one year and must be rotated regularly.

Cocoa and chocolate drink mixes last three to four years if kept dry. Coffee and tea store two to five years in airtight containers.

Bouillon

Bouillon is a good base and flavoring for soups, broths, sauces, rice, and other dishes. It comes in beef, chicken, and vegetarian flavors and in cube, granular, and paste form. It keeps two to five years, granular the longest. Store one to two pounds minimum per person for normal use and up to twelve pounds per person if will be used heavily for meatless dishes.

Chocolate, Cocoa, and Candy

What's life without chocolate? Consider storing two to ten pounds per person for making treats. Keep the chips dry and cool as heat causes a white-gray film on chocolate called "bloom," though bloom does not affect flavor. Storage life is about two years. Cocoa will last several years beyond its expiration date. Other candy, particularly hard candy, stores well.

Cornstarch

Store one to four pounds per person for thickening gravies, sauces, soups, stews, pie fillings, and puddings. Its storage life is five-plus years.

Crackers

If you did not include them under grains, consider some now. Graham, saltine, and soda crackers begin to diminish in quality after a year, so use them in rotation. They will keep longer if protected from moisture and oxygen.

Cream

Sweet and sour cream can add a lot to a survival diet. Cream that does not need refrigeration comes in canned, UHT, and powdered forms. UHT cream has about a one-year shelf life, and powdered cream has a two- to four-year shelf life. Store one to two pounds for each person.

Powdered sour cream makes about four cups per pound and can be used to make cream cheese as well. Powdered sweet cream makes about nine cups per pound and may be used for whipping cream. Nondairy creamers and artificial whipped cream toppings keep about two years.

Flavorings and Extracts

Most flavorings store for two years and one year after opening. But since they are alcohol based, they need to be kept tightly closed to avoid evaporation and additional concentration. Almond, lemon, peppermint, rum, and vanilla are good options for storage. Maple flavoring is important for making pancake syrup.

Lemon and lime juice have a one-year shelf life and, once open, will keep about six months when refrigerated. Lemon and orange-peel granules are another flavoring option.

Food Colorings

These additives enhance appeal to food.

Frostings and Toppings

Dry and canned frosting mixes and toppings such as butterscotch, caramel, chocolate and strawberry add variety. Cake and cookie decorations can also be stored. Most keep one year or more.

Fruits and Fillings

Consider storing coconut, mandarin oranges, cranberry sauce, pie and fruit fillings, maraschino cherries, blackberries, blueberries, dates, figs, and mincemeat if you presently use them in your meal preparations. All keep a year or longer.

Gelatin

Unflavored gelatin keeps for years if kept dry.

Herbs and Spices

Herbs and spices are essential in breaking up monotony. They should be stored in a cool, dark and dry place in airtight containers to prevent evaporation of oils and aromas. Most will keep for two-plus years. Leafy herbs keep less well than seeds. Buy fresh, bright colors with strong characteristic aromas. Do not purchase large quantities that will lose potency before they can be used. Table 18.9 lists the most commonly used herbs and spices.

Table 18.9 Common Herbs and Spices		
· Bay leaves	· Curry powder	· Paprika
· Black peppercorns	· Garlic powder	· Parsley, dried leaf
· Cayenne red pepper	· Garlic granules	· Red-pepper flakes
· Chili powder	· Ginger, ground	· Rosemary, dried leaf
· Cinnamon, ground	· Mustard, dry	· Sage, rubbed
· Cloves, ground	· Nutmeg, ground	· Sesame seed
· Cloves, whole	· Nutmeg, whole	· Thyme, dried leaf
· Cumin, ground	· Onion powder	· Turmeric, ground
	· Oregano, dried leaf	

Ice Cream

Freeze-dried ice cream does not resemble regular ice cream, but it does make an interesting and tasty snack. Even ice cream sandwiches are available. Store rock salt if you're going to make your own ice cream.

Mushrooms

Canned mushrooms can spice up casseroles, meats, and pizza and are available in freeze-dried. One pound equals ten pounds of fresh.

Leavenings

Yeast

Powdered dry yeast in unopened bricks or cans will keep two to three-plus years if kept cool and even longer if frozen. You may want one-half to one-and-a-half pounds per person. It turns brown, slimy, and develops a strong odor with age. You can determine if yeast is still good is by adding to warm tap water and a little sugar. If it doesn't bubble within twenty minutes, it's probably not good. You can also make your own everlasting yeast sponge at home or store freeze-dried sourdough starter, but be sure to have two starts in case you accidentally lose one.

PERSONALLY SPEAKING

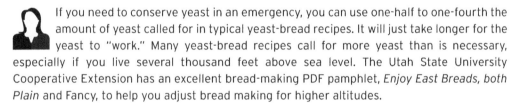 If you need to conserve yeast in an emergency, you can use one-half to one-fourth the amount of yeast called for in typical yeast-bread recipes. It will just take longer for the yeast to "work." Many yeast-bread recipes call for more yeast than is necessary, especially if you live several thousand feet above sea level. The Utah State University Cooperative Extension has an excellent bread-making PDF pamphlet, *Enjoy East Breads, both Plain and Fancy*, to help you adjust bread making for higher altitudes.

Baking Powder

Baking powder stores in its original can for three to five years. You can test a small amount by adding water to see if it still fizzes. Store one to three pounds of baking powder per person. You can make your own from baking soda and cream of tartar or calcium phosphate.

Baking Soda

Baking soda keeps three to five years. Test its potency by adding a half teaspoon of vinegar to a tablespoonful of the baking soda. It should foam and bubble. Besides leavening, it is useful as a dentifrice toothpaste, deodorant, cleaner, first aid, and fire extinguisher. Store two to five pounds per person in an airtight container.

Pectin

Used for making jams and jellies, the liquid version of pectin in brown bottles should be kept in the dark. The powder must be kept cool and dry. Both will keep at least two years if stored properly.

Olives and Pickles

Most olives and pickles will keep for two years. Green and ripe black olives, dill, sweet, bread-and-butter pickles, and relishes are the basic choices.

Rennet

Rennet is used to curdle milk for cheeses and desserts. Twelve rennet tablets will make about ten pounds of cheese. Junket tablets come with directions for making cheeses, custards, and ice cream. You will find them in the pudding and gelatin sections of many supermarkets.

Sauces, Gravies, and Dressings

Storing items your family is used to will make eating more normal in a stressful situation. As most of these types of items have about a two-year shelf life, use and rotate them so you always have some on hand. Table 18.10 lists some of the more common sauces and mixes you may want to have on hand.

Table 18.10 Common Sauces and Mixes		
· A-1 Sauce	· Gravy mixes, assorted	· Soy sauce
· Barbeque sauce	· Hot sauce	· Spaghetti seasoning mix
· Beef-stew seasoning mix	· Ketchup	· Taco seasoning mix
· Chili seasoning mix	· Mustard	· Tabasco
· Cocktail sauce	· Salad-dressing mixes	· Worcestershire
· Enchilada seasoning mix	· Sloppy Joe seasoning mix	

Seafood

Seafood items, like canned anchovies, clams, crab, kipper snacks, oysters, sardines, and shrimp add variety to basic storage.

Specialty Vegetables

Table 18.11 lists some vegetables your family may enjoy that are often additions to recipes and may be stored.

Table 18.11 Common Specialty Vegetables		
· Asparagus, canned · Bamboo shoots · Capers	· Celery, dehydrated · Green chilies, canned · Green peppers, dehydrated	· Mushrooms · Onions, dehydrated · Pimentos · Water chestnuts

Vinegar

Store one to two gallons per person. Cider, distilled, and balsamic vinegars are the most common and will keep two years or longer. Vinegar can evaporate if left open and can take on a slight cloudy appearance, though that does not affect quality.

Vitamin and Mineral Supplements

Vitamin shelf life depends on ingredients and form. Hard-pressed tablets and soft-gelatin capsules keep the longest, while chewable tablets do not keep as long because they readily absorb moisture. Look for expiration dates to determine shelf life, but remember they are usually conservative. If kept cool, dark, dry, and in airtight containers, most store one-and-a-half to five years. Moisture destroys potency, particularly of vitamin C. Dark-brown bottles help keep out light, or you can store vitamins in their original cartons. All supplements with iron should be kept out of reach of children.

NOTES

19 Grain Mills, Bread Mixers, and Kitchen Tools

To properly prepare food in a prolonged crisis, it's likely you'll need specialized equipment and supplies. This is especially true if your preparedness storage food is different from your normal daily food. Some storage foods need processing with specialized tools you'll need to purchase. The right equipment and supplies can make a crisis less stressful.

Since electric power can be nonexistent or only available on an intermittent basis during a crisis, you may need an alternative to the electric appliances found in most kitchens. For those appliances you consider indispensable, it's wise to have a nonelectric choice.

You also may want a backup power source. More information about energy sources in a crisis can be found in chapter 37.

Grain Mills

A grain mill is indispensable when it comes to using stored whole grains. Mills may be electric or manual and may grind grain using stones or metal burr plates. Consider and examine assorted mills. Ask for a demonstration and check reviews. Make sure you get good answers about the mill's capabilities and durability.

> **A grain mill is the single-most-important tool you need to use whole grains.**

PERSONALLY SPEAKING

One of my most important suggestions is that you use your grains consistently, especially wheat! There are lots of compelling reasons for using grains. To begin with, lightly-processed whole grains are healthy, as vitamins are at their peak in freshly ground wheat. Research is showing that a whole-food, plant-based diet lowers cholesterol, blood pressure, and blood sugar and even reverses or prevents heart disease.

FIVE THINGS YOU CAN DO NOW

1 Identify and purchase various hand-held nonelectric kitchen tools that would work in place of electric tools.

2 If you do not own one, compare several grain mills and decide which one would best suit your needs.

3 If you do not own one, compare several bread mixers and decide which one would work for you.

4 Explore making cheese or yogurt and determine what supplies and tools you will need to make them.

5 Make a master list of tools and supplies you would like to acquire. Use the tools and supplies listed in tables 19.7, 19.8, and 19.9 as a guide.

Also, whole-wheat bread made at home, in my opinion, is better tasting than store-bought. Plus, it is a great bargain! Learning to make bread is an important survival skill. It gives you the confidence that you can be self-reliant. To get the most out of whole grains, it really helps if you have the right equipment, and that begins with a grain mill.

You can choose from a variety of high-quality electric and manual mills as well as some dual-function mills. And you'll need to decide what you'll use a grain mill for. Do you want a mill to grind fine flour for baking, or to crack wheat for cereals, or both? Are you going to need it for hard, dry grains like wheat, oily seeds like soybeans, or for making peanut butter? Will you use it regularly, and can it be stored conveniently? Do you want it to be a permanent fixture on the kitchen counter, or will you put it away when not using it? Use the Quick Check to help you decide what qualities you want in a grain mill.

Grinding Burrs or Stones

The most popular burr mills have two grinding burr plates that may be made of hardened cast steel, steel alloy, or corundum— a synthetic aluminum-oxide ceramic. These are all extremely hard and considered to be an improvement over older-model mills with plates made from actual stone. They are used in both electric and manual mills.

In most mills, one burr plate is fixed and the other rotates against it. Some mill designs have cone-shaped grinding burrs. Either way, the grain is pressed and crushed between the burrs to make flour. How closely the plates or cones grind together determines the coarseness of the flour. Depending on what they are made of, some grinding burrs will grind soybeans, oily seeds, and nuts without clogging or absorbing the oil and can be removed and cleaned.

> **QUICK CHECK**
>
> ### What to Look for When Choosing an Electric Grain Mill
>
> ✓ What do you plan to use it for? How often?
> ✓ Do you prefer the speed and efficiency of an electric mill?
> ✓ Or do you prefer the design, flexibility, and the hands-on quality of a well-crafted manual mill?
> ✓ Is the general design efficient and safe?
> ✓ Does it adjust to the variety of settings you need?
> ✓ Does it produce the flour texture you desire?
> ✓ How fast does it grind flour?
> ✓ How messy is it during use, and how easy is it to clean?
> ✓ Will flour dust from the grain mill be a concern?
> ✓ Is it a problem if it is excessively loud?
> ✓ How much counter or storage space does it take up? Can it easily be stored out of sight?
> ✓ Is it built to last a long time, and is it simple and easy to repair?
> ✓ How important is cost?
> ✓ Will it need replacement parts? How available are they?
> ✓ What kind of warranty does it have?

Some well-crafted mills grind with real stones but require more physical effort to get fine flour. Stones also may produce powder when cracking grain or splitting peas. Generally, high-quality stones make finer flour, wear longer, create slightly less heat at finer settings, and are quieter than impact steel burrs. The grind also improves with age as the stones grind themselves to a perfect fit. The tradeoff is the effort needed to grind the grain if using a manual mill.

Impact Grinding Plates

Stainless-steel grinding plates are found in electric mills with powerful motors of one thousand watts or more and are made with two flat, stainless-steel heads with concentric rows of "teeth" that spin within each other at high speeds, bursting the grain kernels at a fast rate. This is called impact grinding, or "micronizing." While they grind grains fast, the process is quite noisy, though some mill brands try to deaden the noise. If you frequently make multiple loaves of bread, this kind of mill is an excellent choice, but it will only grind hard grains and not make cracked wheat. The best way to approximate cracked wheat is to grind the wheat on the coarsest setting and then sift to remove the finer flour.

Electric Mills

For effortless convenience, you can't beat an electric mill. For bread making, look for a heavy-duty mill that will produce fine flour. Some electric mills may have stainless-steel grinding plates. They can be loud and may spit out a fine dust. The housings are usually hard plastic. Their faster grind creates hotter temperatures, but the nutritional value is affected only slightly because of the short grinding time. You can grind enough flour for a bowl of waffle batter or a batch of bread in just minutes.

There are also several electric mills with a synthetic ceramic stone milling action. These mills make exceptionally fine-textured flour and are slower and quieter than the electric mills with stainless-steel grinding plates. And they are attractive enough to sit on a kitchen counter where you can grind flour as needed.

Electric Impact Mills

Blendtec Kitchen Mill

The Blendtec Kitchen Mill (pictured) is an excellent impact mill. It is compact, can be stored in a kitchen cupboard, and is simple to set up, but be sure you attach the cyclone cup and insert the porous "F filter" according to the directions so flour won't blow all over your kitchen. Also take care with the side latches, which are somewhat vulnerable to breaking. This mill can be cleaned with a stiff paintbrush or damp cloth.

It will grind grains, including corn, buckwheat, pinto beans, and soybeans, as well as nonoily legumes, but not moist or oily grains, nuts, seeds, or coffee. It is noisy and takes about two minutes to grind a pound of wheat, but the hopper can be continuously fed and the mill pan will hold twenty-four cups of flour.

NutriMill Classic Grain Mill

The NutriMill made by L'Equip (pictured) is another excellent grain mill that uses an impact chamber with steel burrs to grind fine flour for bread. It also has a unique feature that allows you to slow the RPMs and grind a coarse texture cream of wheat or

cornmeal for cereal. It will also grind grains and nonoily legumes, though it's not suitable for moist or oily grains, nuts, seeds, or coffee. And there's a bagger attachment available where you can easily freeze your flour or share it with friends.

The WonderMill Electric Grain Mill

The WonderMill (pictured) is a first-rate impact grain mill that produces both coarse and pastry flours. It is simple to operate, but one downside is that you must remember to turn the mill on before starting and not turn it off in the middle of grinding. It will grind most grains and nonoily legumes but not oily seeds, nuts, or coffee. It has two main components—the mill and the collection canister, which will hold twelve to thirteen cups of flour.

Electric Stone-Grinding Mills

KoMo Classic Grain Mill

The KoMo Classic Grain Mill is an exceptional electric stone grain mill that produces superbly fine flour at a fast rate as well as cracked wheat for cereal. It is attractive and grinds directly into your own bowl, with virtually no dust. It is the kind of appliance you would like on your countertop to use whenever you need as little as a tablespoon of flour. It grinds all grains and nonoily legumes.

NutriMill Harvest Grain Mill

The beautiful bamboo-covered NutriMill Harvest Grain Mill (pictured) is a stone grain mill designed especially for healthy grain nutrition and gourmet cooking. It grinds fresh flour on demand and adjusts easily from fine bread flour to cracked grains. It will grind both grains and legumes and has an interchangeable insert to prevent cross-contamination for those who desire gluten-free grain.

MockMill

The MockMill, like the KoMo mill, is designed by Wolfgang Mock, and is a top-quality stone-grinding mill whose grinding capability adjusts from finely-ground bread flour to cracked grain. Similar in design to the KoMo and Nutrimill mills, it will grind any hard grain or legume. This mill is notable because it grinds faster and produces a finer flour than similar mills. It is distributed by Breadtopia and is supported by how-to-use tutorials.

Retsel Mil-Rite Mill

The Retsel Mil-Rite Mill is a basic, sturdy mill that uses stones to produce a very fine flour and is excellent for high-quality bread making. It has interchangeable grinding parts, with both synthetic ceramic stones and stainless-steel milling burrs so that it can grind seeds, nuts, and oily grains as well as grains. These mills are sold direct from the manufacturer, Retsel Corporation, in McCammon, Idaho.

Manual Mills

Manual hand mills grind wheat without electricity which makes them perfect for a crisis. They range in price and quality. Several manual mills are very nice quality and are in the price range of electric mills. The less-expensive models are not very energy efficient and can take more than twenty minutes to mill a pound of fine flour. While some people find the manual work required for hand mills satisfying, others find it a tedious, physical strain. Manual grinding may be more successful if grains are ground on a coarse setting first and then reground on a fine setting.

Before selecting a manual grain mill, check the quality of the flour, the amount produced for the effort, and the ease and comfort of the grinding operation.

Also, be aware that manual mills need to be either screwed to a solid surface or clamped onto a counter. If you don't want your wheat grinder permanently attached to your countertop, consider the two-screw clamp designs for stability and distributing the force of the clamp.

Mills with large grinding surfaces, smooth ball bearing action, and flywheels make manual milling easier and more efficient. Good manual mills often come with or offer options such as additional burrs, grain-flaking and meat-grinding attachments, and handle extensions.

Country Living Grain Mill

The Country Living Grain Mill (pictured) is an outstanding durable mill that uses an effortless bearing system with substantial carbon-steel grinding burrs.

Its smooth mechanism is easy to turn, and it will grind pastry quality flour as well as cracked wheat and most other grains.

GrainMaker Grain Mill

The GrainMaker grain mill (pictured) is made in Montana by a family-owned-and-operated company that prides itself on its quality products, craftsmanship, and customer service. The entire mill, including the stones, has a lifetime heirloom guarantee—the best guarantee for any mill. It is an excellent hand mill with large grinding stones and a smooth flywheel mechanism and will make pastry flour for bread making or cracked wheat for cereal. It can be motorized and is available with several attachments that help with that process.

Diamant Grain Mill

The Diamant Grain Mill is a European grain mill that is built to last and is similar in function to the Country Living Grain Mill and the GrainMaker Grain Mill. It will make pastry quality flour as well as cracked wheat. It is sturdy and well designed with a smooth-running flywheel.

Lehman's Own Hand-Cranked Grain Mill

Lehman's motto is "Simple products or a simpler way of life." They offer a range of quality hand kitchen tools for the home and farm. Designed to meet Lehman's specifications, Lehman's Own Hand-Cranked Grain Mill offers both quality and economy and grinds wheat comparable to more expensive mills.

Retsel Uni-Ark Grain Mill

Manufactured and sold by the Retsel Corporation in McCammon, Idaho, the Uni-Ark Grain Mill is known for its durability and flexibility and for being made "without a piece of plastic." It comes with ceramic stones and optional steel wheels so it will grind hard grains as well as oily grains and seeds.

Wonder Junior and Wonder Junior Deluxe Mills

Both the Wonder Junior and Wonder Junior Deluxe (pictured) have stone grinding burrs for grains and dry legumes and a double substantial two-clamp system. The deluxe model comes with additional steel burrs for grinding oily seeds and making nut butters. Both mills can be motorized with a grooved pulley option set up to anything you can attach a belt to. They also have a unique drill attachment.

Roots and Branches Deluxe Grain Mill

The Roots and Branches Deluxe Grain Mill (pictured) is an inexpensive hand mill that grinds coarse flour with a significant amount of effort. You can grind finer flour with a second time through the mill. It can also be fitted with an optional motor that replaces the handle.

Converting Manual Mills

Several top-quality manual mills have conversion kits so that you can operate them more efficiently with a bicycle or electric motor. The Wondermill Junior Deluxe mill even has a fitting that allows you to run the mill with a power drill. Retsel Corporation also has conversion kits for a variety of electrical adaptations.

It is not recommended that you try to adapt an electric mill to a manually turned mill unless it is specifically designed for that purpose.

Family Grain Mill System

The Family Grain Mill (pictured) has a modular system that can fit the stone-burr mill for either a hand-crank or electrical option. It can also be attached to Bosch, Kitchen Aid, and Viking mixers motors. It has a wide range of adjustments and several other kitchen tools can be fitted to its base.

Used manually, it will crack wheat as well as produce medium-fine flour. Double grinding the flour will get you a finer flour. Using the hand crank, it takes about thirty minutes to grind enough grain for one loaf of bread. The flour is coarse, but when ground a second time, it becomes more satisfactory for bread making.

PERSONALLY SPEAKING

You might be hoping I will recommend a grain mill, but you really can't go wrong with any of the mills I have listed. As you'll see from the reviews, there are several fine mills available. Depending on your preference, I would get either one of the excellent electric mills or top-quality manual mills.

I own several, but the one I use most is my Blendtec Kitchen Mill. We bought the first one over thirty years ago. When I gave it to one of my daughters a few years ago, I bought another one because I liked it so well. On the other hand, one of my daughters-in-law loves her Wondermill—especially because there is no dust. But both are great because of their speed and convenience. I recently purchased the Nutrimill Harvest Grain Mill with milling stones, which gives me a few more options for grinding cereals—and I love its handsome bamboo casing. I also have a nice manual mill set aside in case we have a prolonged crisis without electricity.

Bread Mixers

The second-most important kitchen machine for using grains is a good bread mixer. Of course, bread can be made without a bread mixer, but if you plan to make very much bread, this appliance is invaluable. There are several quality mixers on the market, and generally the price point of the mixer will equate to its quality.

Almost all mixers have an assortment of useful attachments that might be included as incentives or offered for purchase. Be aware that the number of watts

---

QUICK CHECK

What to Look for When Choosing a Bread Maker

- ✓ Is the general design efficient and safe?
- ✓ Does it produce adequate bread dough for your use?
- ✓ Can you make small as well as large batches of dough?
- ✓ How important to you are attachments and accessories?
- ✓ How much counter storage or storage space does it take? Can it be stored out of sight easily?
- ✓ Do you want a tilt head, a bowl lift mixer, or set-in bowl with the drive shaft on the bottom?
- ✓ How easy is it to clean?
- ✓ Is it built to last a long time, and is it simple and easy to repair?
- ✓ Does it have parts that will need to be replaced? How available are they?
- ✓ Is the motor made with metal gears and metal housing?
- ✓ Where is the mill made? Is it from a country that has a reputation for superior engineering?
- ✓ What kind of warranty does it have?

of the motor does not necessarily equate with the actual torque power of the machine. Rather, it's often just a marketing gimmick. Do your research and read reviews.

Mixers are loosely divided into three categories: medium-duty, mainstream all-purpose; specialized, heavy-duty designed primarily for bread making; and extra-heavy-duty, near-commercial-grade all-purpose mixers.

All-Purpose Stand Mixers

Breville, Cuisinart, and KitchenAid lead the market in medium-duty, all-purpose stand mixers. Besides adequately performing standard mixing chores, some of the more powerful models also handle kneading bread. Breville, based in Australia, is a relative newcomer to stand mixers, but their mixer is first-rate. Cuisinart's top high-performing mixer has a powerful motor and a three-year warranty. KitchenAid, a staple in kitchen mixers, makes multiple mixer models that vary in power and ability. They make so many models you must be careful as you select a model to be sure you are getting the desired quality and power.

Specialized Bread-Kneading Heavy-Duty Stand Mixers

Some mixers are designed specifically for making large quantities of bread dough. These heavy-duty mixers are powered from below on a drive shaft. They are large capacity and will make six or more loaves of bread. They will also perform other mixing duties and often have attachments for grinding, shredding, and slicing. They can be found in shops that cater to the preparedness and homesteading communities and on the internet sites of similar companies. The most popular bread-making mixers are Anakarsrum, Blendtec, Bosch, and WonderMix.

Ankarsrum

The Ankarsrum mixer (pictured) is exceptionally high quality, Swedish-made, and offers innovative features, such as an infinite variable speed, an automatic shut-off timer, and a wide range of attachments and accessories. It has a six-quart mixing bowl and comes in a variety of colors.

Blendtec Mix 'n Blend

Blendtec, known for its high-capacity blender and powerful grain mill, also makes a heavy-duty mixer that is excellent for bread making. The four-quart mixer comes with a French whisk, cookie whisk, and dough-hook assembly. It also comes with a blender. Look for it on Blendtec's website.

Bosch Mixer

The Bosch Universal Plus Mixer (pictured) is another excellent large-capacity bread maker. It has a 6.5-quart bowl that can handle as many as nine one-and-a-half-pound loaves and as few as one or two. It typically comes with a dough hook and a whisk but has many optional attachments, like shredders, a grater, a citrus juicer, a meat grinder, and even an ice cream maker.

WonderMix Kitchen Mixer

The WonderMix Kitchen Mixer (pictured) is made by the same company that makes the WonderMill. It has a 5.5-quart bowl capacity and strong nine-hundred-watt motor. It has several accessories that make it a complete kitchen tool, including a slicer-shredder, meat grinder, and flaker.

Commercial-Grade Mixers

If you require an exceptionally heavy-duty kitchen mixer, the Hobart Company, maker of many commercial mixers, offers the Hobart N50 five-quart, three-speed, all-purpose bench mixer. It costs about $2,500 and is interchangeable with KitchenAid accessories. It is one-sixth horsepower and has a gear-driven transmission.

Other Food-Processing Tools and Equipment

Some storage foods need processing that requires specialized tools that you may need to purchase.

Grain Roller/Flakers

If you plan to store whole grain oats or desire to have fresh whole-rolled grains or grain flakes, you may want a grain roller. The Margo Mulino Oat Roller (pictured) made in Italy is a stand-alone, hand-cranked flaker that makes nice rolled oats.

Other harder grains need to be soaked first before they can be satisfactorily rolled. The Family Flaker Mill is another option. It is a module of the Family Grain Mills and can be cranked by hand or by using the Family Grain Mill power base. It makes excellent rolled oats and other grain flakes without presoaking. Flaker attachments are available for the Ankarsrum, Bosch, and KitchenAid mixers. KoMo also makes a separate grain flaker.

Blenders

Besides normal kitchen uses, a blender is a useful preparedness tool for mixing noninstant powdered milk, making mayonnaise, blending breakfast drinks, etc. Blenders made by Vitamix (pictured) and Blendtec are top-quality, high-powered machines useful for many food-prep functions. Oster and Kitchen Aid make two well-reviewed, mid-priced blenders. Nutra Ninja Pro Blender is a great value. Of course, all blenders require electricity. A rotary eggbeater and a whisk are good nonelectric alternatives.

Kitchen Knives

In crises, you are more likely to need quality kitchen knives. Look for forged knives with full tangs for high performance and durability. Be sure the knife feels balanced and is easy to manage. Stamped knives are usually not as high quality and do not perform as well. Table 19.1 lists the minimum knives you should have.

Table 19.1 Kitchen Knives	
Minimum Knife Needs	Additional Knife Options
· Chef's knife · Slicer · Utility knives · Paring knives · Serrated bread knife	· Butchering knife set · Chopping knife · Cleavers · Fillet knife · Skinning knives

Whatever knives you have, keeping them sharp ensures they are safe and reliable. You will need a sharpening steel or ceramic rod to straighten the blade edges after each use and a whetstone, diamond-impregnated rod, or other abrasive for occasional sharpening. To help protect knives, store them properly in a countertop or in-drawer knife block.

Table 19.2 General Food Preparation Tools			
· Biscuit cutter · Bottle opener · Breadboard · Butter churn · Can opener, heavy-duty manual · Chopping block · Citrus-juice presser · Coffee grinder, manual · Coffee percolator · Colander	· Dipper · Eggbeater, rotary ball bearing · Flour sifter · Food grater · Food storage bucket-lid opener · Gamma-seal bucket lids · Garlic press · Ice cream freezer, manual	· Ice crusher · Kitchen shears · Kitchen timer · Measuring cups · Measuring spoons · Meat grinder, manual · Meat mallets · Meat slicer, manual · Pasta machines, manual · Potato masher · Potato peelers	· Poultry shears · Rolling pin · Scrapers · Shredders · Spatulas · Spoons · Strainers · Tortilla press · Vegetable slicer, mandolin · Wire whisks · Wooden utensils

Basic Nonelectric Kitchen Tools and Equipment

Many kitchen tools use electricity, but in an emergency, you may need an equivalent manual tool. You may also need to prepare food you usually purchase already prepared and that may require special tools. The list of manual, general food-preparation tools in table 19.2 will help you plan. Check to see which ones you do not already own. Table 19.3 lists basic equipment for cooking. Tools needed specifically for food preservation are not included here but will be discussed in chapters 28–32.

Table 19.3 Basic Cooking Equipment	
· Bread pans · Cake/pie pans · Candy thermometer · Cookie sheets · Corn popper · Dipper · Grills · Hand toaster	· Hot pads · Kitchen scales · Meat thermometer · Muffin tins · Reflective oven · Roasting pans · Waffle iron

Cast-Iron Cookware

If you plan on cooking on woodstoves, in fireplaces, or over campfires, you'll need a set of pots and pans equal to the task. Lightweight steel or aluminum pans with nonstick coatings will develop hot spots and quickly burn through if used directly on or over a fire. The best cooking pots for open fires are heavy cast-iron pots, kettles, skillets, griddles, and Dutch ovens. Cast iron works well for slow cooking over all heats. It is easy to clean and lasts forever if properly seasoned and cared for. It can also be buried under coals and conserves heat. You'll need an iron rod or swinging arm crane if you expect to suspend kettles in a fireplace.

Pressure Cookers

A pressure cooker, smaller than a pressure canner, is an important preparedness kitchen tool. It consumes less fuel for cooking, cuts cooking time for many recipes by two-thirds, and can be used with alternative heat sources. It is indispensable for cooking beans, grains, and many dehydrated foods. Pressure cookers may be stovetop or electric. Look for one that has multiple settings for more cooking options.

Cookbooks

Although you may need to make some substitutions, many of your regular recipes will be suitable for the foods you store. However, you may want some specialized cookbooks for recipes using whole grains, legumes, powdered milk, and dehydrated foods. Also, because they are so different from what most people are used to, a book or two on woodstove and fireplace cooking may be helpful if you plan to use either of these alternative cooking methods.

Kitchen Supplies

Regular household utensils and dishes should be satisfactory (see table 19.4), although a backup set of nonbreakable plastic dishes is advisable. Also consider disposable utensils and paper napkins, plates, and cups for use when water is in low supply or ease of preparation is necessary. Store all paper products in a dry and somewhat cool area and protected from rodents.

Table 19.4 Kitchen Supplies	
· Aluminum foil	· Parchment paper
· Cheesecloth	· Plastic bags
· Dish drain	· Plastic wrap
· Dish pan	· Scouring pads
· Dish rack	· Steel wool
· Dish washing detergent	· Storage lids to re-close dehydrated food cans
· Garbage bags	
· Grocery bags	· Wax paper
· Paper towels	

Making Basic Food Products

Many of the suggested food storage items are the raw ingredients for the basic foods you may use daily. Knowing how to use these raw ingredients will add variety and enhance your diet. You will find recipes for using these basic ingredients on our website: CrisisPreparedness.com.

Making Milk Products

Powdered milk is a versatile food and the foundation for several milk-based products. And, with just a little knowledge and effort, it can enliven a diet. Making yogurt, buttermilk, or cheese is economical and an excellent way to help use and rotate your powdered milk. One cup of liquid milk makes about one cup of yogurt or buttermilk, and one pound of powdered milk makes about one pound of cheese.

Making Yogurt

Yogurt is easy to make, but because it has a sensitive incubation temperature range, it takes some practice to get it to set up properly. You can use whole milk if it is available, but powdered milk also works. You'll also need an active yogurt culture, a nonreactive pot such as stainless steel, a heat source, a cooking thermometer, and a way to keep the culture warm while it incubates, such as an insulated cooler. For creamier Greek yogurt, you'll also need cheesecloth. You can invest in a yogurt maker, but that's not necessary. The best thing to do is experiment with several recipes until you find a method that works for you. You'll find yogurt-making supplies and equipment listed in table 20.1.

Table 20.1
Supplies for Yogurt Making
· Milk, fresh or powdered
· Stainless-steel pan for heating milk
· Bacteria cultures, either powdered or from plain yogurt with live, active cultures
· Clean, sterile containers for culturing yogurt such as Mason jars
· Dairy thermometer, precalibrated
· Warm environment for incubating yogurt cultures (yogurt maker, thermal beverage cooler, heating pad)
· Optional extra powdered milk or liquid pectin for thicker yogurt

---

FIVE THINGS YOU CAN DO NOW

1 Purchase a good cooking thermometer and a kitchen scale.

2 Acquire a yogurt start and experiment with making yogurt.

3 Find recipes for making cheese and gather the supplies you need.

4 Try making homemade crackers. Experiment until you find one you like.

5 Purchase a tortilla press. Try it out. Use a YouTube video to guide you.

PERSONALLY SPEAKING

When my kids were growing up, I would make yogurt in my Salton thermostat-controlled yogurt incubator, and I learned I had to cover the maker with a kitchen towel to help maintain the right temperature so the yogurt would develop. My daughter makes yogurt for her yogurt-loving girls about once a week, but she doesn't use anything fancy to incubate the culture, just recycled plastic sour-cream containers and an insulated cooler she uses to keep the yogurt at a constant temperature during incubation. She usually makes the yogurt in the evening, and by morning the yogurt has developed. Once you get a yogurt strain you like, you can continually refresh and replenish it.

Making Buttermilk

A buttermilk substitute can be made from powdered milk using white vinegar or lemon juice and reconstituted skim powdered milk. Simply place four and a half teaspoons of white vinegar or lemon juice in a glass measuring cup and add enough reconstituted powdered milk to make one cup. Stir and let it stand for ten minutes. It works well in baking and salad dressings and any recipe that calls for buttermilk. "Real" buttermilk is made with a buttermilk culture and whole milk. It is less familiar to Americans and usually only available at health-food stores.

Making Cheese

Cheese making is an ancient culinary art and is one of those survival skills worth developing. If you anticipate prolonged periods of crisis, you might consider acquiring cheese-making supplies and equipment, especially if you have access to cow, goat, or sheep milk.

Making Cheese from Powdered Milk

Powdered milk can also be used to make fresh soft, hard, and cottage cheese. It is especially a good option for making soft cheeses. However, cheese made from skim milk will be drier, have less flavor, and produce a smaller amount. Be aware that hard cheeses take three to sixth months to age, which may be challenging in a crisis. Because of the high heat used in processing, canned milk is not recommended for cheese making.

Depending on the recipe, you may need buttermilk powder or yogurt powder, lemon juice, vinegar, or citric acid crystals. Cultures keep for a couple of years and are available from internet sources and grocery stores. Again, the best thing to do is experiment until you get the results you want.

Table 20.2
Supplies for Cheese Making
• Stainless-steel double boiler
• Slotted spoon
• Thermometer
• One half-gallon or two one-quart jars and caps
• Calcium chloride, especially necessary for heat-treated milk
• Cheese-making starter cultures (store in freezer for two to three years)
• Citric acid and tartaric acid aid in acid development in ricotta, mozzarella, and mascarpone cheeses
• Lipase, an enzyme for enhanced flavor
• Mesophilic culture, requires lower heat (72° F) for soft cheeses made at room temperature
• Rennet in liquid, powder, or tablet form (tablet for longest shelf life), for setting cultures
• Thermophilic culture, requires higher heat (110° F) for hard cheeses

The New England Cheese Making Supply Company is an excellent source for all things relating to cheese making. Besides the milk, for basic cheeses, you'll need the equipment and supplies listed in table 20.2. As cleanliness is critical in cheese making, all equipment must be easy to clean and sterilize.

Making Butter

If you are lucky enough to have cream during a crisis, you may want to make your own butter, which is made by agitating whole milk to separate the butterfat from the liquid and milk solids. Making butter was a weekly task in the nineteenth century and into the twentieth century. Butter was churned in a glass jar fitted with a metal crank and wooden blades. Electric and manual butter churns are available today but are expensive.

Used, vintage Dazey Butter Churns are sold as collectors' items, but practical reproductions, such as the Lehman's Dazey Butter Churn (pictured), are also available. A blender, mixer, eggbeater, or plain glass jar can do the same thing. Butter molds are also useful.

Making Grain Products

Making Bread

The art of making quality bread is also a skill worth cultivating. There are many nuances to making high-quality bread, but with just a little experimentation and persistence, you can achieve fantastic results with just a few simple ingredients from your home food storage. Table 20.3 lists types of flour and the kinds of bread they are used for.

Table 20.3 Protein Content in Flour		
Type of Flour	Percent of Protein	Bread It's Best for
All-purpose white	10.5%–11.7%	Sandwich bread
Bread flour	12.0%–12.7%	Rustic, hearth loaves
Whole-wheat flour	14%	Substantial, wheaty bread

The structure of bread depends on the amount of protein, or gluten, in the flour. Unbleached, all-purpose white flour has less protein and makes a bread with a soft texture. "Bread flour" has a higher percentage of protein, or gluten, and makes a more robust artisan bread. If white bread is your preferred bread, you'll need to store white flour and rotate it since its shelf life is only six months to a year.

Whole-wheat flour has an even higher protein content and makes a hearty, nutty-flavored bread. It's the ideal bread to partner with your long-term wheat food storage because it will help you use and rotate your stored wheat.

QUICK CHECK

Qualities of a Good Bread Pan

✓ Proper size (8.5" x 4.5" x 2.0"
✓ Aluminum or aluminum-coated steel
✓ Substantial feeling
✓ Light in color
✓ Non-stick (optional)

Equipment and Supplies

You will need a grain mill to make whole wheat bread. And if you are going to make bread regularly and rotate your food storage, you'll want a heavy-duty mixer that can handle bread dough. See chapter 19 for a complete discussion of grain mills and mixers.

Table 20.4 Equipment and Supplies for Bread	
· Baker's bench knife	· Bread pans, 4–6
· Baker's peel	· Cooling racks
· Bread knife	· Grain mill
· Bread mixer	· Rising buckets

Table 20.4 on page 215 lists other tools and equipment that make bread making easier. You will need four to six quality loaf pans. Loaf pans come in metal, nonstick, glass, and ceramic. Light-colored metal pans will prevent crusts from getting too dark. Glass and ceramic are attractive but do not conduct heat as well as aluminum.

Breadtopia

Breadtopia.com is a website that is devoted to bread making. The tutorials, recipes, and baking guides will support you in your bread-making endeavors, and their online store offers you a large selection of breadmaking tools and supplies. They specialize in unique heirloom, ancient, and specialty grains.

Making Crackers

Crackers add variety to many meals but are not easy to store because of their short shelf life. But they are simple to make and require just a few supplies and tools. You will need a rolling pin, parchment paper, a pizza cutter, and cookie sheets. Experiment with the basic recipe and decide which toppings you should include in your food storage.

King Arthur Flour has an excellent guide to making crackers on their blog, www. kingarthurflour.com. They offer several recipes and give you helpful tips.

Making Tortillas

Flour tortillas are made with regular white flour, corn tortillas with masa harina. Being able to make them will add variety to your menu during a prolonged emergency.

Flour tortillas are traditionally made from white flour but can also be made from whole wheat or a combination of white and wheat flour. The other main ingredient is solid fat—usually lard, butter, or shortening—which is needed to coat the flour to prevent gluten from developing and making the tortillas hard to roll out.

Corn tortillas are made with masa harina, which has a shelf life similar to flour—about one year. It should be kept in a closed container and, if possible, in the freezer. You will need a pastry blender and a few basic cooking tools. A tortilla press, as opposed to a rolling pin, makes it easier to get thin, flat tortillas. Experiment with tortilla recipes until you find one you are confident with.

PERSONALLY SPEAKING

Tortillas are one of my solutions for a quick meal. I have learned firsthand how good homemade tortillas can be because one of my friends of Mexican heritage generously shares hers with me. Making homemade tortillas is a skill worth developing! I feel fortunate that she was willing to share her recipe. Like other good cooks, she estimates the amounts and just has a feel for when it is right. Her recipe is made with simple ingredients. She starts with 3 cups of flour, then adds about ¼ cup of vegetable shortening or lard (she prefers lard), 1 teaspoon of baking powder, and 1 teaspoon of salt, plus her secret ingredient – 1 teaspoon of sugar. She adds enough water, about 1½ cups, to make a soft ball of dough. The recipe makes a dozen tortillas, depending on size. For a more detailed recipe, see our website: CrisisPreparedness.com.

Making Pasta

The recommended shelf life of pasta is one to two years, but if you can make your own, you won't have to rely only on store-bought pasta. Although pasta is traditionally made with white flour, it can also be made with whole-wheat flour. There are lots of helpful pasta-making tips on YouTube. It is possible to make pasta completely by hand with just a rolling pin and knife, but for ease and more refined pasta, consider a hand-crank pasta machine, such as the Marcato Atlas 150 Pasta Machine (pictured). You may also want a ravioli attachment.

PERSONALLY SPEAKING

My daughter has mastered the art of making ravioli. She really liked the homemade pasta in restaurants and decided she would figure out how to make it herself. Her recipe is basic–120 grams of flour to one egg, with a little less flour if she is making noodles. One of her daughters likes pumpkin ravioli, which has a filling made of half pumpkin, half ricotta. But be warned–it is time intensive. But it's so good it's still worth it.

Cooking with Beans

Dry beans are extremely easy to prepare and cost about one-fifth the price of canned beans.

Some people are intimidated by cooking with dry beans. But there are a few secrets that make cooking beans much easier. The first is to think ahead and start soaking the beans the night before. The second is to cook them with good seasonings. It takes about two hours to cook the beans on your stovetop. Or you can cook them in a crockpot all day long. And using a pressure cooker eliminates the need for soaking and speeds up the process. For more information about cooking with beans, refer to our website, CrisisPreparedness.com.

Although beans are a great addition to your food storage, they can cause gastrointestinal discomfort in some people due to the presence of certain indigestible carbohydrates that ferment in the colon and produce gas. See the Quick Check for tips on how to make bean more digestible.

> **QUICK CHECK**
>
> ### How to Make Beans More Digestible
>
> ✓ Slowly introduce beans into your diet.
> ✓ Soak beans in hot water.
> ✓ Soak beans for at least twenty-four hours.
> ✓ Change the soaking water several times.
> ✓ Rinse beans and discard soaking water at end of soaking period.
> ✓ Rinse beans and discard water after initial cooking.
> ✓ Rinse plain canned beans (those not in a sauce) before using.
> ✓ Sprout beans.

Growing and Preserving Food for a Crisis

Growing Vegetables

No matter how much food you store, it's eventually going to run out if a crisis lasts long enough. Being able to grow your own food will allow you to not only replace the storage food but also to supplement it and make it last longer. This chapter discusses the preparation needed for growing vegetables. The next chapters discuss being prepared to grow fruits, field crops, microgreens, and sprouts.

However, as important as it is to grow your own fruits and vegetables, it does not take the place of your food storage. Living organisms, including plants, are too vulnerable to be your only food source. They require a lengthy period of suitable weather before producing food and are vulnerable to drought, disease, pests, and vandals.

Growing fruits and vegetables is a broad, encompassing process, covering many subtopics. The details of that process will not be covered, this chapter will offer general guidance for gardening, present essential items you should stockpile now, and suggest resources for more information.

Many resources are available to help you navigate the challenge of growing your own fruits and vegetables. One of the best sources of information is your county's cooperative extension service. It works closely with the state land-grant university and offers research-based home-garden information and advice. The information is usually free and will be specific to your area. Other resources about gardening can be found on page 429 in the resource section.

Vegetable Gardening

Successful vegetable gardening comes from learning skills and gaining experience with your unique combination of climate, seasons, soil, water, insects, and disease conditions. Experience also helps you know your personal and family preferences.

FIVE THINGS YOU CAN DO NOW

1 Determine what kind of growing options you have.

2 Gather information at your local cooperative extension service about growing your own vegetables, fruit trees, berries, grapes, etc.

3 Investigate growing vegetables in containers and give it a try

4 Build a grow box and plant a few herbs and vegetables.

5 Stockpile several packages of seeds and vacuum pack them to place in storage.

Becoming an effective gardener does not happen overnight as many of the processes and techniques cannot be accomplished quickly. So, no matter what your circumstances, it is best if you begin to acquire the necessary knowledge and skills now.

For beginning gardeners, this chapter will give basic information and the hope that gardening is something you can manage and even embrace. Many things can be grown when you know a few basic gardening practices.

If you are already a skilled gardener, this chapter will help you evaluate your level of gardening preparedness. It may also give you another point of view to consider.

Start Small

If you are new to gardening, begin small. An easy way to get started is to dedicate a small space in your yard to gardening. You also might consider building a grow box. Try some easy-to-grow things like lettuce, beans, or a few tomato plants. You can also try growing a few herbs or vegetables in a window box or patio pots. Like-minded neighbors may want to join you in creating a small community garden. With time, your practical knowledge will increase and your skills will be refined. Besides consulting many and varied gardening books, you can also gain knowledge from extension pamphlets or university online materials.

> **QUICK LOOK**
>
> ### Easy Vegetables for Beginners
>
> - Radishes
> - Salad greens
> - Green beans
> - Zucchini squash
> - Tomatoes
> - Peppers

At whatever gardening stage you find yourself, you'll want to stockpile important gardening necessities.

Stockpiling Seeds

In a prolonged food shortage, having a good supply of seeds is invaluable. A small amount of seeds represents a large amount of potential food. Seeds will also be valuable for bartering. For best results, purchase seeds from reliable sources, then experiment with various varieties to determine what you like and what works best in your growing environment.

> During a prolonged food shortage, seeds will be worth their weight in gold.

Open-Pollinated Seeds

Seeds can be saved from the present year's crop to be used in future crops if they are from standard, open-pollinated (nonhybrid) plants. Although hybrid plants provide excellent vegetables, the seeds from hybrid plants are unreliable and produce plants that are unlike the parent plant and usually inferior. Be aware that, hybrid or not, some plants will crossbreed and may not produce seeds suitable for saving.

Heirloom Seeds

Heirloom seeds are open-pollinated, nonhybrid seeds that have been passed down from generation to generation. They are most often those used prior to World War II. They represent people's desire to get back to the basics and to avoid high-tech, genetically modified seeds. Some believe they are more flavorful and nutritious than modern hybrids. Heirloom seeds are particularly important in a survival setting because they are usually nonhybrid and the seeds can be collected and replanted from year to year.

On the other hand, many hybrid seeds are developed to improve plant yield, make plants disease resistant, and produce higher-quality, uniform fruits. They may also be bred for improved taste—new varieties of sweet corn are good examples of hybrids with improved taste.

Seed Selection

Ideally, you'll store and rotate the seeds for plants you like that have the characteristics you desire. However, for long-term storage, it's a good idea to store a selection of seeds that are of the open-pollinated, heirloom-type variety. Consider these "seed insurance" and replace them about every five years.

> **Heirloom seeds are important in a survival setting because the seeds can be collected and replanted from year to year.**

Select seeds that grow well in your soil and climate. Consult your county's cooperative extension service for recommended varieties in your area. Choose those your family likes and that will provide balanced nutrition, particularly vitamins A and C. Table 21.1 lists factors to consider as you select seeds.

Table 21.1 What to Consider When You Select Seeds	
How quickly will they produce a crop? · Seed packages list the days to reach maturity. · Know your growing season.	How resistant are they to temperature extremes, drought, and disease? · Seed packages list diseases they are resistant to.
How much do they yield for the space? · Vine crops can be grown on wire frames. · Pole beans take less ground space than bush beans for the yield. · Use grow boxes for a compact garden.	What are some crops that have an extended season? · Broccoli, cabbage, lettuce, and kale grow well in cool weather. · Use succession planting for more productivity.
How well can they be stored or canned? · Potatoes, winter squash, cabbage, and onions all store well. · Tomatoes, green beans, corn, and beets are often preserved by canning or freezing.	Will they add variety, preventing menu boredom? · Grow basil, parsley, chives, and dill. · Grow a variety of different vegetables. · Experiment with unusual vegetables or novelty vegetables.

PERSONALLY SPEAKING

Your choice of seeds for storage will be based on your experience and vegetable preferences. It will also depend on your growing season. Many of my family's favorites are hybrids, which works for us because we are constantly using them up and purchasing new. We also like some of the old open-pollinated standbys.

My garden favorites start with tomatoes, including several varieties for canning, a couple of Italian-paste varieties, and one or two cherry tomatoes. Next comes squash. I could live on summer squash–*costata romanesco* zucchini, regular zucchini, yellow crookneck, and yellow-patty pan. My favorite winter squash is Waltham butternut, which will last until March if stored in a cool place. We also grow sweet corn and have learned that the best varieties for freezing are the tender sweet (se) varieties. We grow Anaheim, jalapeño, and sweet peppers in a variety of colors. We round out our garden with green beans, onions, melons, cucumbers, tomatillos, carrots, beets, and sometimes snow peas if we can get them planted early enough.

A couple of years ago, my husband made three raised beds we fill with a raised-bed planter mix made up of three parts–sand, compost, and peat moss. My favorite things to grow in them are a variety of lettuces, Swiss chard, spinach, cilantro, sweet basil, and a new favorite–arugula.

Recommended Seeds

A good start includes the vegetables listed in "Quick Look—Recommended Seeds." Choose vegetables that your family enjoys and that grow well in your climate and growing conditions. Also include seeds for an herb garden.

QUICK LOOK
Recommended Seeds

• Broccoli	• Kale	• Spinach
• Carrots	• Lettuce	• Summer squash
• Cabbage	• Melons	• Sweet potatoes
• Collards	• Onions	• Swiss chard
• Corn, sweet	• Peas	• Tomatoes
• Cucumbers	• Peppers	• Turnips
• Green beans	• Radishes	• Winter Squash

Storing Seeds

You should store at least a one-year supply of garden seeds; however, storing two years' worth of seeds will give you one to use in the present year, and one to save and use the following year. Constantly replenish seeds so you have a fresh supply. Table 21.2 shows how many seeds are needed for a family of four for one year.

Do not expect any stored seeds to have a shelf life longer than five years.

Look for seeds stored in resealable triple Mylar packages for a longer storage life. Once the packages are opened, the storage life is no different from other packaging. Store all packages of seeds in large, resealable plastic bags or airtight containers

Table 21.2	
Seeds Needed for One Year for a Family of Four	
Bush/pole beans	8 oz.
Beets	1/2 oz.
Cabbage	1/4 oz.
Carrots	1/2 oz.
Corn, sweet	8 oz.
Cucumber	1 oz.
Lettuce	1/4 oz.
Onion	1/4 oz.
Peas	12 oz.
Pepper	1/4 oz.
Radish	1/2 oz.
Spinach	1 oz.
Squash	1/4 oz.
Tomato	1/4 oz.
Turnips	1/4 oz.

Table 21.3 Storage Life for Seeds	
2 Years	Field corn, onions, parsnips, soybeans
3 Years	Asparagus, green beans, carrots, sweet corn, leeks, lettuce, parsley, peas, peppers, hybrid tomatoes
4 Years	Beets, cabbage, cauliflower, chard, okra, pumpkin, radish, spinach, squash
5+ Years	Broccoli, Brussels sprouts, celery, collard, cress, cucumbers, endive, kohlrabi, melons, nonhybrid tomatoes, turnips

Crisis Preparedness Handbook

with oxygen absorbers in cool, dry conditions. Shelf life varies with the type of seed and exact conditions. Table 21.3 gives a general idea of the shelf life of various seeds. Consider storing additional seeds to use for bartering.

Commercially Packaged Storage Seeds

Although limited in choices and amounts, one good way to store seeds long-term is by having the selections specially prepared and sealed in cans. It will increase the shelf life substantially. Be sure to purchase from a seed specialty company that stands behind their seeds. Mountain Valley Seed Company, a part of True Leaf Market in Salt Lake City, Utah, offers excellent storage seeds.

Survival preparedness companies also market seeds packaged in cans or storage buckets for survival use. Be careful when selecting seeds from promoters who encourage you to purchase and not worry about it again. Carefully examine the actual contents and calculate the cost you are paying per package of seeds.

Packaging Your Own Seeds for Storage

You may want to select your own seeds from a reputable seed producer and then store them with oxygen absorbers in a Mylar bag, sealed Mason glass jar, or vacuum-packed bag. If you have access to a dry-pack canner, you may also use it to can your garden seeds. Careful storage of seeds will increase their storage life, and by handling them yourself, you will know exactly which seeds you have stored.

Even if you store seeds in vacuum-packed containers, you should rotate them regularly to keep them as fresh as possible.

Improving the Soil

Good soil is the single-most-important factor in successful gardening. Plants require nutrients to grow well. Nutrients may be added to the soil with commercial fertilizer, but the most significant benefit comes from adding organic material. This includes leaves, grass clippings, plant materials, and composted manures. Adding compost to your soil will benefit it in other ways besides providing nutrients: it will improve drainage, aeration, soil structure, pH balance, and beneficial microorganisms, making your soil viable and healthy.

The addition of organic material to any soil makes it better.

Have Your Soil Tested

If you are concerned about the quality of your soil, take the time to find out its basic characteristics. The important things you need to know are the texture, the pH, the percent of organic matter, and the basic content of macronutrients, like phosphorus and potassium. Nitrogen levels vary widely and are usually not tested for. Contact your county's cooperative extension service, a land-grant university, or a commercial testing service for information about soil testing and recommendations for improving your soil.

Amend Your Soil

Once you know the basic characteristics of your soil, you'll understand what soil amendments and fertilizer you need to add. You can purchase organic or inorganic fertilizer to respond to your specific soil. Purchase enough fertilizer for at least a year.

While a compost pile, pit, or barrel can be used to produce nutrient-rich organic matter, composted animal manure is an even better source of nutrients. Crops like legumes, ryegrass, and buckwheat can be used to increase the nitrogen and organic material in your soil.

PERSONALLY SPEAKING

If you have reasonably good topsoil, the most important thing you can do is take care of it. Almost every year, we add another layer of organic compost to our vegetable garden, and every spring, we broadcast a balanced fertilizer over the garden to replace depleted nutrients. We have found that to get the best quality corn, we need to side-dress it when it's about a foot high.

Controlling Garden Pests

Controlling Insects

In a prolonged crisis where we may need to rely on the crops we've raised, it's important we not allow pests and diseases to reduce our harvest. Keeping a selection of pesticides in your storage can help. The EPA-approved list of pesticides changes frequently, so check with your county's cooperative extension service to see which ones you should store. It's best to store them locked in a ventilated area in a garage or storage shed away from the living areas in your home.

You may also want to grow your own organic pesticide by storing seeds for the chrysanthemum cinerariifolium flower. The dried flower contains the botanical poison pyrethrum and can be ground up and used as a dust. Sprays made from garlic, onions, and other pungent plants also work as repellents.

Learn about beneficial insects and how to protect and encourage them, and practice good horticulture habits to cut down on the need for pesticides. Keep your garden area clean and weed free. Eliminate places for pests to breed and live by promptly and properly disposing of garden refuse. Proper crop rotation will not only reduce the buildup of damaging insects and diseases but also balance the use of different soil nutrients.

Controlling Weeds

Weeds can be the biggest pests because they crowd plants and steal nutrients. Weed early while the weeds are small, and continue to monitor weeds and eliminate them throughout the growing season. This is especially important later in the season when plants go to seed. You can also use a preemergent herbicide after all garden plants are established. Mulching also helps keep weeds down.

QUICK LOOK
Benefits of Mulching
• Control weeds • Retain moisture • Maintain even soil temperatures • Build the soil

Protect your garden with fencing and other precautions. An entire year's harvest can be destroyed in a few minutes by animals or vandals.

PERSONALLY SPEAKING

 The tool pictured is a Corona Clipper Diamond Hoe. It is my favorite garden tool! I love it for getting rid of the small weeds that germinate shortly after planting. It is nice because it has four cutting blades that eliminate weeds as you move it back and forth just below the surface. With a little care, you can maneuver the pointed ends to get in close to corn stalks and tomato vines.

Gardening Tools and Supplies

It's a good idea to at least acquire the basic gardening tools listed in table 21.4. Buy quality tools and maintain them properly, cleaning them after use and storing them where they won't rust or warp.

Rotary tillers are handy for larger gardens, but remember, they need fuel. Clean up the garden and prepare the soil in the fall so you can plant in the spring without cultivating.

Table 21.4 Basic Garden Tools	
· Hoes	· Hand fork
· Shovel	· Gloves
· Rake	· Watering can
· fork	· String or twine
· Wheeled culti-vator	· Hoses
· Extra handles	· Sprinklers
· Trowel	· Wheelbarrow
· Hand weeder	· Hand sprayer (2 gallon)

Getting the Most from Your Garden

Location

A garden site must receive at least four to six hours of direct sunlight daily. Fruit-bearing vegetables need the most light, while leafy vegetables need less, and afternoon sun is better than morning sun. The soil should drain well, and the top six to eight inches should have a loose, crumbly texture. Poor soil can be gradually improved by adding compost, but it can take years. You will also need a source of water, especially in arid climates.

The amount of space you'll need will depend on the climate, soil, and seed varieties you choose. Your gardening skills and methods will also impact yield.

In general, however, you can provide a year's supply of vegetables from a couple of hundred square feet per person with efficient gardening techniques. Although some sources offer tables of projected yields, experience is your best guide.

Planning

The first step in planning a garden is to plan it on paper. Where will you plant crops so they're not in the shade of buildings or other plants? Perennials and fall-bearing crops should be planted in separate plots or at the edges of the garden so they won't be disturbed when the rest is turned under. Make the most of your space by allowing for climbing vegetables, like tomatoes, beans, and even cucumbers and cantaloupe, to grow upward on stakes and trellises.

There are several gardening methods that will produce up to five times as much yield while using less water and allowing for better weed control. For maximum production, you may want to research the following methods as you plan your garden. Resources about gardening methods can be found on page 429 in the resource section.

It's also wise to consider gardening methods specific to your geographic area. Again, your county's cooperative extension service is a great resource. Try container gardening, vertical gardening, hydroponics, or square-foot gardening as viable alternatives when you do not have a lot of space.

Wide-Row or Plot Gardening

Wide-row gardening means that seeds are broadcast in a wide band rather than in single rows. The bands usually consist of two- to four-foot-wide rows or plots that are indented or raised. They often have a six-inch dike around them with one- to two-foot paths in between. Besides utilizing space efficiently, this method conserves moisture and provides a "living mulch" to help with weed control. Leafy crops and small-plant crops, such as beans, carrots, and beets, do well in this arrangement.

Biodynamic or French Intensive Gardening

This gardening method focuses on sustainability and organic-gardening practices and uses well-prepared raised beds or wide rows to more closely space plants, with careful attention to companion and succession planting. The yield may be four times as great as traditional gardening methods.

Raised-Bed or Grow-Boxes

Grow boxes are enclosed structures designed for raising vegetables and other plants. The width of the row should be what you can easily reach from either side, or about three feet. Raised beds allow easy access and offer some of the same benefits as wide-row planting.

Jacob Mittleider developed this gardening method to increase yields. Although the method is somewhat controversial, it deserves consideration if your goal is to grow a high-yielding garden. (See page 429 in resource section.)

Square-Foot Gardening

This gardening method also intends to increase yields within a small space using small grow boxes divided into one-foot squares to minimize watering and fertilizing needs. A drip-irrigation system can also drastically reduce the amount of water needed. Watering deeply two to three times per week helps plants develop a healthy root system, which makes them more drought resistant. Certain varieties are naturally more resistant to drought and, with mulching, they tend to thrive. (See page 429 in resource section.)

Companion Gardening

When you plant crops together for their mutual benefit, you can increase yields, better use nutrients, and lessen care requirements. For example, some plants help shade others, reducing water required and preventing sunburn. Or some plants improve the soil nutrients that benefit other plants. Certain plants, such as marigolds and aromatic herbs, also help deter insects.

Extend the Season

To get the maximum benefit from your garden, learn and use various techniques to lengthen the season. Use succession planting. Replace an early maturing crop with a later one to increase your yield.

Start plants inside in a sunny window or under grow lights for later transplanting. Protect tender plants from frost with hot caps, hotbeds, and cold frames to give them more growing time. Use a greenhouse for year-round gardening.

Investigate hydroponics, which allows year-round gardening in a completely enclosed space without soil. Most important to hydroponic gardening is a suitable light source.

Keep Records

No matter how or where you garden, be sure to keep a detailed record of the varieties planted, how much was planted, when it was planted, and how it was cared for. Note weeding and water and fertilizer used. Include insect and disease problems, time of harvest, and yield. These records will become a valuable planning tool for following years.

NOTES

Growing Fruits, Nuts, Berries, and Grapes

Cultivating orchards, vineyards, and berries is a long-term investment that offers great benefits and is well worth considering. Even if you do not have a dedicated orchard space, you might be able to squeeze some space for a couple of fruit trees or other small fruits varieties into your landscape. Trees are also useful for windbreaks and shade.

Of the many resources available to help you navigate the challenge of growing your own fruits and vegetables, one of the best is your county's cooperative extension service. Each county has an extension office that carefully considers local growing conditions. You will find additional resources about how to grow fruits on page 429 in the resource section.

Tools and Pest Control

To get the most from your trees and vines, you'll need to learn how to prune them properly, and depending on the trees and vines, you may need pruning shears, lopping shears with long handles, or a pruning saw. For tall trees, you'll need a sturdy ladder for both pruning and picking.

Pesticides

You should also store any necessary pesticides, particularly for fruit trees; however, the fruit-tree chemicals available to homeowners have decreased over the past twenty years due to government regulations. Some of the insecticides considered safe for home use include carbaryl or Seven, dormant oil spray, malathion, bacillus thuringiensis (BT), permethrin, pyrethrin, and insecticidal soap. Combination fruit-tree sprays are sold in garden centers.

Most pesticides have a two-year manufacturer's recommended shelf life. Actual shelf life can be longer depending on the storage environment. Label purchase date and store chemicals in a dedicated cabinet, locked if possible, and stay prepared by stockpiling chemicals for at least the next year.

FIVE THINGS YOU CAN DO NOW

1 Decide what options you have for planting fruit trees or other fruiting plants.

2 Get a fruit-tree catalog from a nursery, such as Stark Brothers, and learn about fruit varieties.

3 Study a hardiness zone map to learn about your growing season.

4 Plant a planter with strawberries.

5 Plant a fruit tree.

Apple Bagging

A chemical-free but labor-intensive solution to pest control is apple bagging, where you block common fruit-tree pests by covering the fruit with a physical barrier, usually a bag of some sort. You may purchase bags specially made for this purpose or use resealable sandwich bags or stretchy, disposable nylon foot socks.

To minimize the need for chemicals, remove any fruit left on the trees or that has fallen to the ground to help prevent the next generation of insects.

Selecting Varieties

Table 22.1 gives you an idea of the different kinds of fruits you can grow. Select varieties that will do well in your climate or the hardiness zone of your geographic region. Since fruit blossoms are susceptible to frost, you may want to find or create a microclimate on your property that protects the more tender fruit-tree varieties. If you live in a warm area, you need to know the chill hours required for a particular variety to bear fruit.

Many fruit trees need another tree as a pollinator. Consider what is needed for your variety. If you have a small space, look for fruit trees that are self-fertile. You may also want to consider disease-resistant varieties. A good gardening center or nursery, fruit-tree catalog, or online fruit-tree source can help you decide which trees will work best in your situation.

Dwarf trees are a great choice because they grow to only about 8 or 10 feet and will often bear fruit in 2 to 3 years. Semidwarf trees grow to 10 or 15 feet, whereas standard trees grow 20 feet tall or more. Height is a good indicator of how much space the tree will take up. The larger the tree, the larger the potential crop, but big trees are a challenge to prune, harvest, and maintain.

Table 22.1 Fruits, Nuts, Berries, and Grapes				
Fruit Trees	Tree Nuts	Berries	Grapes	Miscellaneous
· Apples · Peaches · Apricots · Pears · Cherries · Plums · Figs · Pluots · Nectarines	· Almonds · Hazelnuts · Pecans · Pistachios · Walnuts · English walnuts	· Blueberries · Blackberries · Boysenberries · Loganberries · Raspberries · Strawberries	· Table grapes · Juice grapes · Wine grapes	· Currents · Rhubarb

Cultivation Considerations

Fruit Trees

Fruits trees grow best in well-drained soil. A few rocks in the soil is not a problem. Fruit trees also need at least eight hours of sunlight during the growing season.

In colder areas, hardier trees like apple, plum, and sour cherry do well in south-facing areas where they receive more warmth. However, trees that have more cold-sensitive blossoms, like peaches, nectarines, and apricots, may blossom too early in a south-facing spot.

PERSONALLY SPEAKING

 For over twenty years, Jack and I lived in Alpine, Utah, a small town tucked away in a corner of a Wasatch Mountain valley. Many homes in the town were built on the foothills beneath Lone Peak, but our house was situated on one of the lower elevations near the town's center, where the cool air would settle. While this was ideal for cooling off our house at night during the summer and we needed seldom needed air-conditioning until midafternoon, we were more apt to get a late spring or early fall frost. Of course, that was hard on gardens and made the delicate blooms on our fruit trees more susceptible, but we compensated by learning how to extend the harvest and plant less-vulnerable fruit-tree varieties.

Currently, we live on the foothills overlooking Cache Valley, Utah. The temperatures in our valley are renowned for being extremely cold, but we chanced it and planted a couple of peach trees anyway. We hoped that by planting them higher on the hillside and in the shelter of our house, they would be protected from late-spring frosts, and surprisingly, the peach trees have thrived in their perfect little microclimate.

Nut Trees

Most nut trees are large and need a big space in the landscape. Nut trees also grow best in well-drained soil and need long days of sunshine to produce a good crop.

Nut varieties have different requirements. For example, almonds and pecans need a long, hot growing season; chestnuts need acidic soil; English walnuts leaf out early and may be damaged by late frosts; and pistachios need a high, desert climate. Probably the easiest nut to grow in most of the United States is the hazelnut.

Berries

Strawberries

There are two main types of strawberries: June-bearing and ever-bearing. June-bearing strawberries are dependent on day length and bear fruit primarily in early summer. Because of their large concentrated crop, this type is ideal for preserving.

Ever-bearing, or day-neutral, strawberries produce a small crop in June and then continue to produce throughout the summer and into the fall. They are best for a continual supply and grow well in large containers or beds. They need attentive watering and supplemental fertilizer.

Raspberries, Blackberries, and other Brambles

Raspberries and blackberries prefer well-drained, slightly acidic soil (pH 5.2–6.2). By carefully selecting several varieties, you can have a berry crop all summer long. Berries do not suffer from late-spring frosts like other fruits.

Blueberries

Blueberries require the most acidic soil of all the berries, with a pH of 4.5 to 5.5. They also like a well-drained and well-composted soil—much like you would find in the wild woodland. If you have the right soil, blueberries are a joy to grow. Be sure to cover them with a net to protect them from birds.

Grapes

Grapes are not hard to grow if you do your homework and find out which are most suitable for your climate and soil. Also, consider what you'll use your grapes for. They can be grown for table use, and for drying, grape juice, and wine making.

Grapes need well-drained, organic, slightly acidic soil and extensive annual pruning to rejuvenate the vines since grapes grow on the new canes. Plant your grapes from north to south to maximize sunshine. They will also need a trellis or fence to grow on. As with blueberries, use netting to keep the birds from enjoying them before you do.

Rhubarb

Rhubarb is a perennial vegetable used as a fruit in desserts and jams. It prefers cool winters and temperatures below 75° F (24° C) in summer and grows best in fertile, well-drained soil.

Growing Field Crops

Often overlooked in preparation is growing field crops. The most likely field crops to consider growing are wheat and potatoes for people and field corn and alfalfa for animals.

You may be surprised to know that grains like wheat can be grown in every part of the United States. Planting thirty pounds of wheat on an eighth of an acre can yield up to 250 pounds of wheat.

Tools for Growing Field Crops

Growing grains and grasses requires certain tools for planting and harvesting. A grain hand-crank spreader is useful for planting a small plot. For harvesting, you'll need sickles and scythes as well as sharpening tools. A grain cradle is helpful for harvesting more than an acre (see table 23.1).

If you plan to grow potatoes, you'll want a potato fork or scoop.

Table 23.1 Tools for Growing Field Crops
· Grain spreader
· Sickle
· Scythe
· Sharpening tools
· Potato fork
· Potato scoop

Crop Choices

Wheat

Generally, wheat can be divided into winter and spring wheat and soft and hard wheat. The type you grow will depend on your climate and growing season. Hard wheat is best for bread.

Potatoes

There are excellent varieties of potatoes to choose from for the home gardener. Potatoes will yield ten to twenty tons per acre. Potatoes are grown from seed potatoes cut into chunks that contain an eye, where a new plant grows. Historically, a farmer would save and store part of their potato crop to be used for planting the following year. Proper storage must be cool (50° F or 10° C) and dry. Potatoes crops, along with tomatoes and eggplants, should be rotated.

Field Corn

Field corn is grown much like table corn, and harvesting corn to feed pigs is not complicated since pigs will eat all parts of the stalk and ears. Corn seed provides feed for chickens and other fowls.

Alfalfa

Alfalfa is an easy-to-grow, cool-season perennial used to make hay for livestock or as a cover crop and soil conditioner. Growing alfalfa requires more space than the typical urban or suburban home plot. However, you may want to consider it in your preparation if you have enough land, particularly if you have livestock.

Crop Yields

Table 23.2 lists the amount of seed of field crops needed per acre and the yield per acre. For addition resources about growing field crops look on page 429 in the resource section.

Table 23.2 Amount of Seeds and Crops Yields		
Crop	Pounds of Seed per Acre	Yield per Acre
Alfalfa	18–20	3–4 tons
Barley	100	40–50 bushels
Buckwheat	50	25–35 bushels
Field corn	8	65–90 bushels
Grain sorghum	6	55–80 bushels
Oats	8	35–50 bushels
Rye	85	25–35 bushels
Wheat	90	15–35 bushels

24 Growing Microgreens

In a prolonged crisis, another quick way to have access to fresh, living greens is to grow nutrient-dense microgreens. Seeds are germinated in a soil medium or on a felt growing pad where they establish roots and sprout their first leaves, called cotyledons. These initial, tiny plants, ready to harvest in two to three weeks, are known in the horticulture industry as microgreens.

The plant then develops its next set of leaves, called true leaves. True leaves are unique for each plant, and many people harvest the microgreens when these true leaves have grown and the second set is just showing. The plants continue to grow and establish two or three pairs of tender leaves. At this stage, they're called baby greens.

Microgreens can be a delicious addition to your everyday menu. If the idea of growing these tender, nutritious leaves appeals to you, try growing them in a nonemergency setting to develop your skills.

Food Value of Microgreens

The tiny leaves of microgreens are flavorful, colorful, and packed with vitamins and minerals. They also contain powerful phytonutrients that have an important place in a healthy diet. Microgreens get most of their nutritional value from the seeds themselves. As they grow, they also gain nutrition from the soil and from photosynthesis, giving them an added healthful benefit.

> Microgreens are worth experimenting with and are relatively easy to grow.

The young plants contain high concentrations of vitamins and carotenoids and often have more nutritional value than the mature plant.

Plants begin losing nutritional value as soon as they are harvested. Since microgreens are grown nearby, they can be harvested and eaten immediately without losing any of their nutritional benefits.

FIVE THINGS YOU CAN DO NOW

1 Purchase and try microgreens in a couple of different recipes.

2 Gather simple containers and a quality, soil-less potting mixture for growing microgreens.

3 Decide on a microgreen you'd like to try and purchase seeds.

4 Purchase a grow light or find plans for constructing a grow light.

5 Try growing microgreens

Microgreen Seeds

Although any edible plant can be used for microgreens, some varieties are tastier. Seed companies sell seeds specifically developed and tested for microgreens. Table 24.1 lists the most popular.

Table 24.1 Popular Microgreens Seeds		
· Amaranth	· Celery	· Lettuce
· Alfalfa	· Cilantro	· Mizuna
· Arugula	· Chia	· Mustard
· Basil	· Chives	· Parsley
· Beets	· Cress	· Radish
· Bok choy	· Dill	· Scallions
· Broccoli	· Fennel	· Swiss chard
· Carrots	· Kale	

Seed Quality

Use only certified organic or untreated seeds. These range in size from tiny basil and lettuce seeds to larger brassica and radish seeds and even peas. Seed mixes contain seeds that germinate at about the same time.

Amount of Microgreen Seeds to Store

How many seeds you store for will depend on your preferences. Use table 24.2 to help you estimate the amounts. A rule of thumb is to store one-half to one pound of microgreen seeds per person and to store a seed mix or variety of different seeds.

For example, if you plan on two trays a month, plan on two ounces of large seeds, one ounce of medium seeds, and about a half ounce of small seeds for each month. Seed companies can often tell you the yield of the seeds they sell.

Table 24.2 Gauging How Much Seed to Store for Microgreens			
Type of Seed	Ounces of Seeds per 10" x 20" Tray	Two 10" x 20" Trays per Month for One Year	Approximate Yield in Ounces per Tray
Large seeds (radish and cilantro)	1.0 oz.	24 oz.	10–14 oz.
Medium seeds (arugula and kale)	0.5 oz.	12 oz.	8–12 oz.
Small seeds (basil and sorrel)	0.25 oz.	6 oz.	5–6 oz.

Seed Sources

There are many seed companies that offer quality microgreen seeds in bulk. They often show illustrations of the seedlings and categorize them by fast or slow growing. Seeds vary in price from a few dollars per ounce to over twenty dollars per ounce. Look for seeds packaged so that when planted together, they germinate at the same time.

National seed companies like Johnny's Seeds often sell seeds specifically for growing microgreens. Smaller seed companies like True Leaf Market are also good sources for microgreen seeds. Additional companies that sell seeds for growing microgreens or sprouting are listed on page 429 in the resource section.

Storing Seeds

Seeds will last two to five years if stored in a cool, dry, dark environment. Avoid temperature fluctuations, and for best results, vacuum pack or at least keep them in resealable plastic bags. They may also be packed in Mylar bags with oxygen absorbers.

Equipment for Growing Microgreens

Trays

Microgreens are often grown in flat 10- x 20-inch black plastic horticulture trays. If you want to grow smaller quantities, any shallow flowerpot or noncorrosive container will work. Use 10- x 10-inch trays or the plastic clamshells used for carryout that come with their own lids.

Be sure your container has good drainage—punch holes if needed.

Lids

You will need a transparent lid to keep moisture in. You can initially use a second tray as a cover, but as soon as the seeds germinate, you'll need to remove it to allow for light. Clear acrylic dome coverings can be purchased to fit horticulture trays, or you can improvise with clear plastic.

Growing Medium

Biodegradable or compostable felt grow mats made from a variety of natural materials work well because most of the nutrients in microgreens come from the seeds themselves and the roots are not developed enough to pull nutrients from the soil anyway. The felt mats make it easier to harvest, and they eliminate the nuisance of removing bits of soil from the microgreens.

Microgreens can also be grown in a lightweight seed-starting mix, potting soil, or vermiculite. Experiment with growing microgreens in different media to see what works best for you.

Soil Press

A soil press is not essential but can be used to create a flat seedbed if you're using potting soil. A large sheet of cardboard will work, or, for a more permanent press, construct a wooden sheet with an attached handle. You can also smooth the surface with a Popsicle stick or ruler.

Towels

To retain moisture and keep the seeds moist until they germinate, use a cover such as damp paper towels, Chux Cloth, or lightweight cotton towels.

Watering Devices

The device you use will depend on where you are growing the microgreens and the size of the growing containers. The most important thing about a watering device is that it has a gentle showering action. Ideal watering will evenly moisten the soil without stirring up the seeds.

You may use a hose with a sprinkling sprayer attachment outside or a watering can with a sprinkling spout inside. A spray bottle also works for smaller trays.

Heating Mats

Plant-heating mats warm the growing medium from below and help seeds germinate, which is important for warm-weather seeds or if you are growing seeds in a colder climate.

Grow Lights

If you are growing seeds in a low-light environment, you may want a grow light. These may be purchased, or you can build them yourself using a four-foot T5 fluorescent light rod. A sunny south window should also provide enough light.

The grow light pictured is the Jump Start Grow Light System.

Scissors

You will need a good pair of sharp scissors or small hedge trimmers for harvesting. Kitchen shears are usually too dull and will bruise the greens.

Growing and Using Microgreens

Growing Your Own Microgreens

See table 24.3 for step-by-step directions on growing microgreens.

Ideas for Using Microgreens

Ideally, use microgreens immediately after harvesting to receive their full nutritional benefit. But microgreens that are washed and dried and stored in a plastic bag or glass jar will keep for two weeks. Microgreens can be added to salads, scrambled eggs, soups, and casseroles. Put them on salmon, chicken, or tofu. Top a pizza with arugula micros and add kale micros to a sandwich.

	Table 24.3 Growing Your Own Microgreens	
Step 1	If using a grow mat, begin by placing it in the tray. It should be saturated, but not dripping.

Alternatively, if you are using a soil mixture, begin by adding enough water so that the soil just holds together and has a spongy texture. Spread about 1½ to 2 inches of potting soil in your chosen tray to create the seedbed. The soil should be at least a half-inch below the top of the tray. Smooth and flatten the soil, but do not compact it. |
Step 2	Pay attention to seed density when you plant. Generously sow seeds in one layer on top of the mat or soil. Plant seeds so there is adequate room to germinate but close enough to get a good crop. Sprinkle smaller seeds close together and larger seeds farther apart so that each seed has room to germinate. You are aiming for 100 percent germination and close to 100 percent use of the space. Lightly press seeds onto soil using a piece of cardboard or soil press.
Step 3	Covering the seeds is recommended but optional if you can keep the seedbed sufficiently moist. Cover seeds with a thin layer of moist paper towels or Chux Cloth. You can cover seeds with a thin layer of very fine potting soil, but you may end up with bits of potting soil on the microgreen leaves.
Step 4	Gently water the seeds using a spray mister or watering can with a fine spray. Be sure to water thoroughly so that the soil is drenched. In later watering, you'll water just the top layer of soil that may have started to dry out. Seeds must be kept moist in order to germinate.
Step 5	Cover the tray with a clear lid or dome to help keep moisture in. This will create a mini-greenhouse, and the extra warmth and moist conditions will help the seeds germinate. If you are using a warming pad, place the tray on it while the seeds are germinating. Monitor soil moisture and water as needed.
Step 6	Many seeds do not need light to germinate; in fact, some seeds do better if they start in the dark. But once they have sprouted, it's important to give them plenty of light within two to three days. This can be in a south-facing window, outside on a sunlit porch or patio, or directly under grow lights for six to eight hours a day.
Step 7	Continue to monitor progress. Water the microgreens gently as needed, being careful not to overwater and drown the tiny seedlings. Alternatively, water from below by placing the tray in a tub of water and letting the water seep in.
Step 8	Harvest microgreens when the first pair of true leaves has grown by using a sharp pair of scissors to keep from bruising the delicate stems.

NOTES

25 Growing Sprouts

Fresh vegetables may be scarce during a prolonged crisis, particularly if it's not harvest time, and thus, storing a supply of seeds to sprout can help fill that void. Sprouting is a simple process that can be done year-round and provides delicious sprouts in a few days, even in complete darkness. Sprouts supply important nutrients and give variety to a sparse survival diet. Raw or cooked, they add crispness to salads, soups, sandwiches, omelets, and casseroles. or can be eaten by themselves.

Sprouts are the germinated seeds of vegetables, grains, legumes, and nuts. Almost any whole seed can be used, with the notable exception of tomatoes and potatoes—their sprouts are poisonous. Many seeds can be sprouted in combinations.

Nutritious grasses can also be grown by sprouting wheat, rye, or other grains in one inch of soil or on a grow mat and then cutting it when it grows to seven or eight inches.

The Food Value of Sprouts

Sprouts are a nutritious and reliable source of vitamin C and the B vitamins, particularly riboflavin and niacin. Sprouts should be exposed to light for at least a few hours prior to harvesting.

The Whole Grains Council explains that when grains are sprouted, it "is the best way to release all of the vital nutrients stored in whole grains" and that sprouting will "unlock this dormant food energy and maximize nutrition and flavor." Sprouting increases vitamins, minerals, enzymes, and antioxidants, which makes them more digestible.

Bean sprouts contain the protein-binding substances common to all legumes. Cooking the sprouts for at least two minutes inactivates these substances and makes the protein available. This can be done by blanching in boiling water or by stir-frying. Sprouts produce less intestinal gas and discomfort because soaking and rinsing them leaches out the complex carbohydrates that produce gas.

FIVE THINGS YOU CAN DO NOW

1 Gather recipes and ideas for using sprouts.

2 Decide on the sprouts you would like to try and purchase the seeds.

3 Purchase sprouting jar lids for a Mason jar or a seed sprouter of your choice.

4 Learn more about sprouting from one of the resources listed at the end of the book.

5 Experiment with sprouting.

Sprouting Seeds

The Seeds

Sprouting seeds should be certified for sprouting. That means they are untreated, pesticide-free, *E. coli* free, and are likely organic. You can sprout many varieties of seeds. Some popular sprouting seeds you may want to try are alfalfa, broccoli, clover, mung bean, wheat, radish, soybean, mustard, green lentil, and chickpea.

Amount of Sprouting Seeds to Store

Store ten pounds of sprouting seeds per person to provide about a one-half cup serving per day but for sure, store anywhere from five to forty pounds. Twenty-five pounds would provide enough vitamin C to prevent scurvy if no other sources of this vitamin were available.

Sprouting Seed Sources

Seeds can be purchased from either local health food stores or internet sources. You will find sprouting seeds at small companies dedicated to sprouting, regular seed companies, and companies specializing in preparedness products. They should be raw, clean of foreign matter, sorted to insure few broken seeds, and certified safe for sprouting. Often, they will be certified organic and non-GMO. (See page 426 in the resource section for sprouting seed sources.)

Seeds for planting may have been treated for planting with fungicides or pesticides. Treated seeds are required to be so labeled by law and are usually dyed to contrast with their normal color to identify them. Do not use them for sprouting.

Storing Seeds

Store seeds where it is dry, cool, and dark. The number of seeds that will sprout diminishes as seeds grow older, but most have a shelf life of three years or more.

Equipment for Sprouting Seeds

A variety of equipment and tools are sold for sprouting. Some are convenient, but nothing special is necessary. Seeds simply need moisture, warmth, and ventilation to sprout.

You may want to invest in some inexpensive sprouting lids. They may be plastic or metal and often come in sets of three different-sized screens. They screw onto a wide-mouth Mason jar and cost less than five dollars. You can also purchase sprouting jars that come with the lids. The lids pictured are Masontops Bean Screens.

You can also successfully sprout using pans, trays, bowls, strainers, colanders, racks, screen frames, and damp towels.

Growing Sprouts

Following the steps in table 25.1 will get you started growing your own sprouts. See page 430 for resources about sprouting.

	Table 25.1 Growing Your Own Sprouts	
Step 1	Place one to two tablespoons of whole seeds in a wide-mouth jar.	
Step 2	Add two to four tablespoons of lukewarm water to the seeds.	
Step 3	Soak overnight.	
Step 4	Cover the jar with any porous material that has small enough holes the seeds cannot go through, such as nylon mesh, netting gauze, cheesecloth, or specialty lids.	
Step 5	Secure the covering with a canning ring, strong rubber band, or string.	
Step 6	Pour off the soaking water, rinse thoroughly, and then drain completely.	
Step 7	Shake the jar to evenly distribute the seeds and lay it on its side in a location away from direct sunlight. A 65°-80° F (18°-27° C) temperature is best for most seeds. Cress, pea, and rye seeds like it cooler.	
Step 8	Rinse and drain with lukewarm water two to six times each day to keep seeds moist and prevent mold.	

Safe Sprouting

Sprouted seeds have been blamed for a small number of outbreaks of *E. coli* and salmonella poisoning. The source of the contamination seems to be the seeds themselves. A sprouting chamber is essentially an incubator, and the same environment that promotes sprouting also is conducive to bacterial growth, so it's important to take steps to minimize potential contamination. Also, be sure to use sprouts within a few days.

The most important thing you can do for safe sprouting it to buy certified pathogen-free seeds. The sprouting industry is intent on making sure that sprouting is safe. According to the Sprout People, a group advocating for sprouting, "Organic sprouting seed has never been blamed for an outbreak of food-borne illness."

Treating Seeds for Sprouting

If concerned about the safety of sprouting, you may want to sanitize your sprouting seeds. Publication 8151 by University of California at Davis, Division of Agriculture and Natural Resources, explains this process.

They suggest you treat seeds by immersing them for five minutes in a solution of 3 percent hydrogen peroxide heated to 140° F (60° C), and then rinse the seeds under running tap water for one minute.

26 Raising Animals

Though it can be challenging, raising your own domestic animals is a good way to provide an ongoing supply of fresh meat, eggs, milk, fertilizer, fiber, and leather. It is also a satisfying and enjoyable pastime for many people. However, before you consider acquiring animals, you should evaluate the amount of time and effort they require. Daily attention is usually necessary—often two to three times per day with dairy animals—in every conceivable weather condition.

Raising animals requires a long-term commitment to be successful. You also need a suitable location and property that will allow you to raise animals. If you have a suburban homestead as small as half an acre—maybe smaller—you can dedicate space to raising animals.

Domestic animals require some level of protection. They are vulnerable to disease, starvation, and being killed or stolen. They must be taken care of during the stifling heat of summer and the dreary, freezing days of winter. If you are new to raising animals and are serious about it, start with only one or two animals, experiment a bit, and learn and enjoy as you go.

Getting Started Raising Animals

Successfully raising animals requires a good deal of knowledge and effort. A good starting place is to learn from someone raising the animals you are interested in. Homesteading is in full resurgence, so another option is to learn from the firsthand experiences and ample advice of internet bloggers.

Of course, you can always gather information by reading magazines and books dedicated to your specific interest. Your cooperative extension will have good resources, too.

This chapter discusses basic considerations to get you pointed in the right direction.

FIVE THINGS YOU CAN DO NOW

1 Determine if, or which, animals are allowed in your location.

2 Interview people who raise animals and learn from their firsthand experiences.

3 Choose an animal you would like to raise and plan a suitable shelter.

4 Make a list of the supplies you will need to raise that animal.

5 Calculate the initial cost of raising that animal.

4-H

If you have children and desire to help them acquire practical life skills and develop leadership abilities, the 4-H program is a great way to help them in learning to raise steers, pigs, sheep, and goats. Your children get to show them at the county fair, where they are then auctioned.

Selecting Animals

The first thing to consider when selecting animals are zoning laws. And no matter the laws, you're better off if you don't annoy your neighbors. Avoid complaints by keeping the animal area clean and sanitary. If you are just starting out or have limited space, begin with smaller animals. Many backyards can support rabbits kept in hutches or chickens that are kept in pens or allowed to run free within a fenced area. Goats and lambs are another choice. They require minimal space, food, protection, and care. More space and more commitment will allow you to consider larger animals, like sheep, pigs, and cows.

Basic Requirements

Whatever animals you choose to raise, you'll need to know their food requirements and store enough to last them through a crisis. You will also need to provide them with adequate shelter and learn how to keep them healthy and disease-free.

You should also know how to protect your animals from predators and disasters. If you want them to reproduce, of course, you'll need both sexes or at least access to a sire. Keeping records of growth and production for each animal will allow you to know which are the most productive so you can cull the others. Finally, you'll need to know how to slaughter and butcher them or have access to those services if you are raising animals to eat.

Equipment

Each animal requires its own kinds of tools, equipment, and supplies, but no matter what animal, you'll need tools for spreading or distributing feed and tools for mucking manure. For milk animals, you'll need a stool, kick bars or hobbles, milk pails, a milk strainer—stainless steel preferred—and filters. You may also want pasteurization equipment, a cream separator, butter churn, and butter molds. For slaughtering and butchering animals, you'll want to acquire meat saws, cleavers, knives, and scrapers.

Veterinary Supplies

Medicines for animals could be impossible to get during a prolonged crisis, and so you should store a supply of them. To find out which ones might be needed for each animal, you'll have to learn about the diseases that animal is likely to get and accordingly store serums, ointments, sprays, antibiotics, and necessary instruments. Most medicines have an expiration date and will need to be replaced periodically.

Resources

Accurate and knowledgeable information is important when raising animals. In the resource section on pages 429–430 you will find a list of well-regarded general guides along with information sources for each type of animal discussed in this chapter.

Also, as mentioned before, every state has a land-grant university tasked with providing agricultural education to the citizens of the state. It is a valuable resource for information about raising animals as well as other agricultural and horticultural topics. Many of these extensions maintain a useful website.

Specific Animals

The following discussion of popular livestock for small-homestead animal breeders can help you decide whether raising a specific animal is right for you. While modern farms rely on just a few breeds specifically bred for efficiency and high productivity, for the small-homestead animal breeder, there are many different breeds to choose from. Many of these are less-common or endangered heirloom varieties that are good for multiple uses—both for eggs and meat, for example. They live longer, are hardy and often disease resistant, and are just plain colorful, both to look at and in personality.

Raising Rabbits

Rabbits, sometimes known as micro livestock, produce more meat per amount of feed than any other domestic animal. They are able to eat low-protein leafy greens and grain and turn them into high-quality meat protein that tastes much like chicken. Rabbits require only a few minutes of care each day, are relatively odorless if properly cared for, and are quiet and unobtrusive. They require only a small financial investment, need less space than chickens, can be kept indoors, and are easy to keep in good health. They must, however, be protected from predators, including dogs.

Production

Ten does and two bucks will provide one and a half to two pounds of meat per day per year. Rabbits will produce six pounds of meat on the same amount of feed and water takes to produce one pound of beef. A doe will produce two to eight kids each litter. Their productive life is two to four years. They also produce fur and some of the best natural fertilizer (about 40 percent of the total weight of their food).

Space and Shelter Requirements and Equipment

Hutches should have eighteen inches of headroom and are best made of mesh wire that allows droppings to fall through. Be careful that the mesh is small enough to prevent predators from reaching through from below, though. Allow eight square feet for each doe with her liter and four to five square feet for bucks. Special equipment you'll need includes feeders, water bottles, a comb or brush, nail clippers, and nesting boxes.

Food Requirements

Feed should be at least 15 percent to 17 percent protein and can be homemade from corn, grain sorghum, hay, oats, soybeans, and wheat. A dozen rabbits and their young require about three and a half tons of commercial pellets in a year. They also need a salt lick.

Choosing a Rabbit Breed

There are many rabbit breeds to choose from. Your choice will probably depend on what is available in your area. See table 26.1 for varieties of rabbits to raise. Don't let these breed limit you, though, as there are as many as fifty breeds to consider, including some unique endangered heirloom varieties.

Table 26.1 Rabbit Breeds for Different Purposes	
Purpose for Rabbits	Rabbit Breeds
Rabbits raised for meat and fur	American Chinchilla, Silver Fox, Rex, and Champagne d'Argent
Rabbits raised for meat	New Zealand, Californian

Raising Chickens and Other Poultry

Chickens require a small investment in money and only a few minutes each day to feed, water, and gather eggs. They can help keep your vegetable garden free of bugs but can also make a mess and be a nuisance. A rooster is noisy and is not needed for hens to lay eggs, but, of course, is needed for them to lay fertile eggs and reproduce.

Production

Chickens start laying at four and a half months, and their productive life is two to three years. However, they are most productive in the first year. They lay four to five eggs a week, or about one every other day. A dozen hens will provide nearly four-dozen eggs a week. Besides white, chicken eggs may be tan or brown and blue to blue-green, depending on the breed. Colored eggs are just as nutritious and tasty as white eggs. In fact, home-raised chickens will have beautiful, rich, deep-yellow yolks. Chickens also provide meat, feathers, and excellent fertilizer.

Generally, chickens are most productive from spring to late fall. Most breeds need fourteen hours of light per day to keep laying eggs. In winter, placing a light bulb with the chickens will increase production as well as provide some heat.

Space and Shelter Requirements and Equipment

For a dozen chickens, you need a fifty-square-foot coop with nesting boxes and roosts plus a pen at least as big. The coop should be waterproof and provide protection from predators. The pen should have a five- or six-foot poultry wire fence and be covered to keep hawks out.

Besides a safe coop and run, you'll need a waterer and feeder. And for raising chickens, you'll need an incubator, a heat lamp, and a brooder.

Food Requirements

Feed should be at least 15 percent protein. A dozen laying hens requires about twelve hundred pounds of mash a year. A homemade mash of grains and soybean meal should contain a calcium source. Chickens also love old bread and garden waste, like cucumbers, squash, and tomatoes.

Choosing Chicken Breeds

Chicken breeds differ in appearance, temperament, and productivity. Different breeds are suited for different uses (see table 26.2). Chickens may be dual-purpose or used primarily for either egg laying or meat production. Dual-purpose chickens are a good choice for those just beginning to raise chickens or raising them in a backyard setting. They generally have a calm temperament and a desire to forage. Chickens suited primarily for egg production tend to be slender and more nervous. Chickens bred for meat production are larger and plumper than dual-purpose or laying chickens.

Table 26.2 Chicken Breeds for Different Purposes	
Purpose for Chickens	Chicken Breeds
Dual-purpose chickens	Orpington, Plymouth Rock, Rhode Island Red, New Hampshire Red, Sussex, Australorp, and Wyandotte
Chickens raised for egg production	Leghorn, Hamburg, Ancona, Maran, Red Star, and Black Star
Chickens raised for meat	Cornish, Cornish-Rock Cross, Jersey Giant, Fayoumi, Chanticleer, and Freedom Ranger

Look for a qualified breeder who adheres to the American Poultry Association Standards. Also, make sure your chickens have been vaccinated and are free from pullorum and typhoid diseases.

Heritage Breeds

Heritage breeds are heirloom varieties that were raised on country-style farms prior to World War II and before the advent of the industrial farms of today. They have fascinating feathers patterns and colorful eggs. They are good foragers, tend to be healthier, and live longer. They adapt to different environments and climates and produce more eggs. However, heritage chickens grow more slowly than modern breeds.

Backyard breeders often prefer a mixed flock of heritage breeds. Not only are they more interesting, but their unique features help breeders be more attuned to an individual chicken's strengths and weaknesses and to each chicken's specific egg-laying pattern.

PERSONALLY SPEAKING

 Three years ago, our daughter and son-in-law started raising a small flock of beautiful, diverse, heritage chickens. I have been captivated by them and interested as I watch her boys laugh at their antics, take responsibility for feeding and caring for them, and scheme about how to sell their eggs—and we love the super-delicious eggs they share with us. Even as I write this, my husband, Craig, is studying coop designs, and we are trying to figure out how to fit raising chickens into our busy travel schedule.

Other Poultry

If you have a suburban homestead as small as a half acre—maybe smaller—you can dedicate space to raising poultry. Turkeys, geese, and ducks have unique characteristics and offer opportunities if you have space you can dedicate to raising them. Other fowls to consider include guinea hens, Muscovy ducks, peacocks, quail, and emu.

Turkeys

Turkeys are not difficult to raise, are more sociable than chickens, and enjoy human interaction. As with chickens, heritage varieties are worth considering because they have a unique appearance and are good at foraging and insect control.

They need a clean environment, fresh water, and good, high-protein feed. They like to pasture in green grasses and require a larger area for roaming than chickens. You should provide a shelter with five to ten square feet per bird that is well ventilated and shades them from the sun. A seventy-five-square-foot movable turkey coop works well for a dozen turkeys. Turkey manure is hotter than other bird manure and needs to be composted with plant materials before using as fertilizer.

Geese

Geese are extremely hardy, will forage for most of the food they need, and provide feathers and down as well as meat and even eggs. Geese graze on grass and eat standard, untreated poultry feed. They also make excellent "watchdogs" and are exceptionally healthy and disease-resistant. An acre can support about twenty birds. Obviously, they would like a pond, but that's not necessary, although they do need abundant water. They can defend themselves from predators but need a protected place to sleep as they have poor night vision. Especially protect their nests from skunks, raccoons, and similar predators.

Ducks

Domestic ducks are bigger than wild ducks, and their size keeps them from flying off. Ducks need water, and plenty of it! They need a minimum of four cups of water a day but are happier with more. The water needs to be kept clean. Ducks are omnivores, and besides a quality feed, they enjoy rummaging for insects, grubs, and worms. They also love leafy greens, like lettuce and kale. They need company, so plan for more than one, and since they are quite defenseless, ducks must be protected from predators. Ducks are hardy and can sleep outside year-round.

Raising Goats

Goats are social and naturally live in herds. You should have at least two so they will not be lonely—lonely goats are noisy goats and get into trouble! They are primarily browsers, not grazers, which means they prefer to eat leaves, bark, and stems from plants, but they will also graze and clip plants at ground level. They are excellent foragers and ideal for a smaller area.

Production

More people in the world use goat milk than use cow milk. It is mild and easily digested, and two nannies will produce from one to two-and-a-half gallons of milk per day year-round. That will give you cream, butter, and cheese. The average productive life of goats is ten to twelve years, and they produce milk steadily, only stopping the two months before they give birth. Kids can be slaughtered when they are three months old and will provide twenty to twenty-five pounds of meat. Goats can also provide fiber.

Space and Shelter Requirements and Equipment

Goats needs a twenty- to twenty-five-square-foot shed to bed down in and to protect them from the elements. And be sure to protect them against predators. Depending on where you live, these might be wild dogs, cougars, or wolves. Plan for a two-hundred- to three-hundred-square-foot fenced area per goat for roaming and browsing. Electric fencing is also effective.

Besides shelter, to raise goats you'll need a milking stand and stanchion plus stainless-steel buckets and strainers for milking, as well as a hoof trimmer and deworming medicine.

Food Requirements

Goats need two to four pounds of a high-efficiency dairy ration and five pounds of hay per day. That means two goats will require about one-and-a-half tons of the ration and almost two tons of hay per year. A ration can be homemade from grains and soybean meal, and goats also eat root crops. Rock salt is also needed.

Choosing a Goat Breed

There are a lot of things to consider if you decide you want to raise goats. Goats come in dairy, meat, and fiber breeds. Refer to table 26.3. Goats need to be certified free of CAE, CL, and Johne's disease. Also make sure they are free of brucellosis and tuberculosis and are dewormed as needed.

There are good reasons to purchase registered goats. Be sure to get the paperwork at the time of purchase. Each variety of milking goat has its own unique qualities and temperaments. They also vary in the amounts of milk they produce and the levels of butterfat in the milk.

Table 26.3 Goat Breeds for Different Purposes	
Purpose for Goats	Goat Breeds
Goats raised for milk (Swiss/European)	Alpines, Oberhasli, Saanens, and Toggenburgs
Goats raised for higher milk-fat content	Nigerian Dwarfs
Goats for milk production in warm climates	La Mancha and Nubian
Goats raised for meat	Spanish, Tennessee, Boer, and Kiko
Goats raised for fiber	Angora and Cashmere

PERSONALLY SPEAKING

When my children were young, our veterinarian neighbor had a couple of milking goats whose primary job was to keep his weedy backyard under control. He taught our boys how to milk the goats, and when his family went out of town, they milked the goats for him—except for one memorable Fourth of July weekend when my two young daughters and I had the privilege. I learned once again that I could do hard things! After my initial hesitation and knowing it had to be done, I sat down on the stool and figured out how to strip and pull the teats and got a nice bucket of milk.

The most challenging part of it was keeping the thoughtless goat from stepping in the bucket of milk! As payment, we got to bring home all the goat's milk we milked. I was a little squeamish about us drinking the milk, but I strained out the stray hairs and dust and figured out that it made amazingly delicious custard—which also took care of any pasteurization issues. (I have learned since that we could have enjoyed drinking the milk—and I wish I had tried making goats' cheese.) After our experiences with the neighbor's goats, I would recommend raising goats to anyone with a desire for fresh goat's milk—or a weedy backyard.

Raising Sheep

Like goats, sheep are relatively small, have a gentle disposition, and can be trained. And, like goats, they get lonely and need at least one other sheep for company. Good pasture is important, and they will graze on a wide variety of forage. They need protection from predators, including dogs, and sometimes from themselves. They are raised for meat, milk, and fiber. Meat from sheep is called lamb if it is from a sheep younger than a year and mutton if it is older than a year.

Production

A hundred-pound lamb yields thirty-five pounds of meat plus a shearling skin. Many sheep varieties must be sheared every year and provide about eight pounds of wool each time. Hair sheep varieties do not require shearing. Dairy sheep provide milk that is prized for making specialty cheeses like Feta, Roquefort, and Pecorino Romano.

Most sheep need attention to help them stay healthy—they need to be wormed frequently, have their hooves trimmed, their tails docked, and need help with lambing. However, with careful breed selection, you may not need such intense labor.

Space and Shelter Requirements and Equipment

Sheep are very cold-hardy. They need only minimal shelter when lambing and in the most extreme weather. An 8- x 10-foot three-sided shed will house two sheep. Sheep are quite defenseless and need to be fenced in to be protected from dogs and other predators. Hoof trimmers and sheers are basic equipment for raising sheep. You may also want bells to help find any sheep that wander.

Food Requirements

An acre with good grass will graze two to four sheep. And plan on 750–1,000 pounds of hay and 100–200 pounds of grain per sheep. A pregnant or nursing ewe needs a grain and hay supplement as well as salt blocks.

Choosing a Sheep Breed

Sheep have been domesticated for over ten thousand years in many parts of the world, and there are over one thousand breeds. They may be selected for their meat, milk, or fibers

Table 26.4 Sheep Varieties for Different Purposes	
Purpose for Sheep	Breeds of Sheep
Dual-purpose	Corriedale, Columbia, Dorset, Polypay, Romney and Tunis
Sheep raised for meat	Hampshire, Katahdin
Sheep raised for dairy products	East Friesian, Lacaune, and Awassi
Sheep raised for wool	Merino, Cashmere, Shetland, Blue-Faced Leicester

(refer to table 26.4). Because sheep are sensitive to climate, it's a good idea to check with your cooperative extension for advice on varieties that do well in your area. If you want to breed your sheep, you'll also want to consider what variety of rams you have access to. You may be able to get lambs from a local sheep grower in the spring when lambs are born. Crossbred animals can also be beneficial for vigor. All should be vaccinated for tetanus.

PERSONALLY SPEAKING

 One of the most important events of the year for my husband, Craig, is the Cache County Fair 4-H livestock auction. This is where the kids in our community bring the livestock they have raised for their 4H projects to be auctioned with the hopes of making a little money for their efforts. Craig loves the thrill of bidding on the animals and the satisfaction of helping the kids. He "boosts" most of the animals he buys—meaning he just raises the price per pound the kids get but doesn't have them slaughtered for personal use. He also usually buys a steer or two, several pigs, and a half dozen lambs to be butchered and distributed primarily to the people who work for his landscape company and to our extended family. At first, lamb was not very popular, but as we have learned how to properly season it and roast or grill it, we've found it to be a tender, juicy, flavorful meat.

Raising Pigs

Pigs are intelligent foragers and are cleaner than commonly believed, although they do produce offensive odors. Pigs do best if they are not raised alone—when raised with other pigs, they become competitive and will eat more. Pigs are more challenging to raise than the other animals we've discussed, but they might be just the right animal for you.

Production

A good sow will birth two litters a year, averaging ten pigs per litter. The number and quality of piglets will depend on how well the mother pig is fed and cared for. A 200- to 250-pound hog dresses out to about 140 pounds of bacon, ham, sausage, chops, and roasts, as well as valuable lard. Pig hide is some of the strongest yet softest leather around. It takes from five and a half to six and a half months to raise a pig to slaughter size, and its productive life averages eight to nine years.

Space and Shelter Requirements

For the most success in raising pigs, you need a lot of room. The larger the area you can give pigs, the more diluted the odor will be—away from your house is best! And if you can provide pigs a pasture, you will not need as much feed. Pigs need a twenty- to fifty-square-foot shed for shade. Cover the floor with four to six inches of straw to help minimize odor. Pigs are strong and love to root. To keep them contained, you'll need a strong, dig-proof fence. You'll also need to provide them a wallow, where they can cool off the summer.

Food Requirements

Pigs need a *lot* of food. An acre will support five to six pigs if they are supplemented with two to four pounds of coarsely ground grains daily. If you figure three pounds of grains per pound gained, you'll need six to seven hundred pounds of grain per mature hog. Excess milk, eggs, vegetables, and table scraps can be fed to pigs, and you may be able to supplement your pigs' diet if you can find a day-old bread store or bakery; a dairy with old milk; a cheese factory; or a grocery store with old produce where you can get food for your pigs for free or inexpensively.

Like all animals, pigs need clean water every day. You will need a sturdy watering container so the pigs will not knock it over.

Choosing a Pig Breed

Like the other animals we've discussed, there are heritage pig breeds the small farmer should consider. The advantage is that these pigs have specific qualities you may be looking for. If you are considering breeding pigs, you would be helping keep the breed from becoming extinct. Questions to ask: What is the breed's temperament? Is it easy to handle? What are its grazing and feeding habits? How much meat will it produce? How fast does it grow? What is the quality of its meat? What are its maternal instincts and mothering skills? Some common pig breeds are Duroc, Chester White, Hampshire, Yorkshire, but look for heirloom varieties as well. Pigs should be inoculated against hog cholera.

PERSONALLY SPEAKING

 My husband was raised on a small dairy farm—the ideal place to raise a few pigs to feed the family. The pigs foraged and grubbed on three or four acres and ate the excess colostrum from the freshened dairy cows; the leftovers from the vegetable garden; the windfall summer apples; pumpkins; cornstalks; and anything else his dad could think of. He supplemented their diet with whey and cheese ends from a local cheese factory and grain from the local farmers' co-op. Because it was a constant effort to satisfy their enormous appetites, his dad had to be very inventive!

Raising Cows

These large animals may be raised for beef or milk. Cows require a fenced pasture or a corral, as well as daily feeding and watering. If you have enough pasture land, they are not difficult to raise, but be prepared for when your best plans go wrong. Plan for drought, parasites, and unforeseen veterinary costs.

Production

A good dairy cow will average two to three gallons of milk per day for ten months out of the year. Most steers yield between nine hundred and twelve hundred pounds of beef.

Space and Shelter Requirements and Equipment

If possible, separate the acreage into several pastures so that grass and forage have time to recover after cows graze. A shed with one open side with be adequate. Each cow needs about one hundred square feet of space and each calf at least thirty. You'll need forty to fifty bales of straw yearly for bedding and sturdy fencing to protect your cows and corral them. You will need a stanchion for milking, a manger for feeding, and a trough for watering.

Food Requirements

Each cow needs one to three acres to graze unless supplemented with hay, and that is expensive. With adequate summer pasture, you'll need about a ton of grain and two to four tons of hay per cow for the winter. Raising one beef cow requires about five hundred pounds of grain and a like amount of hay.

Choosing a Cow Breed

Jersey and Guernsey and other smaller breeds are best for milking. Herefords and Angus are bred for beef. Angus beef have few problems and are one of the best for meat.

All should be free from brucellosis and tuberculosis.

Keeping Bees

Obtaining Bees

Bees may be gathered in the wild from a swarm, but these bees may carry disease or have weak genetic material. Bees obtained by package or nucleus will need careful introduction of the queen to the hive. Italian bees are best for beginners.

Production

Bees produce honey and beeswax and help pollinate fruit trees and other plants. They require minimal attention and, after the first year, will produce thirty to one hundred pounds of surplus honey per hive. Start a hive in the spring when blossoms appear. It is best to have at least two hives, and you should have ample supers for the bees to store the honey in.

Space and Shelter Requirements and Equipment

Hives need to be placed where they get adequate sunlight and can be protected from winter winds and strong southern exposures. You will also need to protect them from insecticides and proximity to humans. Bee gloves, a smoker, hive tool, veil, uncapping knife, and a honey extractor are important tools for successful beekeeping.

Special Considerations

Bees require special care during winter, particularly if temperatures fall below 50°F (10°C). Bees need supplementary feeding during the winter, and their hives need to be insulated and protected from rodents.

Aquaculture

Raising fish offers another healthy food source. How you approach raising fish will depend largely on the amount of fresh water you have available, its quality and oxygen content, and its temperature.

Space and Shelter Requirements

On a small scale, you can raise fish in backyard ponds, pools, or tanks. If you are fortunate to have a farm pond, you've got the ideal setup. The pond should be deep enough that it does not freeze in winter and remains somewhat cool in summer. A pond will also attract migratory waterfowl, especially if you plant wild rice and other food for them.

Production

If you're going to raise fish in tanks, you'll need a filtering system and heater. A solar-powered, twelve-feet-diameter by three-foot-deep, vinyl-lined, aboveground swimming pool with a dome cover and filter-aerator can produce forty to one hundred pounds of fish per year. Ponds will produce from one hundred pounds per year naturally, to ten to twenty times that if supplemented. Figure on two pounds of high-protein supplementary ration per pound of fish harvested.

Varieties

Be sure to grow regionally appropriate fish. Check local regulations. Blue Tilapia grows fast from fry to fillet in eight months. It needs water at 80° F (21° C) to thrive but can survive temperatures as low as 48° F (9°C). Crappies, bass, and bluegill are good choices for ponds in the Midwest. Trout are harder to raise and require cold, well-aerated water.

Foraging for Food

Some people believe they can just "live off the land" in the event of a crisis. In fact, there are a few who seem to be looking forward to the possibility. They are convinced that if they ever had to, they could gather all their food from wild sources.

Foraging food from the wild would provide bare subsistence at best.

They imagine digging up a few wild onions, picking a handful of berries, perhaps catching and eating a fish, and lying back under a tree, full and contented. In reality, foraging enough food in the wild to live off is not a realistic option.

Consider the amount of wild food that would likely be available. In any major crisis, hordes of unprepared people with the same idea would converge to scavenge the wild areas, quickly gathering, killing, and eating everything in sight. It would be life-threatening just to be there!

And that does not even account for herd migration or winter when game is scarce and edible plants lie dormant. Even in the backcountry it would be a desperate struggle to keep from starving to death.

Nevertheless, in some circumstances, foraging could be used to stretch and add some variety to the survival diet or to keep you alive *temporarily*. Keep in mind that the first rule of obtaining food from the wild is that it should not take more energy to acquire food than it will give you in return.

Wild Edible Plants

There is a large variety of plants humans can eat, but there are limitations. First, most contain fewer than one hundred calories per pound. You would need to eat from fifteen to more than fifty pounds per day to

---

QUICK CHECK

Wild Edible-Plant Safety

✓ Most green plants are not poisonous.

✓ Just because an animal safely eats a plant does not prove it is safe for humans.

✓ White berries are almost always poisonous.

✓ Red berries are sometimes poisonous.

✓ Avoid unknown plants resembling cucumbers, melons, or parsnips.

✓ Assume all mushrooms are poisonous unless you know the difference.

FIVE THINGS YOU CAN DO NOW

1 Collect resources about food sources that could be foraged in your locale.

2 Interview an expert in edible plants and learn how to locate them.

3 Learn how to use edible plants. Explore recipes and try them.

4 Learn and practice hunting and fishing skills. Connect with others who can teach you in the field.

5 Learn how to clean and prepare animals that have been killed in the wild.

---

meet your energy needs! Even berries only average about 250 calories per pound. Nuts provide more calories but are not always available.

Second, if you are not used to them, wild plants may challenge your taste buds and digestive system. Your body may need a period of adjustment to be able to handle the increased fiber. This all assumes you know the difference between what is poisonous and what is not! Some poisonous plants look quite similar to nonpoisonous ones, and some parts of a plant are nontoxic, while other parts are toxic. Follow the guidelines in the Quick Check, "Wild Edible-Plant Safety" to help determine the safety of a plant.

Learning to Distinguish Edible Plants

Although some edible plants are easily distinguished, others take great skill to identify. The best way to develop this skill is by learning in the field from someone already knowledgeable. If that isn't possible, consult guidebooks. Each area of the country has its own indigenous edible plants, and you'll need a guide to the plants common to your local area.

Guide Books

Use the criteria in the Quick Check "Criteria for Selecting Guide Books" to help select a good guidebook. Refer to pages 430–431 in the resource section for specific examples of guidebooks.

> **QUICK CHECK**
>
> ### Criteria for Selecting Guidebooks
>
> ✔ Look for numerous clear illustrations of the plants and their edible parts at harvest time. Color pictures are best.
> ✔ Descriptions should include the plant's range, habitat, and when and how to collect and use it.
> ✔ Poisonous species that might be confused with nonpoisonous species should also be mentioned, with details on how to tell them apart.
> ✔ There should be enough species for your locale to make it a valuable reference source.
> ✔ Recipes telling you how to prepare the plants to make them palatable are also very helpful.

Fishing, Hunting, and Trapping

Normally, fishing and hunting are sports with accepted methods and are governed by laws and regulations to maintain a healthy population of sport animals. I don't recommend that you disregard hunting and fishing regulations.

Fish

Fish are a low-calorie food, averaging four hundred calories per pound for whitefish and seven hundred calories for salmon. Unless you fish in the ocean, much of what you can catch today is stocked from hatcheries that probably will not be raising fish in a prolonged crisis.

It's never a bad idea to have a stockpile of line, leaders, sinkers, lures, and especially hooks; however, the best way to catch the most fish quickly is with nets and traps.

You will want a good fillet knife for preparing and filleting fish.

Ocean and bay shorelines can supply shellfish, like mussels, clams, and oysters, and crustaceans, like crab and lobster. If this is an option for you, you'll need appropriate gear, such as shovels, rakes, and traps.

Small Animals

Although survival courses teach how to live on rodents, lizards, snakes, frogs, turtles, bird eggs, and even insects, and although rabbits, squirrels, porcupines, birds, and other small animals are edible and considered tasty in different regions of the country, this kind of diet would be a radical change for most Americans. Like most wild things, small animals are low in calories, and you would need quite a few to stay alive. Most are also low in fat content. Rabbit meat, for instance, has about two and a half percent fat.

When it comes to killing small animals, you have several options. Slingshots with #9 buckshot will work but require practice to perfect your aim. Pellet guns are inexpensive to shoot and are quiet, accurate, and effective at short ranges. Shotguns are very useful but far from quiet. Select #6 shot or game load for an all-around ammunition.

Traps are an efficient way of obtaining small animals. They work quietly twenty-four hours a day with little attention. Also, you may want to store wire of different sizes for snares.

Game Animals

Venison has 550 to 700 calories per pound yet is only 4 percent fat. I recommend a rifle unless you are a proficient bowhunter. Store plenty of ammunition, a hunting knife, and a skinning knife. It is best to learn the basic skills now from an experienced hunter. Using a red filter over a light will allow you to see the game at night without them seeing you. Close observation of their feeding and watering habits will help you have success. A guide to animal tracks will help you identify what it is you're following and what it's doing.

Wild-Game Recipes

Once you've got whatever game you were after, you'll need to know how to dress it and prepare it for the table. Check out the resource section on page 431 for where to find more information.

Fur Animals

If you plan to trap fur animals like muskrat, beaver, otter, and mink, store six to twelve steel traps. Sizes #1 and #1½ are the best all-around with #4 for beaver. Acquire a book or two on trapping to tell you about animal habits, how to trap them, and how to properly prepare the traps to eliminate human scent.

Cold Storage to Prolong Food Life

Many fresh, unprocessed fruits and vegetables can be preserved using cold-storage methods. These methods were used long before refrigeration and are the easiest and least expensive ways of prolonging the harvest where there is not a year-round growing season. This allows more variety in the survival diet. Also, some fruits and vegetables are more suited to cold storage than other methods of preservation.

What Foods to Store in Cold Storage

Produce you might consider for cold storage includes apples, pears, beets, cabbage, carrots, garlic, horseradish, Jerusalem artichokes, onions, parsnips, potatoes, pumpkins, rutabagas, sweet potatoes, turnips, and winter squash.

Fruits and vegetables can be either homegrown or purchased during the harvest season, and foods in cold storage can be kept up to eight or more months in the right conditions.

Cold Storage Conditions

Optimum shelf life is achieved by maintaining proper temperature and humidity. Fruits and vegetables can be divided into two groups. The first needs cool, dry conditions: 45° to 60° F (7° to 18° C) with 70 percent or less humidity. The second needs cold, moist conditions: 32° to 45° F (0° to 7° C) with 80 percent or above humidity. Adequate ventilation is needed to properly control the temperatures and moisture levels and to remove the gases given off by ripening fruit. Table 28.2 on page 268 lists the ideal temperature and humidity ranges for storing common vegetables.

Managing Temperature

Higher temperatures cause sprouting, ripening, and deterioration. Wide fluctuations in temperature should be avoided, as alternating freezing and thawing cycles are particularly damaging.

FIVE THINGS YOU CAN DO NOW

1 Consider your options for creating a cold storage facility and take steps to make it.

2 Make a list of vegetables that you might want to store in cold storage. Consider how you will store them.

3 In the fall, purchase food such as bags of onions, garlic, potatoes, winter squash, etc. to store in cold storage.

4 Plant winter squash or other vegetables in your garden that can be kept in cold storage.

5 Plant root crops such as carrots and beets. In the fall, cover them with bags of leaves or other insulation to keep them in the ground and available throughout the winter.

Daily monitoring is recommended during the coldest periods of winter so that you can adjust and correct for temperature extremes. Use a minimum/maximum type of thermometer to monitor temperature. When the temperature approaches freezing inside, a low-wattage light bulb or small kerosene lantern can be used to keep the space above freezing. Electric heaters may produce too much heat.

Managing Moisture

Produce will rot with too much moisture and shrivel and dry out with too little. Moisture levels can be accurately checked with a hygrometer. High humidity can be maintained by placing large pans of water near the air-intake vents and by sprinkling water directly on the produce or the floor. A three-inch layer of clean sand, sawdust, or coarse, well-washed gravel covering on the floor will help retain moisture. Normally, this will not be enough for root crops. It is best to store them in perforated polyethylene bags or box liners.

Windows or other vents help regulate the temperature and humidity. Outside air that is colder or warmer can be allowed in or shut out to cool or warm the space. Excess moisture can be vented, or high outside humidity can be used as needed. Two vents are needed for each separate space to allow for proper airflow. Daily adjustments may be necessary. Ideal conditions are not always available, but try to come as close as you can with the space and temperature ranges you have.

Preparing Food for Cold Storage

Shelf Life

The shelf life estimates for various varieties of fruits and vegetables are listed in tables 28.1 and 28.2. The exact shelf life will depend on your facilities, temperature and moisture fluctuations, and the variety of the produce. Late-maturing varieties usually store best. Select varieties that are the best keepers whether you grow them at home or purchase them in late fall or early winter.

Shelf life is also dependent on the condition of the produce prior to storage. The better the quality to begin with, the longer the shelf life.

Harvesting

The ideal time to harvest produce is in the morning when it is cool and dry. If you must pick during the warm part of the day, allow your produce to sit overnight to cool. Cut root-crop leafy tops off one inch above the crown right after digging. Leave stems on winter squash and pumpkin. To prevent moisture loss and spoilage, do not trim roots. Root crops should also not be exposed to sun or wind.

Except for tomatoes, all produce should be picked when it is mature but not overripe. Pears should be fully mature but still hard and light green.

Preparing for Storage

Wipe off excess soil, but do not wash prior to storage, then gently rub off any remaining soil with a soft, dry cloth. The produce should be handled as little as possible, and carefully, to prevent cutting and bruising. Any produce that is bruised or shows signs of disease or spoilage should be culled. Inspect produce on a regular basis once it is stored. Spoilage will quickly spread to adjacent produce, so any spoiled produce needs to be removed. Mold on squash can be removed by rubbing softly with a cloth that has a little vegetable oil on it.

Cure Prior to Storing

Garlic, onions, potatoes, pumpkins, and squash should be cured prior to storing to harden their skins and prolong their shelf life. Cure pumpkins and squash by leaving them in the field for two weeks after picking, exposing them to the sun and air. If it is rainy, they can be placed in a room at 80° to 85° F (27° to 30° C) for ten days (a furnace room may be ideal).

Acorn squash does not need curing and will store at 45° to 50° F (7° to 10° C) for one to one and a half months. Sweet potatoes are cured by placing them in a humid room at 80° to 85° F (27° to 30° C) for ten days to two weeks. White potatoes should be cured for seven to fourteen days at 60° to 75° F (16° to 24° C), and onions and garlic should be placed in the sun for five to seven days.

Packaging

Produce will last longer if properly packaged for storage. Tomatoes should not touch each other and are best individually wrapped in wax paper or paper towels. Apples can be wrapped in shredded paper. Cabbage, pears, pumpkins, squash, and sweet potatoes can be individually wrapped in newspapers or dry burlap. Separate layers of produce with newspaper, straw, peat moss, or grass. This may all seem like a lot of effort, but it will extend the time you have food available.

Storage Containers

Recycle containers. Be sure they have smooth inner surfaces without protruding staples. Use open containers in dry areas to promote air circulation. Stack crates and similar containers. There should also be about four inches of space beneath containers. Containers for humid storage should be closed. You can cover them with moist burlap or old kitchen rugs.

Where to Locate Cold Storage

The local climate will dictate what facilities will be adequate for cold storage in your area. A cool, dry area can be found in many homes, but if not, you may need to construct an outside root cellar or other special structure to provide a suitable cold, moist area. All storage areas must be dark, and any windows should be screened. Large slatted louvers over windows will shade them while allowing air circulation.

Basement Storage

Basement storage rooms work best located in the coolest part of the basement, usually the northeast or northwest corner. They should be away from the furnace, chimneys, heating ducts, and hot-water pipes. The ceiling and interior walls should be insulated if possible. The door should be tight and secure. Concrete outside walls should not be insulated. The room must be rodent proof.

Fifteen to twenty square feet per person will be enough for average needs. Partition the room if two areas are needed. To make air circulation and cleaning easier, use removable, heavy-duty slatted shelving and flooring. Three-foot deep shelving allows enough room to store pumpkins and squash without them touching.

If you close the heat vents to storage areas in heated basements, that space can be used for ripening tomatoes and for short-term storage of onions and potatoes.

Outside Root Cellar

Root cellars work well for keeping a large amount of produce. They take time and money to build but require minimal upkeep. They can be built entirely underground, dug into a slightly sloping hillside, or bermed on three sides. Like basement storage areas, they can do double duty as a storm or shelter if planned for ahead of time.

The best root cellars are made of reinforced concrete, but they can also be made from cinder blocks, bricks, stones, and even wood. A firmly tamped dirt floor helps maintain proper humidity levels. The roof and walls should be waterproof. If possible, locate any doors on the north.

Free-Standing Storage

Barrels, crates, boxes, garbage cans, and discarded refrigerators or freezers can all be used for storage. They may be smaller and less expensive than a permanent storage area but are generally less effective. Fill these containers with alternating layers of straw, leaves, or similar packing along with the produce. When filled to the top, cover the container with straw and then earth.

Outdoor Mounds

Cabbage and root crops can be stored in an outdoor mound if winter temperatures are not too cold. Make a bottom layer of straw or leaves and then cover it with a cone-shaped pile of produce. Cover with more straw and then place about three to four inches of soil over the straw. Place a cap of boards, sheet metal, or plastic on top to keep the water out.

Cover the mound with one-inch mesh chicken wire or hardware cloth to help discourage rodents. If the area is not well draining, dig a shallow drainage ditch around the perimeter. Because the entire mound should be used upon opening, it is preferable to make several smaller mounds rather than one large one, storing a variety of vegetables in each mound.

Covered Pits or Trenches

These can be handier than mounds because you only need to use the portion uncovered at one time. They are about two feet deep and as wide as necessary to store the produce without crowding. It is best to line the bottom of the hole or trench with rocks to aid in proper drainage. Cover the rocks with alternating layers of straw or a similar dry material with produce. The top can be covered with bales of straw or hay for easy access. Use a slanted roof to help prevent rain and snow from getting inside. All mounds, pits, and trenches should be made in a different location the following year to prevent contamination.

Other Locations

Other areas that can be used for storage include insulated crawlspaces under porches, covered window wells, outside basement entry steps, enclosed porches, balconies, fire escapes, and unheated pantries, halls, attics, garages, or spare rooms.

Storing in Place

Hardy root crops like carrots and beets can often be left right in the ground. They can be covered with an eighteen-inch to two-foot blanket of mulch topped with chicken wire and weighted down with a rock or covered with bales of hay or straw. How much mulch you need will depend on your climate. The vegetables can be dug up until the ground is too frozen, the remainder harvested in the spring.

PERSONALLY SPEAKING

 We are always trying to figure out ways to extend our harvest. Last year we planted several rows of carrots in our raised beds. In the fall we covered the carrots with large bags of dry leaves to insulate the ground and protect the carrots. It worked perfectly! All winter long we harvested fresh, sweet carrots, digging up the last of them in March. The airy soil mix in the raised beds made them easier to harvest than if they had been planted in regular garden soil. Our next plan is to plant beets and carrots side by side to make it easier to harvest both. The challenge will be to figure out how much to plant.

Maintaining the Storage Facility

Your storage facility should receive a thorough cleaning every summer. All containers, bins, and shelves should be removed, and the entire area swept and disinfected, if possible, and aired out to eliminate mold and disease. Any necessary repairs can be made at this time. It is best not to reuse packing materials and cardboard boxes.

Keeping Fruits and Vegetables in Cold Storage

Fruit

A few fruits can be successfully stored up to several months, ideally in a root cellar dedicated to fruit storage. Fruits give off ethylene gas that causes vegetables to spoil, so they should be stored separately from vegetables. All fruits require temperatures between 32° F and 40° F and a humidity of 80 to 90 percent. Apples and pears have the longest storage potential. Citrus fruits, including oranges, lemons, limes, and grapefruit, can be stored for shorter periods. Grapes can also be stored for a few months. See table 28.1 for the storage life of common fruits.

Table 28.1 Storage Life of Common Fruits	
Fruit	Approximate Storage Life
Apples	4–6 months
Grapefruit	1–2 months
Grapes	1–2 months
Oranges	1–3 months
Pears	2–5 months

Vegetables

Vegetables are more variable than fruits in the conditions required for successful storage. The storage requirements for common vegetables are described in this section, and table 28.2 gives you the storage life for them at a glance.

Potatoes

Brush off excess soil and store in a cool (35° to 40° F or 2° to 3° C), dark place. Plastic bins lined with damp sand are a good container choice.

Onion and Garlic

Store onions and garlic in a dry, cool place. Do not store near potatoes because potatoes give off moisture that will spoil onions and garlic.

Root Vegetables

Carrots, beets, parsnips, and turnips can be stored in tubs layered in peat moss or sand. Brush off excess soil, clip off tops, and store tubs in a cool, dark place. Vegetables like cabbages, rutabagas, and turnips have strong odors and should not be stored with potatoes or fruits. Most people prefer to store these vegetables outside the house.

Pumpkins and Winter Squash

Pumpkins and winter squash will store up to six months. They benefit from curing in a warm place for a couple of weeks prior to storing. Apply a thin layer of oil to pumpkins and winter squash to protect their skins, then store in a moderately cool place, between 50° and 65° F. A cool, unused and unheated bedroom is a good storage location.

Peppers

Store peppers in the traditional way by stringing them on a line to dry. Avoid letting individual peppers touch each other.

Tomatoes

Store both green and unblemished red tomatoes in a shallow box lined with paper. Keep tomatoes from touching or wrap them individually in paper. Crisscross the boxes and store them in a cool, dark area. The cooler the temperature, the longer it will take for green tomatoes to ripen and the longer they can be stored.

Table 28.2 Storage Life of Common Vegetables	
Vegetables Needing *Cold Moist* **Conditions**	**Approximate Storage Life**
Beets, carrots	3–5 months
Brussel sprouts, cauliflower	1–2 months
Cabbage	3–4 months
Celery	2–3 months
Leeks	1–3 months
Vegetables Needing *Cool Moist* **Conditions**	**Approximate Storage Life**
Eggplant, green peppers, tomatoes	1–2 months
Potatoes, white	4–9 months
Vegetables Needing *Cool Dry* **Conditions**	**Approximate Storage Life**
Garlic	6–7 months
Onions	6–7 months
Vegetables Needing *Warm Dry* **Conditions**	**Approximate Storage Life**
Peppers, hot chili	6 months
Potatoes, sweet	4–6 months
Pumpkins	2–3 months
Squash, winter	2–6 months
Based on information from Cornell Cooperative Extension, "Storage Guidelines for Fruit & Vegetables"	

You will learn where to find more information about cold storage and root cellaring on page 431 in the resource section.

PERSONALLY SPEAKING

 We make it a yearly challenge to see how long we can store the tomatoes we've grown in our garden. Our first frost is usually around October 1, so we have learned to use good storage practices to make our tomatoes last—and we usually enjoy tomatoes from our garden all through the fall and even in our Christmas Eve salad. The secret is to select blemish-free tomatoes, then place them individually on newspaper-covered cardboard trays on shelves in a cool basement storage room. We select firm red, orange, pale-orange, yellow-green, and even green tomatoes without any cracks or spots and use the ripest ones first. As time passes, the quality is not quite as good as summer-ripe tomatoes, but the flavor is better than that of the bland tomatoes from the grocery store.

Keeping Other Items Cool

Dairy products and other produce can be kept cool for short periods of time using iceboxes, evaporative coolers, spring houses, and cooling cabinets. Historically, before refrigeration was widespread and where the climate allowed, ice for the iceboxes was harvested during the winter from ponds and streams and then stored in cellars or insulated icehouses for summer use.

NOTES

29 Home Canning to Preserve Food

Food can be preserved at home in a variety of ways, the characteristics of a particular food determining the best method. It will also depend on climate and storage conditions as well as space and equipment. A combination of several methods is usually best. Table 29.1 gives comparisons of the relative costs, energy needs, effort, time required, and vulnerability during a crisis.

Table 29.1 Comparison of Home Food-Preservation Methods					
Method	Cost	Energy	Effort	Time	Vulnerability
Cold Storage	Low	Low	Moderate	Low	Moderate
Canning	Moderate	Moderate	High	Moderate	Moderate
Freezing	High	High	Moderate	Moderate	High
Dehydrating	Moderate	Moderate	Moderate	High	Low
Pickling	Moderate	Low	Moderate	High	Moderate
Smoking	Moderate	Low	Moderate	Moderate	Moderate

This chapter gives you an overview of canning and a list of the equipment and supplies needed. Each of the references at the end of this chapter will give you detailed information necessary to safely can a large variety of foods at home. Subsequent chapters provide an overview of other preservation methods.

Home canning is the most common way to preserve food at home. It's inexpensive whether you are canning produce you have grown or animals you raise, hunt, or fish, and it requires a moderate amount of labor and skill.

FIVE THINGS YOU CAN DO NOW

1 Determine what types of food you would like to preserve through home canning.

2 Purchase a *Ball Blue Book: Guide to Preserving*.

3 Collect canning jars at thrift shops.

4 Purchase a boiling bath canner or steam canner and other canning tools.

5 Locate local farmers' markets and farms where you can purchase fruits for canning.

Home Canned Products

Whole Fruits and Vegetables

Canning is an excellent method for preserving fruits and vegetables that are normally thoroughly cooked. It is ideal for preserving whole fruits, like peaches, pears, apples, cherries, plums, figs, and berries. Vegetables suited for canning are tomatoes, asparagus, green beans, lima beans, beets, carrots, corn, peas, potatoes, pumpkin, and winter spinach.

Jellies, Jams, and Preserves

Fruit that is too large, too small, or irregularly shaped can be made into jellies, jams, preserves, conserves, and marmalades that add delicious variety to your diet.

Proper amounts of pectin and acid are necessary to make jellied fruit products. Some fruits, such as apples and plums, are high in natural pectin, but you'll have to add it to others. Lemon juice or citric acid is added for flavor and to aid in gel formation. There is more natural pectin in slightly underripe fruit than in ripe fruit. To get enough pectin when making jams and jellies, use about one-fourth underripe fruit to three-fourths ripe.

Jellies, jams, and preserves should all be canned in jars with rings. They require five to ten minutes of processing in a water bath or steam canner. Sealing the jar with paraffin is not recommended.

PERSONALLY SPEAKING

Growing up in Oregon gave us easy access to all kinds of berries. We picked strawberries at the local patches and grew our own boysenberries, blackberries, and blueberries. So, at an early age, I was introduced to the art of jam making. My mother believed in making things simple, so she used commercial pectin packages. I learned that if you followed the directions exactly, you ended up with close-to-perfect jam. In the Intermountain West, where I live now, we have more access to fruit grown on trees than we do berries. My three favorite jams to make now are apricot, peach, and plum. I've learned that peaches have enough natural pectin that packaged pectin isn't necessary. When we eat peach jam made from peaches grown on our own peach trees, it is like eating a bite of captured sunshine.

I encourage you to try making homemade jam! It is not hard, and it is so, so much better than the mass-produced national brands! I know you'll be proud of your own gourmet jam—it will rival those sold in gift shops and boutiques. And this is not a gender-specific skill—my brothers have been making jam since they were teenagers.

You will find several of my favorite jam recipes on our website, CrisisPreparedness.com.

Juices and Ciders

Juices are a good way to use less-than-perfect fruits and vegetables after completely removing any bruised, spoiled, or overripe portions. Juices may be canned and stored as concentrates or made into ciders, vinegars, and wines.

Apple Cider

The best apple cider is made from blending apple varieties, including sweet, sharp, and bittersharp. If you have apples you would like made into cider, you may be able to find someone in your area who presses small batches. Alternatively, you may purchase the presses.

Pickles and Relishes

Pickles and relishes can be made from many different fruits and vegetables and will add variety in your diet. Pickling can be accomplished through fermentation in a crock or by freshpack in canning jars. Both use brine and vinegar as preservatives. Sugar, herbs, and spices are added for flavor. Pickles are usually canned using the water-bath method unless they will be eaten soon after being made.

Meat, Poultry, and Fish

You will need a pressure cooker to safely can beef, pork, poultry, game, fish, and seafood at home in either glass jars or tin cans. Combinations like stews, soups, and chili may also be canned.

Home-Canning Safety

Although some home-nutrition experts caution about the safety of canning at home, it is completely safe if the directions followed are from a reliable source, such as the USDA canning guide, National Center for Home Food Preservation, or a current extension guide (see page 431 in resource section). Like most of the survival skills described in this book, it's best if you practice this skill before a crisis. Home-canned foods are often superior in quality to purchased canned foods, and home canning is a great way to enjoy your homegrown food and save money.

Home-Canning Methods

Boiling-Water-Bath Canning

Water-bath canning is suitable for high-acid produce. This includes all fruits, tomatoes, and pickled vegetables. The food can be put in the jars uncooked, then cooked during processing. This is called raw pack. Or the food can be cooked and placed in jars hot and then processed. This is called hot-pack canning.

Water-Bath Canner

Boiling-water-bath canners are large pots that come with a jar rack to securely hold jars and keep them off the bottom of the pot. The most common, smaller canner size is 21.5 quarts; it will hold seven pints or quarts. The typical large thirty-three-quart canner will hold nine quart jars and fourteen pints. Traditionally, canners are made from granite ware—an enamel-covered steel, but aluminum and stainless-steel canners are also available. Make sure the canner is deep enough to allow two to four inches of space over the jars.

Steam Canning

Steam canning is also suitable for high-acid produce. The products may be raw packed or hot packed. Steam canning uses about half the energy and takes less time because you do not have to wait for a large amount of water to boil. It also eliminates most boil-over messes. Best of all, product quality is superior. For details about how to use a steam canner and further information, consult the two extension bulletins from Utah State University and the University of Wisconsin listed in the resource section.

Steam Canner

Most steam canners consist of a shallow pan with a tall cover that fits over pint or quart jars. Jars are placed on a rack that sits above the boiling water, and hot steam circulates to process the product. Excess steam vents through small holes on the cover. The canner pictured is the Victorio 7-Quart Aluminum Steam Canner.

Dual Purpose Canner

Victorio makes a dual-purpose canner with a unique double-sided rack system that sits flat for water-bath canning or may be inverted for steam canning. Its built-in temperature gauge allows you to monitor the temperature, and a glass lid lets you see the contents. It is also stainless-steel and can be used on induction-stove surfaces.

Pressure Canning

Pressure canning is required for all low-acid vegetables and meat. This is because the dangerous bacteria found in low-acid foods are only killed completely at 240° F (116° C), and that temperature can be attained only under pressure. As altitude increases, so must the pressure necessary to attain the correct temperature. Consult a canning book for adjustments needed. Pressure canning can be done with either glass jars or tin cans.

Pressure canning is not recommended for high-acid produce. It will result in a lower-quality product due to the longer total time spent at high temperatures.

Pressure Canners

One of the most important pieces of canning equipment is a pressure canner, especially if you intend to can vegetables and meats. There are several brands and all will work well if properly used. They are larger than a pressure cooker and need a rack to place the jars on. Check the capacity of the canner. Some are large enough that you can double-stack pint jars.

All-American Pressure Canners (pictured) are unmatched in quality and use a clamp system that requires no gaskets. They use a gauge to measure pressure and are durable enough to last a lifetime. They are also more expensive than other brands.

Presto and Mirror also make good pressure canners. Both have locking handles and a rubber gasket seal. Presto pressure canners use a gauge, and Mirror canners use five-, ten-, and fifteen-PSI pressure controls. The Granite Ware Pressure Canner is relatively new on the market and includes a steamer insert

If your pressure canner needs them, be sure to store extra pressure-release valves or gaskets. No matter which brand you have, make sure the pressure gauge is checked regularly to verify its accuracy. If it uses fixed weights instead of a gauge, have it properly adjusted to the correct altitude. A canner that uses fixed weights may be better for long-term survival use because it does not require periodic calibration like the gauge type does. Maintain the pressure canner in working order by periodically cleaning the petcock and safety-valve openings.

Home Canning with Cans

Canning with cans requires heating the contents and exhausting the air before sealing the cans and processing them in the canner. To achieve the necessary vacuum, the food in the can must be at 170° F (77° C) or higher. The can is then sealed, processed in the canner, and cooled in cold water. The quicker cooling results in slightly better produce texture and less heat damage to vitamins.

Home-Canning Equipment

Not only will you need the canners described in the previous section, you'll need a variety of canning tools to increase the number of products you are able to can and to make canning easier. Even if you do not presently can food, it is a good idea to have the necessary equipment and supplies stored. Not only might you use them in the future, but they could be valuable for bartering. Tables 29.2 and 29.3 list canning supplies you might find useful.

Table 29.2
Basic Canning Equipment

· Canning funnel	· Food grater	· Manual food mill
· Canning tongs	· Food grinder	· Measuring spoons
· Cutting boards	· Jar lifter	· Measuring cups
· Colander	· Jar wrench	· Pot holders
· Cooling racks	· Kitchen shears	· Silicone spoons
· Dipper or ladle	· Kitchen scale	· Slotted spoons
· Food chopper	· Knives	· Timer

Table 29.3
Specialized Canning Equipment

Canning Fruits	Canning Vegetables	Jelly Making	Pickle Making
· Apple corer · Apple peeler · Cherry pitter · Pear corer	· Bean slicer · Cabbage shredder · Colander · Corn cutter · Pea sheller	· Candy thermometer · Cheesecloth · Jelly-bag strainer · Jelmeter · Flat bottomed kettle: 8 to 10-quart	· Nonreactive metal utensils · Crock or food-grade bucket and covers · Weights

Food Strainer and Sauce Maker

A food strainer is a valuable piece of canning equipment because it extends the versatility of the products you can. A food strainer makes it easy to process apples and tomatoes into sauce. Two well-known

brands are VKP and Squeezo. The one pictured on page 275 is the VKP Food Strainer. You may also want to purchase the berry and pumpkin screens.

Steam Juicer

A steam juicer is a three-part pan that steams the fruit to extract and collect the juice. It will effortlessly extract the juice from grapes, berries, and apples with no need to strain them. Steam juicers may be aluminum or stainless steel.

Equipment for Canning Jams and Jellies

Unless you are using a steam juicer exclusively to make juice, you'll need a muslin or canvas jelly-bag strainer. Or you can make your own from cheesecloth. For jellies, jams, and other preserves, you may want a jelmeter—a graduated glass tube that helps you determine the amount of pectin in the fruit. You will also need tools for mashing the fruit and an eight to ten-quart flat-bottomed kettle for cooking. I recommend a candy thermometer to take the guesswork out.

Equipment for Canning Pickles

Pickles require utensils and containers made of stainless steel, aluminum, or glass to avoid reactions with acids or salts. Fermenting is done in a crock or stone jar, plastic food storage bucket, large glass jar, or an unchipped, enamel-lined pan. You will need covers and a weight to hold the produce below the brine's surface. A plastic bag filled with water placed on top of a plate will work.

Equipment for Canning in Cans

Home canning with cans is also an option. Tin cans are more durable than glass jars and can be stored in the light, but they cannot be reused and require an extra step to seal.

Wet-pack canning requires specialized cans and a can sealer. Sealers are set to seal a specific size can, but additional sized chucks for sealing may be purchased. Dry-pack canning requires a sealer that seals #10 cans. The sealer will also allow you to seal seeds, ammunition, and other things in cans. All American and Ives-Way make several hand-operated can sealers. Ives-Way Model 900 can sealer is pictured. (See page 85.)

All American and Ives-Way make several hand-operated can sealers. Expect to pay from $300 to $1,500 for a sealer. Wells Cans and Ives-Way make hand-operated can sealers. Ives-Way Model 900 can sealer is pictured.

You will also need a glass food thermometer. And you will need a bottle capper if you anticipate capping bottles.

Home-Canning Supplies

Jars and Lids

Stockpile all the glass jars you'll need for your family. A good time to stock up is at the end of the growing season when the canning season is over. Stores mostly sell half-pints, pints, and quarts. Use the jar size your family can eat at one meal. Store extra jars to replace any that might get broken. Wide mouth are a little more expensive but are easier to clean and are best for tomatoes and larger fruits. Narrow-mouth are fine for smaller produce, sauces, and juices.

Store at least two seasons' worth of lids and enough rings for your jars. Jar lids could be in extremely short supply in a prolonged crisis. Keep them cool and dry, and they'll keep five years or longer.

Reusing Jar Lids

Tattler makes a two-piece reusable lid that includes a nitrile rubber seal and a food-grade plastic lid. A regular jar ring is needed. They are warrantied for five years, but in practice may be used longer.

Gasket Seals

Luminarc and Bormioli Rocca are two companies that make canning jars with their own attached, reusable lids. They are made of heavy glass with a hinged glass lid. The seal is a rubber-ring gasket. They are more expensive than regular jars, but the rings can be reused.

Cans

Cans come in a variety of sizes, ranging from three- to twenty-eight-ounce. The handiest sizes are normally the #303, #2, and #2½. Cans are available with plain lids or pull-tab lids and in a range of sizes. Large #10 cans are frequently used for dry-pack canning, and cans cost from one to two dollars, depending on quantity and size. They are expensive for a one-time use, so use them only for high-value products, like meat or fish. Cans come in different qualities. Select better cans for longer storage.

Pectin

Include pectin, either powder or liquid, in your storage. You can make your own pectin from tart apples if necessary.

Ascorbic and Citric Acid

Ascorbic or citric acid crystals prevent discoloring of fruit. Powdered ascorbic acid compounds should be stored in a cool place where they will not absorb moisture. They will keep at least two years.

Sugar, Vinegar, and Salt

Store extra sugar for canning and jam making. For pickles and relishes, store vinegar with a 5 to 6 percent acidity. Also store pickling or canning salt, sugar, and spices.

NOTES

Freezing to Preserve Food

Despite the high quality of most frozen foods, you should not rely on freezing as your main method of preservation. Its primary drawback is its susceptibility to power failure. You will find recommendations later in the chapter about what to do if this happens. It is also expensive when compared with other food-preservation methods.

The Basics of Freezing Food

The Pros and Cons

Freezing is an excellent and easy way to preserve many fruits, vegetables, and meats. However, because of the ongoing energy requirements, it is expensive when compared with canning and dehydrating. Because of its vulnerability, during a prolonged power outage, you would need an alternate power source to keep the freezer running or the means for preserving the food that defrosts. Another risk to freezers is that they can malfunction, putting your frozen food at risk at any time.

Equipment for Freezing Food

Of course, to freeze food you need a freezer. A chest freezer is better at retaining cold when opened. Look for one with sliding baskets that help you organize. Also use small removable tubs to organize smaller items. A lock will prevent accidental opening and theft. Plan on about six cubic feet per person.

Table 30.1
Freezing Equipment and Supplies

· Ascorbic acid	· Funnel, wide-mouth
· 6- to 8-quart pan for blanching	· Heavy-duty resealable bags
· Colander	· Tongs
· Cooler or ice chest	· Vacuum bags
· Freezer containers, assorted	· Vacuum sealer
· Freezer wrap and tape	· Wire basket for scalding

A cooler or ice chest is useful during a power outage. Items stay frozen longer if they are tightly packed. Coolers also allow for portability of frozen items.

FIVE THINGS YOU CAN DO NOW

1 Purchase a freezer if you don't already own one.

2 Defrost and organize your freezer.

3 Purchase berries or other produce on sale, flash freeze it, and then store in containers in your freezer.

4 Set the freezer thermostat at zero degrees or below, especially if you anticipate power outages.

5 Keep your freezer full. If necessary, fill clean milk jugs with water, freeze, and then use to maintain temperature in a full freezer.

Vacuum Sealing Food for Freezing

Proper wrapping prevents freezer burn. The ideal method is to use a vacuum sealer. Food Saver makes a line of vacuum sealing systems (pictured) that work well for meats and many fruits and vegetables. It can be a challenge to use if the item has excess liquid because the vacuum sucks the liquid and prevents a seal from forming. You will need a supply of bags in various sizes to go with the sealer. Table 30.1 on page 279 lists the equipment and supplies needed for packaging frozen foods.

Ascorbic acid is used as an anti-oxidant and to prevent discoloration of fruits. Use sugar to wet-pack fruits for added flavor.

Flash Freezing

Flash freezing prevents the formation of large ice crystals that rupture cells and harm texture and flavor. It works best if your freezer is set at -20° F (-29° C) or below. Place a single layer of produce or pieces of meat on a precooled tray, then place the tray in the coldest part of the freezer for twenty-four hours. Repackage the frozen items in a vacuum-sealed bag or other container.

Freezer Backup

One way around part of the vulnerability of frozen food is to have an alternative energy source so that outages and shortages will not affect you as much. If you live where you are frequently affected by power outages, a backup generator is a good idea. Small portable generators are another option. Refrigerators and freezers that run on propane or kerosene can also be used as a backup to your regular freezer. And, of course, you'll need to store fuel for the generator or for the alternative refrigerators and freezers mentioned. (See chapter 36)

Quantities of Frozen Food

Meats and seafood freeze well, especially if vacuum-sealed. Many kinds of produce also freeze well, including applesauce, asparagus,

Table 30.2
Approximate Yield from Fresh Fruits

Fruit	Fresh Quantity	Frozen Quantity in Pints
Apples	1 bushel (48 lb.)	32–40
Apricots	1 lug or crate (22 lb.)	10–12
Berries	1 crate (24 qt.)	32–36
Cantaloupe	1 dozen (28 lb.)	20–24
Cherries	1 lug (25 lb.)	8–10
Cranberries	1 box (25 lb.)	46–54
Peaches	1 half bushel (24 lb.)	16–24
Pears	1 bushel (50 lb.)	40–50
Pineapple	5 lb.	3–5
Plums	1 bushel (56 lb.)	38–56
Strawberries	1 crate (24 qt.)	36–40

Table 30.3
Approximate Yield from Fresh Vegetables

Vegetable	Fresh Quantity	Frozen Quantity in Pints
Asparagus	1 crate (30 lbs)	15-22
Beans	1 bushel (30 lbs)	30-45
Beets	1 bushel (52 lbs)	35-42
Broccoli	1 crate (25 lbs)	22-26
Carrots	1 bushel (50 lbs)	32-40
Corn	1 bushel (35 lbs)	14-17
Peas	1 bushel (30 lbs)	12-15
Peppers	3 peppers	1
Tomatoes	1 bushel (53 lbs)	15-20

berries, green beans, lima beans, broccoli, carrots, cauliflower, corn, peas, peaches, peppers, potatoes, and spinach.

Use tables 30.2 and 30.3 to help you determine how much fresh fruit and vegetables are needed to produce a specific yield. The actual number of pints will depend on the variety, size, maturity, and quality of the produce. It will especially depend on the way the food is prepared.

PERSONALLY SPEAKING

 Freezer space is kind of like real estate. Not all things that can go in the freezer are of equal value, and they do not get the same amount of freezer "real estate." I carefully consider what gets space in the freezer. They are usually expensive, convenient, or rare.

Seafood, meat, and nuts get priority because they are more expensive than other items and the cost to replace them if they spoil is greater. The peppers we have at the end of the summer get space because they are convenient. We slice them and seal them in vacuum-packed bags for fajitas and stir-fry. Since we like our fresh frozen corn much better than any store-bought varieties, it gets a fair amount of space because it is rare—we can only get it when we harvest our garden. The same goes for our frozen pesto sauce made from the basil we grow.

I recommend you use freezer space economically and prepare produce so that it takes up the least amount of space. Since I make jam from peaches, apricots, and berries, I puree them in jam-sized batches, label them, and freeze them. They take up less space than whole or sliced fruit and are ready go when I want to make fresh jam.

I also like to use the freezer to take advantage of grocery-store sales. I have learned that butter freezes well and is usually cheapest right before the fall holidays when grocery stores are trying to get your business. So that is when we stock up and keep it in our freezer to use all year. When chicken breasts go on sale in bulk, we buy a package or two and vacuum pack individual breasts to freeze and use later.

One of my favorite conveniences is to keep nuts and seeds in the freezer. Again, I buy them in bulk and keep them secure in their own bag inside a second, resealable bag. I dedicate a storage basket to all nuts and seeds so they are easy to find when we need them. Freezing nuts keeps them from becoming rancid. It also keeps out annoying pests. Though I have kept nuts in the freezer for at least two years, they are usually eaten long before that.

Lastly, I use the freezer to make life easier. I like to cook in big batches, so if there is room in the freezer, I make batches of cooked beans, specialty sauces, soups, and baked dishes and freeze them in meal-size containers, then thaw them when we need a quick meal.

All the things in our freezer are a comforting part of our short-term food storage. We realize that a prolonged power outage would have us scrambling to take care of the thawing food, but we are willing to live with that risk.

Remember, in an extended power outage, you'll have a few days to preserve the food using another method.

Shelf Life of Frozen Foods

The biggest risk to frozen foods is the freezer burn caused by dehydration and oxidation. It occurs over time when frozen food is exposed to air. You can lessen this with proper packaging. A vacuum sealer will give you the highest quality frozen goods. Repackage items like fish fillets and frozen shrimp in vacuum packages.

Freeze purees and soups in plastic containers with a layer of plastic wrap securely fitted on top of the liquid.

The other factor in shelf life is temperature. The best temperature is 0° F (-18° C) or lower. Although food kept at or below this temperature will store indefinitely and can be safely eaten, the quality will decline over time. Table 30.4 lists the optimal storage life at 0° F (-18° C) or below. Practically speaking, you may be able to store frozen foods longer.

What to Do with the Freezer in a Power Outage

If your freezer loses power for whatever reason, there are things you can do to protect or salvage your food. In warm weather, you'll have to act fast. In cooler weather, you'll have a few days to solve the problem.

	Table 30.4 Storage Life for Frozen Foods		
Meat	Months	Vegetables	Months
Ground beef	2–3 months	Asparagus	6–8 months
Stew meat	2–3 months	Beans	8–12 months
Beef roasts	8–12 months	Beets	12 months
Fish, fat varieties	2–3 months	Broccoli	12 months
Fish, lean varieties	6–9 months	Brussels sprouts	8–12 months
Lamb chops	3–4 months	Carrots	12 months
Lamb, ground	2–3 months	Cauliflower	12 months
Lamb, roasts	8–12 months	Corn	8–12 months
Pork chops	3–4 months	Okra	12 months
Ham and sausage	1–2 months	Peas	12 months
Pork roasts	3–8 months	Peppers	8–12 months
Organ meats	3–4 months	Pumpkin	12 months
Miscellaneous	Months	Winter squash	12 months
Baked goods	2–3 months	Spinach and greens	12 months
Butter and lard	6–8 month	Fruits	Months
Cheese	6–8 months	Citrus fruits and juices	4–6 months
Coconut, shredded	8–12 months	Mixed fruit	6–8 months
Cookies	8–12 months	Other fruits	12 months
Cream	3–4 months	Berries	12 months
Eggs (not in shell)	8–12 months	Nuts	6–12 months

Protecting the Food in a Nonworking Freezer

First, only open the freezer if it's absolutely necessary. If possible, do not open it at all. Even a nonworking freezer will keep food frozen for some time. How long it will keep food depends on how full it is, the types of foods, the size of the food packages—bigger is better—

During a power outage it is best not to open the freezer at all.

and the temperature of the room where the freezer is located. A fully stocked freezer can keep temperatures at satisfactory levels for two to three days providing it is kept closed. A partially filled freezer may keep food frozen only half as long. Conserve the cold by insulating the freezer with blankets and newspapers.

Using Dry Ice to Keep Things Frozen

If you can get it, dry ice will extend the time you can keep food frozen. A twenty-five to fifty-pound block will keep the temperature of a half-full freezer below freezing for two to three days. Put the food close together and then, using tongs or gloves, place the dry ice on a layer of heavy cardboard over the center of the food. A large block of dry ice will last longer than several small ones.

Because dry ice evaporates and produces tremendous amounts of carbon dioxide, the area should be ventilated and the freezer door left slightly ajar to prevent a dangerous buildup of pressure.

Anticipated Power Outages

If you are anticipating a power outage, transfer the foods you'll use right away to a quality camping cooler with ice. You may also want to prepare gallon jugs of ice to place in the freezer to fill the air space and help keep it cold. With advanced warning or if you are subject to frequent power failures, set the freezer temperature to its lowest setting. The colder the temperature inside the freezer when the outage begins, the longer the food will stay frozen.

You may be interested in a power failure alarm to alert you whenever the power goes off. Should your freezer be inoperable beyond these measures, you can take the frozen food to a commercial locker, assuming the outage has not affected them, or immediately preserve the food by canning, dehydrating, or smoking it.

When to Refreeze

As a rule, if a food is safe to eat, it is safe to refreeze. In practice, with most foods, that means they have not thawed completely and are still under 40° F (5° C). How can you tell? Some ice crystals will still be present. Their existence means the food is between 32° F and 40° F and can be refrozen. Exceptions to this are deli meats, fish, and other seafood that spoil very quickly. Expect refrozen foods to have a lower quality and to not keep as long. They should be used as soon as possible.

If foods have completely thawed, do not refreeze, and use at once. Uncooked vegetables, meats, and poultry can be cooked and then refrozen. Fruits can be canned or made into preserves. Fruit-juice concentrates ferment when spoiled and can cause the cans to explode. Throw out any off-flavor fruits.

Home Dehydrating to Preserve Food

Drying food is the oldest method of food preservation, dating back thousands of years. Even as late as the nineteenth century, it was one of the main ways foods were preserved and stored.

Dehydrating is an inexpensive way to preserve food. Most fruits and vegetables, such as corn, onions, peas, peppers, tomatoes, and herbs, are good for drying.

Solar dehydrators and electric dehydrators are the main methods of drying food. Sun drying is the least-expensive method since it takes advantage of sunlight. Although it is possible to dehydrate in a kitchen oven, it is expensive and hard to control—many ovens do not get to a low enough temperature to properly dehydrate food.

A recent development in dehydrating is freeze-drying for home use. Freeze-dryers are expensive but can be made cost-effective if several families split the cost.

Superior Quality of Home-Dehydrated Produce

When produce is harvested, the very best grade is used for fresh produce in markets. Large canning companies use the next best grades. What is left over is used for dehydrating. Commercial dehydrating may use higher temperatures to speed up the dehydration process, which can reduce nutritional value.

To assure the best quality and nutritional value, a great option is to dehydrate foods at home. By using a good dehydrator and lower temperatures (110° to 145° F or 40° to 60° C), along with selecting quality produce, you can have an excellent product. You can

---

QUICK CHECK

What to Look for in a Dehydrator

✓ Does it have enough total surface for your needs?
✓ Does it have enough trays?
✓ Does it have an adjustable thermostat?
✓ Does it have an auto shutoff?
✓ Does it have a timer and alarm?
✓ How easy are the trays to clean?
✓ Can trays be washed in the dishwasher?
✓ Does it have trays or sheets suitable for fruit roll-ups?
✓ Is the heat evenly distributed?
✓ Is the airflow evenly distributed?
✓ How noisy is it?

FIVE THINGS YOU CAN DO NOW

1 Read about dehydrating in the *Ball Blue Book: Guide to Preserving*, an extension-service pamphlet, or an online source.

2 Look at plans for DIY solar dehydrators and choose one you would like to make.

3 Purchase a dehydrator.

4 Choose a fruit you would like to dry, such as apples, and experiment with dehydrating them.

5 Try making some jerky.

---

save money by growing your own produce, purchasing fruits and vegetables in season, or buying in bulk during special sales.

If food is properly dried, packaged, and stored, its shelf life will be from six months to more than two years.

Types of Dehydrators

An electric dehydrator is easy to use and does the most consistent job. There are two basic types: vertical flow (usually round) and horizontal flow (usually rectangular). Although sun drying is least expensive, it is not as consistent as other methods because of air and temperature variations. See table 31.1 for descriptions of three popular dehydrators.

Table 31.1 Advantages and Disadvantages of Vertical and Horizontal Dehydrators		
Dehydrator Style	Advantages	Disadvantages
Vertical Dehydrator (heat source on bottom)	· Small and compact taking up less counter space · Often expandable, with ability to add more trays · Less expensive · Satisfactory for fruits and vegetables · Lower exposure to airflow better for delicate foods	· Less-effective airflow · Uneven heat distribution · Less-efficient use of space · Often require tray redistribution during drying for best results
Vertical Dehydrator (heat source on top)	· Small and compact, taking up less counter space · Often expandable, with ability to add more trays · Less expensive · Designed for optimum vertical and horizontal airflow · Ability to dehydrate jerky · No dripping on the heating element	· Round trays with less-efficient use of space · Limited space for dehydrating large quantities
Horizontal Dehydrator (heat source at back)	· Even and efficient airflow · Efficient use of space meats as well fruits and vegetables · Dries meats as well as fruits and vegetables · No need to rotate trays · Good for dehydrating large quantities of food · Less likely to mix flavors in mixed batches	· Larger in size, requiring more counter space · Requires significant storage space · Heavy to move · Usually more expensive
Dehydrators shown are Cuisinart Food Dehydrator, Nesco GardenMaster Food Dehydrator, and Excalibur 9-Tray Hydrator		

Vertical-Flow Dehydrators

Vertical-flow dehydrators have a heat source that may be either at the base or at the top of the dehydrator. A fan helps airflow vertically throughout the dehydrator. Some designs force air over the food trays for better circulation. The challenge of this design is the tendency for the food-filled trays to block airflow. The round design also limits the amount of space available for food.

Horizontal-Flow Dehydrators

Horizontal-flow dehydrators have a heat source and fan located at the back. Airflow is more evenly distributed than in a vertical model because it flows across the drying trays. Heat is well distributed in most models, making them excellent for dehydrating meat. They also require less tending and rearranging of trays.

Solar Dehydrators

Commercial solar dehydrators are hard to come by, but you can make your own simply with screens that have an insect-proof covering. Both simple and elaborate DIY solar dehydrator plans can be found with an internet search.

Tools and Supplies for Dehydrating

Most of the work in dehydrating fruits and vegetables is in preparing them for dehydrating. To shorten prep time, use time-saving tools. The list of specialized kitchen tools for canning in chapter 29 is useful for dehydrating as well. The equipment and supplies listed in table 31.2 will also help with dehydrating.

Table 31.2 Equipment and Supplies for Dehydrating	
• Ascorbic acid • Cheesecloth to cover trays • Manual food slicer • Salad spinner • Oxygen absorbers • Silicone or other BPA-free plastic sheets for fruit roll-ups	• Vacuum-packing equipment and bags • Waxed paper or plastic wrap for separating fruit roll-ups • Wire basket for scalding and blanching produce

Dehydrating Vegetables

Preparing Vegetables for Dehydrating

Vegetables are easy to prepare for dehydrating. Begin by paring and slicing uniform pieces. The most important step is blanching, and almost all vegetables must be blanched. Blanching deactivates enzymes, speeds up drying by relaxing the tissue, preserves vitamins and flavors, and sets colors. Arrange in a single layer on dehydrating trays.

How Dry Should Vegetables Be?

The best temperature for dehydrating most produce is 125° to 140° F (57° to 60° C). Drying times will vary widely according to the individual item, the size of the pieces, the humidity in the air, and the dehydrator used and drying in the sun takes longer than drying in an electric dehydrator. Most vegetables are dry when they are brittle or crisp, although tomatoes and mushrooms should be pliable. Dehydrated vegetables should have about 95 percent of their moisture removed.

Dehydrating Fruits

Preparing Fruit for Dehydrating

Prepare fruits by paring and slicing uniformly. Pretreat light-colored fruit to help prevent darkening by dipping them in a solution of ascorbic acid or Fruit Fresh (one tablespoon per quart). Blanch fruits with skins like grapes, plums, and cherries in boiling water for thirty to sixty seconds.

How Dry Should Fruit Be?

Fruit is dry when it is leathery and pliable and retains its characteristic color. Dehydrated fruit should have about 80 percent of its moisture removed. Produce should be allowed to cool before you test its dryness. Although not ideal, it is also better to over dry than under dry. Long-term storage requires a drier product.

Storing Dehydrated Fruits and Vegetables

Produce should be cooled after dehydrating and before being placed in storage containers. Seal the dried food in vacuum-packed plastic bags, in a resealable plastic bag, or other airtight container. Add an oxygen absorber. Place packages in a #10 can with a plastic lid or in a glass jar. Store all dehydrated products in a cool, dry location. Jars are best if kept in the dark.

Dehydrating Meat

Dehydrated meat is usually marinated and known as jerky. But it is possible to dehydrate untreated meat as well.

Use lean cuts of beef, other red-game meat, salmon, and turkey for jerky. Do not use pork because the dehydration temperatures are not high enough to kill the trichinosis parasite.

Preparing Meat for Making Jerky

It is a personal preference whether to cut meat with the grain or against the grain. Slicing with the grain will make it chewier, and slicing against the grain will make it more prone to falling apart. Slightly defrosted meat is easier to slice. See table 31.3 for complete instructions on making jerky.

Marinating

To add flavor, marinate the meat for ten to twenty-four hours, depending on how seasoned you want it to be. Refrigerate it during marinating. Many jerky recipes use soy sauce and Worcestershire sauce, along with sugar and spices like pepper, garlic, and onion powder. Test out several recipes and sure to stockpile the ingredients you like.

Testing for Doneness

Periodically cut into jerky to check for doneness. How fast it dries will depend on how thick the jerky is and how moist it is from the marinade. Jerky is ready when you can bend and crack it, but it should not be brittle. It should be a dark brownish-red in color.

Table 31.3 Making Jerky	
Step 1	Slice the meat into very thin strips, removing as much fat as possible.
Step 2	Coat the jerky with salt. Add spices if desired. Salt helps with the curing and speeds dehydrating.
Step 3	Spray trays with nonstick cooking spray.
Step 4	Lay out the meat strips, leaving space around each piece to allow for good airflow.
Step 5	Set drying temperature to 165° F (80° C). Check every couple of hours. It should take about eight hours to thoroughly dry.

Storing Dehydrated Meat

Jerky was not originally intended for long-term storage. The idea is that you make jerky as meat becomes available, either after a good sale at the grocery store or after slaughtering a steer or game.

Store jerky in a refrigerator or freezer if possible. Homemade jerky should be eaten within one to two months. Commercial jerky can last one to two years because it's cured with sodium nitrate, a preservative.

Freeze-Drying at Home

The latest contribution to food preservation at home is the Harvest Right home freeze dryer—a home appliance that will freeze-dry food. Food that is made using this dryer is excellent and rivals commercial freeze-dried foods.

It uses a refrigeration condenser that freezes food to a very low temperature (-40°F). A vacuum pump creates a vacuum inside the drying chamber, which pulls off the ice crystals, which then change to vapor and evaporate. It will handle both fresh and cooked foods, including combination meals. The biggest limitation is cost, but look for prices to continue to come down.

Crisis Preparedness Handbook

Salt Curing, Smoke Curing, and Pickling to Preserve Food

Salt curing, smoke curing, and pickling are additional ways to preserve food. This chapter will give you a description of the basic processes as well as the supplies and equipment you'll need.

Salt Curing

Salt curing is an ancient preservative method. Pork is the most traditional cured meat, although game meat, beef, veal, lamb, chicken, and fish can also be cured. Ham, bacon, and sausage are the most common cured meats in the United States. Many cultures have developed their own unique cured meats and sausages, such as Italian prosciutto and salami.

Curing Salts	
Sodium nitrite and salt:	Prague Powder #1, known as Insta Cure #1, or pink curing salt
Sodium nitrate and salt:	Prague Powder #2, known as Insta Cure #2, or Morton Tender Quick

Curing salts are special blends of table salt and sodium nitrite or sodium nitrate used to preserve meats, poultry, and fish. Curing salts work by osmosis, drawing out the moisture to create equilibrium between salt in the cure and salt in the meat. They are necessary for curing meat because they prevent botulism. Other bacterial and mold growth are also inhibited in this environment. In addition, curing salts keep the fat in meat from going rancid and give it an appetizing pinkish color.

Wet Curing or Brining

Wet curing uses salts dissolved in water to make a brine solution, in which the meat is submerged so that the salts penetrate it. Wet curing is more manageable than dry curing because there is less guesswork about whether the salt solution has permeated the meat. The thickness of the meat will determine how long it takes the salt solution to permeate the meat. Salt moves through the muscle tissue at the rate of one inch per seven days. Modern recipes require that the end product be refrigerated.

FIVE THINGS YOU CAN DO NOW

1 Look for a pickling or sauerkraut recipe. Purchase a crock, pickling salt, vinegar, and spices, and give making pickles or sauerkraut a try. Use the freshest cucumbers or cabbage you can find.

2 Purchase a good book about sausage making, along with basic sausage-making equipment.

3 Research meat smokers and purchase one. Be sure it has a reliable thermostat.

4 Try smoking salmon or other fish. This is an especially good idea if you catch the fish yourself, but salmon is also available at many grocery stores year-round.

5 Perfect a rib recipe and smoke precooked ribs as a final step for added flavor.

Dry Curing Meat

Dry curing uses curing salts either to coat whole pieces of meat or to mix in with ground meat formed into sausage. Spices and other ingredients may be added for flavor.

Treated meat must be cured in a carefully maintained environment, where both temperature and humidity are controlled. Sanitation and the quality of the meat are also important. It is a long, slow process with many variables that have to be managed.

Although the ingredients are basic, the craft of dry curing meat, known as charcuterie, is sophisticated. It is both a science and an art. The key to successful meat preservation is knowledge and experience. If possible, find a master meat curer who can teach you the craft. Additionally, you may want a good reference book to guide you. (See pages 431–432 in resource section.)

Country-Cured Hams

Country ham is the dry-cured hind end of a pig. Typically, the meat is salt-cured and then hardwood smoked to give it a unique flavor. Different regions of the United States have their own distinctive recipes. Extension services will have recipes for the region they serve. Country-cured smoked hams will keep at least a year.

Safety Precautions

To avoid botulism poisoning from improperly-prepared dry-cured meats, it's critical you gain the appropriate knowledge and skills. Be sure to use the correct curing salts, proper tools, and correct curing methods, and use recipes from reputable sources, such as extensions. Look for recipes that use weights and percentages to calculate the amount of salt cure in any given recipe.

Smoking

Smoking is a type of drying that adds a unique smoky flavor to meats that a smoker or dedicated smokehouse. Historically, it was used to preserve meat, usually in conjunction with curing. Today it's primarily used to flavor meat, and virtually all smoked meats need to be refrigerated.

The meat or fish is salt cured before it is smoked. Hams, bacon, meat, and fish preserved this way and should be stored below 40° F (4° C).

Cold Smoking

Meat smoked at temperatures less than 140° F (60° C), or cold smoked, can be a potential source of botulism because the food does not get to a high enough temperature to kill pathogens.

Cold smoking should only be used on foods that have been previously salted or cured. It requires exactness in order to be a safe method of food preservation. If you desire to use this method for preservation, you'll need top-quality, accurate digital thermometers and precise scales. You will also need expert guidance and tested recipes that you follow with precision. All meat that has been salted and cold smoked is still raw and should be cooked prior to eating.

Hot Smoking

During hot smoking, the meat is cooked at temperatures between 200° to 225° F (91° to 107° C) until the meat reaches an internal temperature between 145° to 165° F (63° to 73° C). It is used to flavor and slow-cook meat but not to preserve it. Meat prepared this way should be refrigerated.

PERSONALLY SPEAKING

Smoked salmon is a family favorite, and both my husband, Craig, and my brothers have developed expertise. It's an easy process that consists of slicing the fillets into strips, brining them in the refrigerator overnight in a simple salt solution, and then smoking them for a couple of hours. My brother uses a salt-and-sugar solution, but Craig usually uses just salt. It's not necessary to buy an expensive brining salt. The secret to smoking it is to do it slowly and at low a temperature. Be sure to set a timer so you don't overcook it!

Another of Craig's specialties is smoked ribs. Each summer, Craig puts on a rib fest to thank the employees of his landscaping company. He cooks a crazy amount of ribs, and one way he finishes them is by smoking them for an hour and a half. That wonderful smoked flavor is a favorite—there are very few leftovers!

Sausage Making

Sausage making is another method of preservation. Dry-cured sausage such as salami requires curing salts, a lactic-acid bacteria starter culture, and a controlled curing environment. Meats preserved this way require expertise and care.

Meat, poultry, and game scraps can be ground and made into sausage. You will need a meat grinder. A manual, cast-iron meat grinder with cutter plates and an attachment for stuffing sausages is best. A funnel can also be used for stuffing. An attachment on an electric appliance will also work but is susceptible to loss of electricity.

Artificial casings may be stored, or you can use the small intestine from a hog or sheep. Lightweight muslin is used for sausages over one and a half inches in diameter. A selection of spices can help make sausage to your taste preferences.

Dry-cured sausage will keep for six months to more than a year if properly cured. Otherwise, it must be canned or frozen for long-term storage.

Safety Precautions

All meat is susceptible to pathogens and bacterial growth. Although fresh internal muscle is sterile, when sausage is ground, the pathogens are incorporated into it. When a piece of meat is cooked, the pathogens on the surface are killed, but to be safe, the internal temperature must reach 165 F (74 C).

Pickling

Pickling is a process for preserving food in an acid solution. One common pickling process uses an acid solution made from a vinegar solution of at least 5 percent acidity. Another uses brining and fermentation to create an acidic solution made of lactic acid.

Pickled Vegetables

You can pickle vegetables with either a vinegar solution or by fermentation. Vegetables made with a vinegar solution are processed, placed in jars, and canned in a water-bath canner.

Brined pickles and sauerkraut are made with a brining process that uses fermentation to create an acidic environment from lactic acid. Once the vegetables have reached a desirable stage of pickling, they may be refrigerated or processed in a water bath.

Safety Precaution

Be sure to use a recipe from a reputable, lab-tested source, and make sure the vinegar-to-water ratio is at least half vinegar. Use produce and meat that is as fresh as possible. Take precautions to sanitize work surfaces and preparation tools. And use crockery, glass, food-grade plastic, or stainless steel for containers and utensils.

Equipment for Curing, Drying, and Pickling

Temperature-Humidity Sensor

This sensor needs to be accurate, and so you should be able to calibrate the hygrometer part of it with a hygrometer calibration kit.

Meat Thermometer

This is important for monitoring the internal temperature of smoked meat.

Kitchen Scales

It is important to accurately weigh your meat and curing salt.

Curing Chamber

To properly cure meat, you need a curing chamber that can maintain a temperature of around 50° to 60° F (10° to 13° C) and high relative humidity, around 65 to 80 percent. Curing chambers are often built from old frost-free refrigerators with a temperature regulator specifically calibrated for dry curing.

Smoker

Smokers come in multiple sizes with a variety of features. Look for one that has a good temperature control system. Electric smokers are convenient but may offer an inferior taste. Charcoal and gas smokers usually require more adjustment and may or may not come with a thermostat. A good smoker is made of steel, is reasonably heavy, and closes tightly. A grill can double as a smoker.

Fermentation Crock

There are two kinds of fermentation crocks—open and water-sealed. The water-sealed crocks have the advantage of being airtight, with less mess and odor, but they are more expensive. A five-liter crock will hold about ten pounds of cabbage and is fairly easy to lift and clean. For larger amounts, consider a ten-liter crock.

Meat Grinder

You'll need a basic hand meat grinder or meat-grinder attachment if you plan to use ground meats to make sausages.

Sausage Stuffer

A hand-crank sausage stuffer is superior to add-on sausage-stuffer attachments. Make sure the piston that pushes the meat into the stuffer has a plastic or rubber gasket that doesn't allow the ground meat to squeeze out.

Curing Salt-Sodium Nitrite

This is the most common curing salt. It contains 94 percent plain salt and 6 percent sodium nitrite, as well as a tiny bit of red dye so that it doesn't get confused with regular salt. The sodium nitrite helps prevent botulism growth and is the top-recommended curing salt.

Curing Salt-Sodium Nitrate

This curing salt is used for slow-cured meats. Over time, it slowly breaks down into sodium nitrite and is called for in some recipes for dry curing meats.

Morton's Kosher Salt

This salt does not have additives that may affect the quality of cured products, but be aware that it is larger grained than regular table salt, and because the grains are larger, a teaspoon of kosher salt contains about half the amount of salt as table salt. Adjust recipes for weight as needed.

Preparing Your Shelter for a Crisis

Emergency Evacuation

A disaster like a wildfire, flood, landslide, earthquake, tornado, hurricane, or chemical or nuclear accident can force you and your family to evacuate your home without warning. More than likely, you'll have to rely on yourselves until the civil authorities can effectively respond. If you are prepared with an emergency evacuation kit, often referred to as a 72-hour kit or bug-out bag, you'll have the bare necessities and a minimal level of comfort during that period. A survival kit will not only give you more control of the situation but also help boost morale.

Planning for Emergency Evacuation

When it comes to an emergency evacuation kit, there is a lot of advice about what you need. However, your first task is to decide what you expect to use it for. Then you'll plan for the items that will help you get through an immediate crisis of about three days.

Possible Reasons to Evacuate

What kind of crises are you likely to face? Assess the likelihood of various scenarios. Do you live in an earthquake or active volcano area? Could tsunamis impact you? Are hurricanes likely or probable? Or tornados? Are flooding or landslides a possibility? What about wildfires? Are blizzards and winter storms a problem for you? Do you live downwind from a nuclear reactor? Do you live near a chemical plant or an interstate highway or railroad where chemical spills might occur? What are the chances you'll be forced from your home? Where do you anticipate going? Refer to worksheet 2.1, "Rating Your Likelihood of Experiencing Disasters" in chapter 2.

Because what we can store in our emergency evacuation kits is limited by cost, space, and weight, our emergency evacuation kit should be prepared to handle high-probability, high-risk events.

FIVE THINGS YOU CAN DO NOW

1 Decide the three most likely reasons you may need to evacuate.

2 Purchase containers for your emergency evacuation kit.

3 Make a list of the items you want in your emergency evacuation kit and prioritize them.

4 Take an extra pair of athletic shoes to work in case you can't drive and have a long walk home.

5 Give gifts that can be used in an emergency evacuation kit (space blankets, water straws, flashlights, etc.)

Container Options

Portability is essential, and so your kit needs to be just large enough to contain the most useful and critical supplies. The container should be sturdy, waterproof, and easily organized. It should fit in a place that is accessible at a moment's notice.

Each person should have their own container for individual gear, food, and water. Having individual kits will spread the load and lessen the impact in case of separation. The container options listed will hold roughly what a person can carry., though it is more efficient to share certain items, distributing them among family members or keeping them in a group or family container.

Five-Gallon Buckets

These large white buckets with handles and tight-fitting lids are waterproof and inexpensive. They are often used for commercial 72-hour kits. Although they are portable, they cannot easily be carried long distances. An alternative is to strap the bucket on a pack frame. Buckets have the added advantage that they can be used to sit on, to carry water, or for emergency sanitation.

Duffel Bag

Large duffel bags work best if you expect to be using a vehicle for transportation. They are inexpensive and come with handles. A disadvantage is that they are hard to carry for long distances. But like buckets, they can also be strapped to a pack frame.

Backpack

A backpack is an excellent option and may be something you already have lying around the house. It can be transported either by vehicle or on foot. Backpacks also keep your survival gear close and allow you to have your hands free.

You can carry more on your back than you can tote in a handheld container. A person in good physical condition can carry about one-fourth of their body weight, but 20 percent of your body weight is more realistic and comfortable.

One downside is that high-quality, comfortable backpacks are usually expensive, especially if dedicated solely to a survival kit. If you are a backpacker, you'll already have made this investment. If so, one option is to fill zip-top bags with your evacuation items and store these bags in the pack. Remove and then replace them when you use the backpack for recreation. You will also find that there is an overlap because a lot of backpacking gear is also ideal for survival situations.

Other Choices

Improvised containers might be suitcases, ice chests, or garbage cans. A large fishing tackle box can help organize items but is awkward to carry for long distances. Remember, the container must be small enough to fit in your vehicle.

Putting Your Own Emergency Evacuation Kit Together

There's no reason to purchase a commercial kit when you've got a little know-how. Chances are a commercial kit will not fit your unique needs anyway. Buying one means relying on someone else's selection and quality.

When you put your own kit together, you know exactly what's in it. You will have taken the time to think about each item and how you'll use it. That thought process alone will give you an advantage in a crisis. You will not be spending your time rummaging through an unfamiliar kit to figure out what's in it and how to use it. It may initially take more time to put your own together, but it'll be worth it in the long run.

Water and Food

Water

Water is essential. Unfortunately, it's bulky and heavy, weighing over eight pounds per gallon. best practice is to include two quarts of water per person and a personal water-purification bottle or straw. (See chapter 9 for filter recommendations.) You may also want to have a heavy-duty water filter and purification tablets in a communal kit. Add a collapsible bucket or jug with a spigot to carry water.

Store a case or two of half-liter water bottles or a five-gallon container of water near the vehicle you'd likely use in an evacuation. Recycle this water so you always have fresh water ready to go.

QUICK LOOK
Water Essentials
For Each Individual:
• Two quarts of water
• Personal water filter or straw
• Water bottle or canteen
For Each Group:
• Collapsible bucket or jug
• Large-capacity water filter
• Purification tablets
• Solar still (optional)

Food

Store foods in your emergency evacuation kit that do not need refrigeration and can be eaten without cooking. Do not worry about them being nutritionally balanced—for the short-term, it's enough to have palatable calories. Plan on at least 2,000 to 2,400 calories a day.

Freeze-dried foods are my first choice for dinner entrées. They taste good and are lightweight, for easy portability. However, they require boiling water to reconstitute, so you'll need a small stove and fuel, or at least a way to get very hot water. Many other foods are available in pull-tab cans or small packets. See table 33.1 for suggested food items. Many of the foods listed do not have a long storage life, so you'll need to replace them periodically.

Each kit should have a spoon, fork, and sharp knife. A bowl or plate and a stainless-steel camp-

Table 33.1 Suggested Food Items		
• Bouillon cubes	• Fruit juice	• Pork and beans
• Cheese	• Milk, powdered	• Protein bars
• Chocolate bars	• Meat, canned	• Spaghetti, canned
• Crackers	• Hard candy	• Soup
• Freeze-dried entrées	• Hot chocolate	• Stew
• Fruit, canned	• Jerky	• Tuna fish
• Fruit, dehydrated	• Peanut butter	

ing cup that can be used as a small cooking pot would also be useful. Include a P-38 G.I. can opener—they are handy and inexpensive. Keep detergent and a small dishpan to clean the utensils after use.

Sanitation, Hygiene, and First Aid

The lack of proper sanitation can quickly become a crisis of its own.

Waste and Sewage

You should have a supply of heavy-duty plastic bags with twist ties. Individual commercial kits, like the Travel John Disposable Urinal and the Biffy Bag Pocket Size Disposable Toilet, are useful waste containers.

If you are using a large bucket as a container, it can be turned into a makeshift latrine. Emergency toilet seats are made specially to fit buckets. You will also need plastic garbage-bag liners, disinfectant, deodorizer, and toilet paper. A good disinfectant is a 1:10 ratio of liquid bleach to water.

Personal Hygiene

Personal hygiene is important in a crisis. Not only will it help contain the spread of disease, it will improve morale and relieve stress. Consider the items listed in table 33.2.

Bleach

Household liquid bleach is effective for purifying water as well as

Table 33.2 Personal Hygiene Items		
· Chapstick	· Hand lotion	· Shampoo
· Comb and brush	· Hand sanitizer	· Shaving supplies
· Disposable diapers	· Hand soap	· Sunscreen
· Facial tissue	· Hand towels	· Toilet paper
· Feminine-hygiene products	· Insect repellant	· Toothpaste and toothbrush
· Fingernail clippers	· Mirror	· Tweezers
· Fingernail file	· Moist towelettes	· Washcloths
· Glasses, contact solution	· Personal wipes	
	· Petroleum jelly	

disinfecting human waste. Two pints is the minimum recommended by Homeland Security. It loses its potency after a year and should be recycled and replaced.

First Aid Kit

Be sure to include a basic first-aid kit with a good instruction booklet, pads, gauze, adhesive tape, and plenty of large sterile bandages. Include a supply of special medications with prescriptions anyone in your group or family requires and moleskin or Spenco Second Skin to protect feet from blisters.

Clothing and Bedding

Suitable Clothing

Include a change of warm, durable work or outdoor clothing suitable to the coldest weather you expect. Include a rain poncho, work gloves, a hat for shade, and a wool cap for warmth. Sturdy, comfortable shoes with socks are also important. Do not forget to have backup survival clothing and footwear in your car and where you work.

PERSONALLY SPEAKING

Not all clothing or bedding will fit in an evacuation kit. Besides that, you may want your best work jacket or work shoes and hiking boots available for other uses. A good compromise is to make a short list on a 3" x 5" card of things you can quickly grab and add to the evacuation kit. If you tape it to the lid of your bucket or pin it to the top of your bag or pack, you won't have to think about what to grab when you need it.

Space Blankets

Almost every emergency list recommends a space blanket, also known as a Mylar, or solar, blanket. Originally developed by NASA, these blankets are made of a reflective aluminum vapor and polyester and will reflect and retain up to 80 percent of your body heat. They can also protect you against rain and wind. They take up little space, weigh only a few ounces, and cost less than five dollars. Be sure to include several in your kit, but also check the quality because it varies.

Be aware that space blankets cannot really be used as a regular blanket. They are very thin and are susceptible to tearing, and they retain body heat only if there is air space between the blanket and your body. It works better when used along with a blanket or other insulating layer.

Space blankets can also be used for collecting rainwater, as a signal, or as a ground cloth.

Wool Blankets

Wool blankets sustain warmth even when wet. They are also naturally fire-resistant. Compare the weight and thickness of blankets. You may want to wash a wool blanket several times to improve softness and remove odor.

Down

Though less durable and not waterproof, down-filled throws are another blanket option. They are lightweight and compress into small stuff sacks.

Woobie

First introduced during the Vietnam War, it is also known as "US Military All Weather Poncho Liner Blanket." This all-purpose blanket is loved and recommended by many members of the military and is essentially two pieces of sturdy nylon cloth with a layer of polyester filling. Designed to keep you warm and dry, it is about 82" x 62" and costs close to fifty dollars. If tightly compressed, it will roll to about a three-inch diameter.

Sleeping Bags

There are countless sleeping bags to choose from. Consider your expected needs. Will you be able to dedicate a sleeping bag solely for your evacuation kit?

One sleeping-bag system to consider is the military Modular Sleep System manufactured by Tennier Industries. This four-piece system includes two sleeping bags that may be used together or separately. It also includes a Gortex outer covering and a compression stuff sack.

Shelter

A shelter system is designed to protect you from exposure to the elements. An emergency shelter can be improvised with a cloth or tarp and ropes. Look for a heavy-duty variety with grommets in the corners. The Grabber All Weather Blanket and similar tarps have a waterproof plastic side and a reflective side that can aid heat reflection. Other options are tube tents and backpacking tents.

Communication

Radio

A portable radio can help you keep connected to the news. You can get important information about relief efforts and weather forecasts. Get either an AC/DC radio and regularly check the batteries, or a solar-powered radio with rechargeable batteries. A hand crank feature is also useful. The one pictured is the RunningSnail 090P Solar Crank Radio.

A GPS (global positioning system) can give you your exact position on a map or, with a compass, help plan a travel route. But a GPS relies on satellites and only works if they do.

For more detailed information about communications, see chapter 40.

Cell Phones

Just about everybody relies on a cell phone today. But don't count on them in an emergency. They have serious limitations. When the power goes down, so do cell towers. If the towers happen to stay functional, they usually receive a huge increase in demand during a crisis, making it difficult to get through.

If you do hope to use a cell phone, plan for a way to recharge it. A charger that uses alternative energy, such as battery, crank, or solar power, is essential.

Lighting, Heating, and Cooking

Flashlights

The variety of flashlights to choose from is almost limitless. Flashlights can cost as little as two dollars and as much as two hundred dollars. For more flexibility, look for flashlights that use traditional alkaline batteries and rechargeable li-ion (lithium) batteries interchangeably. Determine how long the batteries will last. Look for LED lights and a variety of output modes.

Alternative-Energy Flashlights

If you want to forgo traditional power sources, look for flashlights that generate energy through solar cells, by squeezing a lever, by shaking, or by turning a crank. Even better, find one that's also a cell-phone charger. The one pictured is the ThorFire solar-powered, hand-cranked flashlight.

Light Sticks

Cyalume Technologies makes six-inch light sticks that chemically create a bright fluorescent light that lasts eight hours. They have a five-year storage life. Be sure to get industrial or military grade, not cheap glow-stick toys. UV Paqlite (pictured) makes a reusable light that provides over ten hours of illumination. Simply recharge it with sunlight.

Solar-Powered Lanterns

The inflatable emergency solar lantern made by MPOWERD Luci (pictured) will run for twelve hours after an eight-hour charge. They are flat, lightweight, and inexpensive enough that you could have several.

PERSONALLY SPEAKING

My favorite backup light source is the Luci inflatable lamp by MPOWERED. It is cylinder shaped, about six inches in diameter, and six inches tall, and makes a great little ball of glowing light. I love to give them as gifts. They also offer a variety of lighting effects, from crystal-clear to soft frosted and candle glow to fun, glowing pastel colors.

Flares and Survival Candles

Flares, typically used in a roadside emergency, will burn for about twenty minutes. Their primary use is for signaling, but they also are an excellent fire starter. Survival candles have been recommended in the past, but they have the inherent risks of an open flame. Still, they will burn for a long time. And, surprisingly, they also offer some heat in an enclosed area.

Survival Candles

Survival candles have been recommended in the past, but they have the inherent risks of an open flame. Still, they will burn for a long time. And, surprisingly, they also offer some heat in an enclosed area.

Emergency Heating

For moderate amounts of heat, you can store a clean-burning catalytic heater with fuel, but be aware that they take up substantial space and weigh quite a bit. Follow the manufacturer's directions.

Matches

Be sure to have plenty of waterproof and windproof matches stored in a waterproof match safe. You may also want a metal match and a magnesium fire starter. A disposable lighter is small and cheap and may come in handy too. Steel wool and a battery are another fire-starting source.

Emergency Cooking

You won't need to cook if you select precooked foods for your emergency evacuation kit. But if you like to eat certain things warm, a small backpacking stove and fuel won't add a whole lot of weight. The versatile Omnifuel Multi Fuel Backpacking Stove by Primus (pictured) works with several types of liquid fuel and is a good choice if you're already a backpacking enthusiast.

Alcohol stoves are small, compact, efficient stoves that use denatured alcohol as fuel and can be used to boil water for warm beverages and reconstitute freeze-dried meals. They work best with a windscreen. Although flammable, if properly stored, denatured alcohol is shelf stable and can last three years. The one pictured is the Vargo Decagon Alcohol Stove.

Security, Money, and Transportation

Securing Personal Documents

What you do with personal documents and valuables will depend on the kind of crisis you expect, but it's always important to safeguard personal documents and valuables, such as credit cards, securities, deeds, contracts, and insurance policies. Protect these items with a heavy-duty metal and watertight military ammo can or a fire safe. Determine which items you'll need to have with you. Make copies of documents and put them in your emergency evacuation kit.

Weapons

A small handgun, such as a .22 or 9mm pistol, may serve you well. There are many choices. Be sure to keep a good supply of ammunition on hand.

Money

Do not overlook having money in your emergency evacuation kit. In a crisis, cash or coins may be the only way to purchase things. Credit-card machines and ATMs may not work if the power or other systems are not operating. Keep a supply of cash in small denominations or rolls of coins because making change might not be an option.

Transportation

Any vehicle you plan to use should contain a small kit with common repair parts and tools (See table 41.2 in chapter 41 for a list of items for a basic, onboard tool kit). Also, get in the habit of always keeping your gas tank at least half full.

Miscellaneous Items

Table 33.3 lists other items you may want to include in your emergency evacuation kit.

Table 33.3 Miscellaneous Items		
· All-purpose knife	· Folding shovel with serrated edge	· Razor blades
· Boots/shoes	· Food for three days	· Resealable bags
· Cable saw	· GI can opener	· Salt tablets
· Change of clothing	· Hat/cap	· Scissors
· Canteen or bottle	· Knife, spoon, fork,	· Sewing-repair kit
· Coat	· Map of evacuation routes	· Sleeping bag
· Compass	· Multi-tool	· Snare wire
· Detergent	· Paper/pens	· Steel camping cup, plate
· Dishpan	· Paper plates and cups	· Survival manual
· Documents and valuables	· Parachute cord, fifty to one	· Water
· Duct tape	hundred feet	· Water-purification tablets
· Emergency blanket	· Plastic cutlery	· Work gloves
· Extra socks	· Plastic drop cloth	
· Folding stove, fuel	· Rainwear	

Where to Keep Your Emergency Evacuation Kit

Again, this is going to depend on your situation, but it's a good idea to have several types of emergency kits you keep in different places

At Home

The most important thing about where you store your emergency evacuation kit is that it's easily accessible. Although a basement storage area may seem like a logical storage place, it may not be accessible in a disaster, such as an earthquake or fire. If you have just a few seconds to evacuate, you may not want to take the time to go to the basement to grab it.

The garage is another common place to store your kit. It will be close to the vehicle you'll be using to evacuate. But be aware that widely fluctuating temperatures will likely affect the quality of the items you store, so you'll want to frequently rotate heat-sensitive items. Keep your kit clean by placing it in a closed cupboard or bin.

In a Vehicle

If you spend a good share of your time away from home, it makes sense to have an emergency evacuation kit in your vehicles. It is especially true if you live where winter storms, earthquakes, or hurricanes can potentially strand you.

At Work

Think about what you'll do if an emergency happens while you are at work. Will it be an option to shelter in place? If so, keep a small, self-reliant emergency kit in your work space. It should include basics that would help you survive for at least twenty-four hours. Also, keep backup survival clothing and durable footwear where you work. (See chapter 35.)

Get-Back-Home Survival Kit

A get-back-home kit can help you get home should you need to evacuate an urban building or other location away from home. You will probably be on foot unless you are a bike rider. Transportation is likely to be disrupted, with streets in gridlock, roads impassable, and mass transit shut down.

If it becomes necessary for you to shelter in a government facility, your get-back-home survival kit will help sustain you.

Basic Items in a Get-Back-Home Survival Kit

Use a small backpack with reflective tape. Table 33.4 lists a few items to consider for your get-back-home survival kit.

Table 33.4 Get-Back-Home Survival Kit		
· Comfortable athletic shoes, socks · Dust mask or bandana · Feminine-hygiene items · Hand lotion · High-energy nutrition bars · Laminated map of routes home · Lip balm	· Package of tissue · Personal medications · Personal wipes · Personal toilet bag · Pocket knife or multipurpose tool · Poncho · Small denominations of cash, and coins	· Small plastic garbage bags with twist ties · Space blanket or another lightweight blanket · Sunscreen · Water

Urban Dangers

Urban survival conditions can be dangerous. There may be dust and fire to contend with. Law enforcement may be stretched thin or nonexistent. You will need to make a calculated decision about your best option if you evacuate. You should have a map of safe routes. In your kit, also include a whistle on a lanyard and mace or pepper spray to deter dogs. Do not display valuables as you evacuate.

Plan Ahead

Plan for which route you would take if you were stranded in a city. Laminate a map of your local area with your route and alternative routes outlined. For safety, plan to travel in groups. Find others who would be evacuating with you who are going your direction, and team up. Make family plans of where to meet, with alternative second and third options in case the first option is not available.

Home Security

There are many things you can do to better prepare your home for times of crisis. Budget limitations may not allow you to do all of them, but the more you do, the better prepared you'll be.

Home Maintenance and Security

There may be extended periods of time when you're the only repair person available should something break down. Acquire a basic home repair book—there are many available. Also, supplies and replacement parts may not be easily available, so plan for that.

Take Care of What You Have

The first thing you can do is to keep what you have now in good repair. Do preventative maintenance on all items possible. This especially includes your home. Paint it when needed and have any roofing, plumbing, heating, electrical, or structural problems corrected.

Maintaining Appliances

Over the last twenty years, circuit boards and electronic controls have replaced mechanical controls on appliances. When the electronic parts fail, they are expensive to replace and not easily repaired. In a prolonged crisis, they would likely be unrepairable and irreplaceable.

If you're concerned about the possibility of needing to be self-reliant for a long period of time, you can take preventative steps to lessen the inconvenience when modern appliances break down.

The solution is to look for and purchase appliances with simple mechanical features that are easy to repair and that require minimal maintenance. Also, purchase spare parts and any special tools needed for repairing mechanical appliances. Appliances are usually not difficult to repair if you have the spare belts, bulbs, cords, switches, etc., they require.

FIVE THINGS YOU CAN DO NOW

1 Conduct a walk-through of your home and list any needed maintenance.

2 Identify one appliance or item that needs to be replaced and make plans to replace it.

3 Identify hand tools that would be useful and acquire one or more.

4 Identify any fire hazards on your property and take steps to remove them.

5 Take precautions to secure utilities and teach all family members the process for proper shutoff.

PERSONALLY SPEAKING

We recently replaced an eleven-year-old, top-brand dishwasher because the electronic part to fix it was half the cost of a new dishwasher. On the other hand, we still own a Maytag washing machine we bought when my youngest daughter was born over thirty years ago. It has no fancy electronic parts and is not very water efficient, but it has only been repaired once and it keeps on washing clothes. In the earlier edition of CPH, Jack recommended purchasing the spare parts and tools needed to repair appliances yourself, but I think you can do that only if you have mechanical appliances rather than electronic ones. Manufacturers still make mechanical appliances. You may not find them on the display floor. Just ask for them.

What to Store

Secondly, besides maintaining what you have, store tools, parts, and supplies for basic home repairs. Also, keep spare parts and repair manuals on hand. Repair manuals can often be found online, along with directions for repairing. If you keep them in electronic format, make sure you'll have a way to view them in a time of crisis.

Hand Tools

Consider investing in hand tools that do not require electricity. Buy quality tools that will last. Although dependent on a power source, power tools are also important because of their efficiency.

Tables 34.1 lists basic tools that will allow you to handle simple maintenance, minor repairs, and modest building projects. You need to decide if you'll want heavier tools, such as axes, picks, shovels, wood mauls, and wedges.

Depending on the extent of your desire for self-reliance or if you have a skill or want to develop one, you may want to acquire additional tools for woodworking, carpentry, blacksmithing, metalworking, masonry, welding, small-motor repair, etc. Determine your priorities and consider those that would be useful during an extended crisis.

Maintaining Tools

Tools must be properly stored to prevent rust and other problems that could destroy their usefulness. First, control the humidity if you can. Second, keep your tools and shop clean. And third, you may want to coat tools with protective coverings, such as mineral oil or a synthetic conditioner. Store tools in a dark, dry location where they can remain straight, or hang them by their handles.

Home Security

It's logical that if you go to the effort to acquire and store items that will ensure your physical well-being, you'll want to maintain the security of your home, protecting them from fire or theft. Of course, you'll want to take precautions to protect your life and the lives of your family. There are several safeguards you should take to prepare for the security of your home.

Table 34.1		
Basic Tools for Home Maintenance and Repair		
Hand Tools		

· Block plane, 7"
 · Bolt cutters
 · Bubble level, 2'–4'
 · Chisels, assorted cold and wood
 · C clamps, 2 pairs of 4" and 10"
 · Claw hammer
 · Files, assorted
 · Framing square, 16" x 24"
 · Hand drill with bits
 · Hex-key sets
 · Measuring tapes, 12', 25', and 50' steel
 · Metal tin snips
 · Paintbrushes, assorted
 · Pliers
     Groove-joint, 10"
     Lineman's, 7"
     Needle-nose, 6"
     Slip-joint, 6"
	· Propane torch
	 · Protective goggles
	 · Putty knife
	 · Sawhorse brackets
	 · Saws
	     Crosscut carpenter saw, 26"
	     Rip saw, 26"
	     Hacksaw with assorted blades
	 · Screwdrivers
	     Flat-head, assorted
	     Philips, assorted
	 · Sharpening stone, combination
	 · Sledgehammer, small
	 · Staple gun
	 · Utility knife with spare blades
	 · Vise grips, 7" and 10"
		· Wall scraper, 6"
		 · Wheelbarrow
		 · Wire brush
		 · Work apron
		 · Wrecking bar, 30"
		 · Wrenches, assorted
		     Adjustable, 8" and 12"
		     Combination, ⅜"–1¼"
		     Pipe, 10" and 14"
		     Socket, ⅜"–1¼"
		
		 Power Tools
		
		 · Circular saw
		 · Corded or battery-operated drill
		 · Impact drill
		 · Reciprocating saw

Fire Protection

First, make your home as fireproof as possible, then install several smoke alarms, both electric-powered photoelectric and battery-powered ionization. The key to the proper location of smoke alarms is to place them between hazards and people. That usually means in central hallways outside bedrooms. Test all units after installation and at regular intervals. Also, install carbon monoxide detectors.

Firefighting Equipment

Next, you should invest in firefighting equipment. Start with some ABC multipurpose, rechargeable fire extinguishers with the highest ratings at the heaviest size you can handle (probably ten to twenty pounds of agent). Put one near the kitchen, one in a major hallway, one in the garage, and perhaps a smaller one in your automobile.

A home-rechargeable water unit would be a good backup. Keep a garden hose and nozzle hooked up, ready for use. Have fire ladders available for all locations higher than the ground floor. Emergency escape masks and Water-Jel Fire Blankets are additional precautions. These blankets work quickly to dissipate heat and protect much longer than a regular blanket soaked with water.

Without ready access to the services of a functioning fire department, you will need more extensive firefighting capabilities to handle larger fires, both accidental and intentionally set.

Fire Management in Forest and Brush Areas

Every year we are reminded of the devastation caused by forest and brush fires to rural areas and even large suburban lots. The risk is real in susceptible areas. Prevention is key to your home surviving a raging fire.

Fire Prevention

Homes built in low-density developments, on steep slopes, in wind corridors, and surrounded by wildlands are more at risk for fire. Surprisingly, homes are more susceptible to fire where the native trees and shrubs have been cleared and exotic grasses and weeds have invaded.

Windblown embers are the most likely source of fires. Again, a large, cleared perimeter is not a protection. Instead, your home becomes an ember catcher. It is better to clear a relatively small thirty-foot perimeter, with an additional seventy feet partially cleared, where embers can be "caught."

It is critical to invest in class-A roofing and ember-resistant vents. It is especially important to remove debris from gutters and locate wood stacks away from structures. Avoid locating wood fencing and overhanging trees near structures.

Earthquake Preparation

Most people, even in earthquake-prone areas, are not prepared for them. Thankfully, there are several simple, low-cost measures that can be taken now to decrease the possibility of damage and injury from a major earthquake.

Protect Yourself During an Earthquake

First, learn what to do in an earthquake to best protect yourself. If indoors, do not run outside but get under a doorway or sturdy piece of furniture, against an inside wall in a central hallway or bathroom, or in a corner. Move away from heavy furniture that could fall on you.

One of the most common injuries caused by an earthquake is from breaking glass. Flying glass shards can be deadly. Stay away from windows. At the very least, face away from them. Keep slippers near your bed so that if an earthquake occurs at night, you can walk without cutting your feet on glass shards.

Think about and discuss the effects of an earthquake with your family and friends. Consider how you would react in your situation. What can you do to minimize the risk at home, at work, and while driving in an automobile? Make your home and its contents as resistant to earthquakes as is practical.

PERSONALLY SPEAKING

Living a good share of my life in the Pacific Northwest and along the Wasatch Front in Utah, the possibility of an earthquake has always been a cause for uneasiness. Not long ago, there were a few small-scale earthquakes in the Salt Lake City area that heightened awareness. People are now suggesting not just having a 72-hour kit, but having a *72-second kit*. The items in this kit will help you respond immediately to the mayhem following a large-magnitude earthquake. A good place to store this kit is in a tote under your bed so you have easy access to it in the middle of the night.

72-Second Disaster Kit

What will you need immediately following an earthquake? How will you respond if an earthquake hits in the middle of the night? See the Quick Check for ideas of items to include in a *72-second* disaster kit.

Reinforcing Structures

What type of structure do you live in? A carefully engineered and properly constructed wood-frame home is among the safest due to its built-in give. An unreinforced masonry home can suffer much greater damage. Windows shatter from bending unless frames have been braced with plywood sheathing. You can bolt your home to its foundations with steel sill anchors either during construction or by retrofitting. Brace masonry chimneys by tying them to the frame with steel straps and nails. Place safety film on one side of large windows.

QUICK CHECK
72-Second Earthquake Disaster Kit
✓ Sturdy shoes
✓ Socks
✓ Dust mask
✓ Goggles
✓ Work gloves
✓ Light stick on lanyard
✓ Flashlight
✓ Pry bar
✓ Portable kinetic radio
✓ Whistle or signaling device
✓ Bottles of water
✓ Contacts information
✓ Money in small denominations

Securing Gas Appliances

Gas meters are especially dangerous because the connecting pipes may rupture and cause gas leaks. Begin by securing your gas water heater by bolting it down or securing it to the wall with heavy-gauge steel strapping. This may prevent it from tumbling over and breaking water and gas lines, which may be ignited by the pilot light. Do the same with gas furnaces and other gas appliances. Also, install an earthquake-proof natural-gas shut-off valve and flexible tubing and connections on gas appliances.

Securing Loose Objects

Use camper-type latches on cupboard doors, store heavy items on lower shelves, and secure top-heavy furniture to wall studs. Secure storage shelves with lips or bars to help prevent cans and jars from falling. Properly hang heavy mirrors, paintings, and light fixtures. Locate beds away from windows and heavy furniture.

Disaster Preparation

There are many things you can do to protect yourself and better prepare your home from the physical effects of natural and manmade disasters.

Utility Precautions

One of the first things to do in preparing for any disaster is to learn how to shut off your gas and water. Place a natural gas shut-off tool or wrench where all responsible members of the family can find and use it quickly. Ruptured gas lines can explode and cause fires. Broken water mains can flood and introduce contaminants. Teach members of the family how to shut off water to prevent contamination and how to preserve the water in your lines. Severed electrical wires can electrocute and start fires. Teach all family members to treat any downed wire as a live wire.

Disaster Materials to Store

Keep a good supply of sandbags and pumps if flooding is a possibility. Store sheets of plywood and heavy-duty plastic for tornadoes, hurricanes, and the aftermath of earthquakes.

Emergency Shelter

If your home is destroyed or made uninhabitable by a disaster, make sure you have an alternative shelter. This could be a trailer, mobile home, camper, tent, or even just a canvas or nylon tarp. Another suitable temporary shelter is a tipi, like the Plains Indians used. Even in the dead of winter in a cold climate, it can be kept tolerable inside with a small fire. If you are forced to live for any length of time in a tent, you will appreciate cots or ground cloths with air mattresses or pads.

Securing Your Home

Sadly, through necessity, our homes are becoming more like fortresses. Although it is impossible to make your home impregnable, you can make it a lot harder to get into.

Doors

For the most security, exterior doors should be heavy-gauge steel with no glass, mounted tightly in a steel frame with inside or pinned hinges. Install quality door locks and dead bolts and wide-angle door viewers. Secure sliding glass doors with locks made especially for that purpose. Especially pay attention to outside basement doors and weak garage doors.

Windows

Windows are a vulnerable part of a home. Windows made of tempered glass are four or five times stronger than annealed or standard glass windows. Tempered windows are required by code in some places in a home, like in bathrooms or near doors. Tempered glass is more expensive but considerably harder to break. If tempered glass breaks, it crumbles into small pieces, whereas standard glass shatters into dangerous shards.

Hurricane-resistant or stormproof windows are made with a covering of polyvinyl butral (PVB) or ethylene-vinyl acetate (EVA). They can still shatter if struck by a flying object but will break in a spider-web pattern rather than flying shards. In a hurricane, these windows not only protect against water damage but also help maintain a home's pressure and keep it from collapsing.

Windows can also be protected with plywood—five-eighths-inch is recommended. The plywood can be attached with hurricane clips.

Windows can be protected from intrusions with bars and grids. These do not have to be fortresslike but come in many decorative options.

Alarms

Install burglar alarms and other intruder-detection devices. Discourage intruders by properly locating and trimming trees and shrubs, using blinds on windows, and providing nighttime illumination of your entrances and yard

Engrave or otherwise mark valuables for identification.

Hidden Rooms

Somewhat whimsical, these rooms have become popular in high-end homes. They are expensive. But hidden rooms also have a place if you want to hide valuables, food, and emergency supplies in your existing home. For survival purposes, there are advantages to making it a DIY project.

Safe Rooms

Depending on your survival philosophy, you may want to consider a safe room or panic room. Safe rooms are designed to withstand hurricanes, tornadoes, and chemical or biological contamination. Plan on ten square feet of space per person. The amount of time you expect to be in a safe room will determine the supplies you'll store there.

In theory, a safe room is also a place you can retreat to if your home is invaded. It supposes a serious breakdown in society. If not used wisely, however, it can become a trap. Thoughtfully research and consider your options.

Protecting Valuables

Hiding Valuables

There is nothing like an ostentatious display of goods to make you a target of looters during a crisis. The old cliché "Out of sight, out of mind" is a good guide for protecting your stockpile and other valuables. You can use hidden rooms and disguised closets, and safes and vaults.

Entire books have been written on how to safely hide items. If you have a lot to hide, get one of them. One book you might check out is *DIY Secret Hiding Places* by Steve Plant (2015). Burying things in PVC pipe "safes" is an option. When they are buried six feet deep or so, even metal detectors have a hard time locating them.

Safes and Vaults

Safes are tested and rated for both fire resistance and burglar resistance. Look for a safe that is fire-resistant at least up to an hour. For more details about selecting a quality safe refer to "Top 10 Things We Recommend before Buying a Safe" by The Safe House or "Top 10 Things You Must Know before Buying a Safe" by Safe and Vault.

QuickSafes makes two ingenious hiding places that use RFID (radio frequency identification technology). To open the safes, you simply touch the RFID key to the hidden access point. The QuickVent Safe looks like an ordinary heating vent. The QuickShelf Safe is "hidden in plain sight" as a decorative wall shelf.

Guard Dogs

A well-trained guard dog is a substantial deterrent against intruders. Whichever dog you choose, you must invest the time, effort, and money to properly train it. Each breed has unique characteristics, so it is critical to educate yourself and find the breed that will meet your needs.

Larger dogs are usually more intimidating. German shepherds, Belgian Malinois, Doberman pinschers, Rottweilers, Airedales, bullmastiffs, boxers, Rhodesian Ridgebacks, and Bouvier des Flandres are breeds to consider.

If you're going to spend money to buy a dog, take the time and effort needed to make sure it's properly trained and poison-proofed—training it to only take food from its own dish or when commanded. Most problems with dogs are caused by pet owners who do not properly train their animals. Also, get it vaccinated for rabies, distemper, and parvovirus. Store proper food and medications.

Survival Homes, Homesteading, and Retreating

Building a Survival Home

As you can see, preparing your home for crisis can be quite involved. If a survival home or retreat is of interest to you, refer to *The Secure Home* by Joel M. Skousen for more information. Joel covers not only how to remodel your present home to make it suitable for survival but also how to build a survival home from scratch. Sections cover planning, integration of security and self-sufficiency systems, construction and implementation, and recommended equipment and suppliers.

Homesteading

There is quite a bit of crossover between preparedness and survival and the homesteading movement. Extensive research by Brigham Young University has shown that a family of six can be almost completely self-sufficient on two and one-half acres of decent farmland with a dozen chickens, a similar number of rabbits, and two milk goats.

Retreating

Some locations will be better than others during a crisis. Yet, determining the best areas cannot always be determined from maps of population density, earthquake zones, and fallout patterns. Certainly, large population centers must be among the worst under any scenario. Small towns away from large urban areas are a better option in all but the worst cases.

However, most people desiring to prepare for a crisis cannot or do not want to move from their present location. That leads to the option of having a retreat located away from the normal habitation.

To Retreat or Not to Retreat

The subject of having a retreat is controversial. Some feel it is the only viable alternative, while others question its value. You can research further on various internet sites to help make your decision.

If the retreat is in a small rural community far enough away from your present home to do any good, what is the probability you'll safely reach it in a crisis? Will you be able to drive through roads jammed with frightened, angry, and most likely violent people? Is your vehicle capable of traveling off-road? Will there be enough fuel to get to your destination? And will you find someone else already there, willing to fight for what they have claimed possession of?

So, getting to a distant retreat in a time of crisis could be difficult at best and depend on a great deal of luck. And making the decision to leave for the retreat would be a most difficult one. On top of that, a working retreat takes years to adequately develop. Storing all your survival gear in your current home, packing it all up, and taking it with you when the crisis develops would be a daunting task.

Relocation

Some survivalists believe the only safe, sane solution for those living in major metropolitan areas is to decide now, no matter what the resulting difficulties, to relocate to a suitable location as soon as possible. This would most likely require substantial changes in lifestyle and a reevaluation of personal values. Lots of people have made this kind of change and have established a simpler lifestyle and learned unique ways of supporting their families. Based on your expectations for the future, you'll have to decide if relocating is a realistic move for you.

Survival Communities

Survival communities or compounds present an option for surviving societal collapse with a like-minded group of people. If this interests you, look for one whose philosophy aligns with your own. Some require a yearly membership per person, almost like a time-share.

Different Point of View

The book *The Survival Retreat* by Ragnar Benson will give you information about the retreat option. He's also written *The Modern Survival Retreat* from a different perspective.

See page 432 in the resource section for more sources of information about preparing and securing your home for a crisis.

Clothing and Bedding

Clothing and bedding are necessary to shelter and protect the body from cold, heat, wind, rain, sun, and injury. In a survival situation, without the civilized comforts we depend on, there will be a greater burden placed on the clothes we wear and the bedding we sleep in.

Survival Clothing

Most typical American adults probably have enough clothes in their closet to last the rest of their lives, literally. Although you may already have a good supply of clothes, there is a good chance you will need different types of clothes for survival conditions. While it's always a good idea to have an extra supply of the clothing you normally wear, the focus of this chapter is basic survival clothing.

Survival clothing should be durable, warm, and comfortable. You will likely need several options to suit your situation, climate, and seasons, and you should have enough survival clothes to last a year. Function obviously will be more important than style and appearance.

Traditional Outdoor Clothing

Good-quality outdoor clothing is often the best to stockpile. Natural fabrics, like wool and cotton, are more absorbent, and wool retains its insulating properties even when wet. Tightly woven fabric is more snag and tear resistant and wears longer, but loosely woven fabric can offer more insulation. A few quality brands of traditional clothing are Filson, Pendleton, and Woolrich.

FIVE THINGS YOU CAN DO NOW

1 Inventory cold-weather clothing for your family, making sure each person has warm and water-resistant clothes for the following winter.

2 Check that you have durable, rugged, waterproof footwear. Look for ways to dovetail recreation and preparedness uses. Children need decent footwear too.

3 Be on the lookout for a variety of gloves and hats for multiple uses in both warm and cold weather.

4 Inventory bedding. Acquire enough that your family could keep warm if there were little or no heat in your home. Consider purchasing several sleeping bags at discounted prices to keep on hand as extras.

5 Review the sewing tools and supplies in tables 35.3 and 35.4. Select those you think would be valuable to you and store them in a plastic container.

Cold-Weather Outerwear

Consider what special adaptations you will need for a cold climate. You will most likely want more warm clothing added to your list. Approach these two ways. You can choose heavily insulated garments or use the principle of layering.

Down Insulation

Down, an excellent insulator when dry, comes primarily from geese and ducks and is the fluffy, wispy filaments found underneath their feathers. Down is rated by its loft—the higher the loft, the higher the insulating ability. Goose down has a higher loft than duck down, but it is more expensive. Down is very compressible, and a large jacket or sleeping bag can be compacted into a small stuff bag. Down is also expensive and loses its insulating ability when wet. However, an alternative is hydrophobic down which is treated to shed water.

Synthetic Insulation

Synthetic insulation, like down, traps the body's warmth in its soft, billowy fibers. And it insulates even if wet. It is less expensive than down but weighs more, for the same insulating capability. Surprisingly, it is not as durable as down and loses its insulating ability each time it is compressed in a stuff sack.

Recreational Outdoor Performance Apparel

Many outdoor recreational companies use high-performance fabrics to produce quality apparel. They are often made of high-tech synthetics and offer higher strength, resist abrasion and mildew, wick away moisture, wash better, dry more quickly, and provide excellent warmth. Clothing made from fabric with 50+ UPF sun protection should also be considered for some survival uses.

Gore-Tex

Gore-Tex is a microporous fabric that is both waterproof and breathable. It is preferred for outerwear and footwear.

Fleece

Polyester fleece is an insulating fabric that is lightweight, retains warmth, and can be inexpensive. It comes in a variety of weights and is used for lightweight jackets and often layered with other clothing. Polar Fleece is the original brand name.

PrimaLoft

PrimaLoft is a microfiber thermal insulation used in jackets and sleeping bags. Unlike natural down, PrimaLoft provides warmth when wet. Products using PrimaLoft are found in high-end brand names but also in company brands like Cabela's and L.L. Bean.

SmartWool

SmartWool is a product made from merino wool and treated to make it itch free and resistant to shrinking. It makes a comfortable, warm underlayer and is also used to make socks for multiple uses.

Thinsulate

Thinsulate is another synthetic thermal insulation that provides warmth without bulk. It retains warmth when wet and is less expensive than down. It commonly comes in 100-, 150-, and 200-gram weights and in a product called high-loft. It is used as insulation in gloves, jackets, and boots.

Basic Clothing for a Year

Stockpile clothing like you would any other commodity, using it and replenishing it. Always have one or two of each important item in reserve. If outdoor recreation is part of your life, you should have little trouble mindfully stocking up on the clothing you would need in a crisis.

Table 35.1 is a generic list of the minimum suggested amounts for basic clothing items. It is intended as a starting point. Use it as a checklist for items you may need to acquire. Clothing is very personal, so be sure to adapt it to your situation.

Table 35.1 Basic Survival Clothing per Person for One Year (Adapt to Age, Gender, Climate, and Lifestyle)	
· 8 sets underwear · 2 sets long underwear · 1-2 pajamas or nightgown · 1 warm robe · 2 T-shirts · 2 cotton turtle-necks · 2-4 work shirts (chambray, etc.) · 2 cotton flannel or chamois shirts · 2 heavy wool shirts · 1-2 sweaters · 4 pairs jeans, pants, overalls · 1 pair heavy wool pants or snow pants · 1 water-repellent windbreaker or slicker	· 1-2 lightweight jackets · 1-2 winter work coats · 1 heavy-duty winter parka · 1 straw hat · 1 knit cap or balaclava · 1 heavy-duty work belt or suspenders · 2 pairs leather work gloves · 2 pairs winter gloves or mittens, inserts · 2 pairs work shoes or boots · 1 pair waterproof boots/overshoes · 2 pairs shoelaces for each pair of boots/shoes · 12 pairs socks (8 light, 4 heavy)

Survival Footwear

A good case can be made that the single-most-important survival item, after food and clean water, is good footwear. Shoes are almost always in short supply during times of crisis. For example, George Washington's Continental Army suffered greatly, among other things, due to a lack of shoes.

Durable Shoes and Lightweight Boots

The everyday survival shoes or boots you select must be both comfortable and tough. A new shoe or boot should fit comfortably and not have to be broken in. You may need to go up a size to give your feet plenty of room and to accommodate heavy-duty socks.

Buy the best quality shoes or boots you can afford. Have two pairs and alternate between them to let them dry out and prolong their life. If you will deal with inclement weather, the upper part of a shoe or boot should be insulated and waterproof. They are best made from sturdy leather or a padded, heavy-duty breathable synthetic material, like Gore-Tex. For long wear, look for rugged construction and durable soles, such as Vibram. Hiking shoes or cross-trainers are both good, lightweight choices.

Heavy-Duty Boots

You may also want a pair of heavy-duty work boots or extra-sturdy hiking boots. Look for work boots with soles that repel water and oil and are slip and electric-shock resistant. Although very protective, these boots are also less flexible and heavy to walk in.

Other features to look for are gusseted or full-bellows tongues to keep dirt out and steel or composite shanks to protect the foot.

Winter Boots

If you expect to work in the cold and snow, you'll want insulated boots or rubber-bottomed boots with spare felt liners. Foam-rubber insoles will cushion the foot, and thick felt insoles will help insulate it from the cold.

Socks

A good pair of socks should offer extra padding and keep the feet dry. Socks should also dry quickly and work equally well in cold or hot climates. Socks made from merino wool are an excellent choice. They are durable, antibacterial, and wick away moisture.

Shoe Accessories

You can purchase ready-made shoelaces or nylon or leather shoelace yardage and cut it to fit. The length of shoelace needed for a pair of shoes can be estimated by counting the number of eyelets or hooks on one side and multiplying by six inches. Also store foot powder and moleskin to help prevent blisters.

Gloves and Headwear

Gloves

Do not overlook gloves. Have a good supply and a variety of leather work gloves and flexible nitrile gardening gloves to protect your hands. They must have a comfortable, snug fit. For cold weather, you'll need gloves and mittens designed to keep your hands warm.

Select reusable rubber or synthetic gloves to protect your hands when using caustic chemicals or skinning and dressing animals that may be infected with disease. Also, have a supply of gloves in case you need to clean or handle waste materials. They should be tough but not too expensive since you may not want to reuse them.

PERSONALLY SPEAKING

I think you should pay as much attention to quality, proper-fitting work gloves as you do the right pair of work boots. I have learned that I prefer a softer, supple leather with smooth seams to protect my hands against blisters, splinters, cuts, and scrapes. I also like them to fit my hands closely—a sloppy fit makes it hard to use my fingers, and I end up frustrated and removing the gloves. My favorite gloves are made from deerskin. In winter, if I just need to keep my hands warm, I prefer quality mittens—my fingers need each other to stay warm. I also want stretchy wristbands that are long enough so there is no gap between my mittens and jacket sleeve.

Hats

Be sure to have hats for protection from the heat extremes as well as from excess exposure to the sun and rain. Your head should especially be protected in cold because about 10 percent of body heat leaves through the head.

Balaclava

A balaclava is a knit hat that extends to cover the face, protecting it from wind and cold. Be aware that some jurisdictions prohibit the use of face coverings in public places.

Bandana

Bandanas are versatile and inexpensive squares of cotton fabric that have limitless uses. They protect from dust and sun and act as an evaporative cooler when wet. They can be used as a sweatband, face mask, tourniquet, pot holder, sling, pouch, neck gaiter, etc.

Scarves

A scarf, larger than a bandana, is likewise versatile and may be made from cotton, wool, silk, or synthetics and protects from sun, wind, and rain. The best fabrics are comfortable against the skin, such as North x North's large 40" x 40" merino wool kerchief. Coolibar's 24" x 24" scarf, designed for high-performance, gives UPF 50+ sun protection. The traditional Middle-Eastern shemagh, made of cotton and traditionally wrapped around the head and neck to keep out sand dust, can be used as a towel, ground cloth, blanket, sling, etc.

Clothing for Infants and Children

Infants

Plan for disposable diapers in your emergency storage supplies and evacuation kit. They can also be used for bandages and when someone experiences intestinal distress.

In a crisis, you may eventually run out of disposable diapers. As a backup, store three dozen cloth diapers and sixteen to twenty-four plastic pants in assorted sizes through toddler. In lieu of diapers, store twenty-five yards of twenty-seven-inch diaper flannel. You will need pins for fastening cloth diapers.

A minimum of ten yards of cotton flannel will make an adequate supply of blankets and baby clothes. Baby sleepers take about one yard of sixty-inch stretch terry or cotton knit.

You might also want to store heavy blankets, blanket sleepers, and a snowsuit for colder climates. These will also help in unheated homes.

Children

It is difficult to store enough sizes to properly match growing children. Buy loose-fitting clothing that can span several sizes. Children should have comfortable walking shoes and socks.

Caring for and Storing Clothing

Washing Clothes

If the power is out, your electric and gas washers and dryers won't do you much good. Of course, a clothesline and clothespins can replace the dryer. Washing options will take you back to the nineteenth century. The first option is washboards and washtubs, sometimes called utility tubs. A clothes wringer will also be helpful. A second option is a wringer washer, which is a half barrel on legs with an attached handle and wringer. Lehman sells one for about $600.

Estimate the amount of laundry detergent needed based on your current usage. You will probably want to store bleach and perhaps some bluing to keep whites white.

Caring for Footwear

Footwear will last years if kept clean and dry, and shoe trees will help footwear retain its shape. Store leather preservatives and waterproofing as well as paste wax for polish. Most footwear is made from chrome-tanned leather and needs a protective, wax-based treatment, like Sno-Seal and Outback Leather Seal, or a protective silicone treatment. Oil-based compounds are used for the oil-tanned leather found in some boots and belts, but be careful because they will weaken the glue on footwear with cemented soles. NeatsFoot Oil and Mink Oil keep leather soft and pliable. Saddle soap cleans and softens leather. For simple repairs, store Barge Contact Cement and thinner.

Storing Clothing and Footwear

Clothing, fabrics, boots, and bedding should be stored clean and in a cool, dry, and dark environment. Heat will crack leather, too much moisture can cause mildew, and light deteriorates some materials and fades colors. Mothballs or crystals can be used to protect woolens from moths and other insects, and desiccants can be used to reduce moisture.

Use acid-free tissue paper for wrapping, and plastic bags to keep dust out. A cedar chest with a tight-fitting top is among the very best storage containers. Sleeping bags and clothing made with down should be stored loose and not in stuff sacks. Label containers with contents.

Sewing Clothing

As an alternative to stockpiling clothing, you might store the fabric and other necessary supplies to sew clothes.

Ready-to-Wear versus Home-Sewn

The advantage of storing ready-to-wear clothing, of course, is that it can be used immediately. The advantage of storing fabrics and the necessary supplies is that fabrics can cost one-third to one-half as much and can be made as needed into the items and sizes required. Sewing clothes is especially advantageous for families with growing children if you are unable to purchase clothes during a prolonged crisis. On the other hand, sewing clothes requires a sewing machine, supplies, time, and a reasonable skill level.

Estimating Fabric Yardage Requirements

Woven cotton fabrics are usually 45-inch wide. Woven wool and home decorating fabric is usually 54-inch wide. Cotton and synthetic knit fabrics are generally 60-inch wide. Patterns can often be arranged with less waste on wider widths of cloth.

An estimate of fabric requirements can be determined based on a person's body measurements. For example, calculate the yardage requirement for an adult shirt by adding as follows: (length of front of shirt) + (length of back of shirt) + (2 x the length of arms) + (collar and cuffs). For pants, calculate by measuring the length from the waist to the floor, adding a hem allowance, and doubling that measurement.

For more exact requirements refer to table 35.2. Most of the fabric requirements in the table are based on 45-inch wide fabric. If fabric is 54-inch to 60-inch wide, you will need about two-thirds as much. Actual patterns will give you more precise amounts.

Table 35.2 Yardage Requirements for Basic Clothing						
Items of Clothing	Infants	Children	Youth	Women	Men	Suggested fabrics
*Underpants		1/4	1/2	1/2–1	1/2–1	Cotton knit, tricot
Nightgown (long)	1	2–3	3 3/4	3 3/4–4 1/4	3 3/4–4 1/4	Flannel, cotton knit
Pajamas		1 3/4–2 3/4	3 1/4	3 3/4–4 3/4	4 1/2–5	Flannel, cotton broadcloth
Robe (long)		1 1/2–2	3 1/4	4–4 3/4	4–4 3/4	Flannel, terry, quilted fabric
*T-shirt (short-sleeved)	1/2	1/2–3/4	1	1–1 1/2	1 1/4–2	Cotton, cotton-polyester knits
Shirt (long-sleeved)		1 1/2–2	2 1/4	2–2 1/4	2 1/4–2 3/4	Cotton and wool flannel, broadcloth, chambray
Pants	3/4	1–1 1/2	2 1/4	2 1/2–2 3/4	2 3/4–3	Denim, corduroy, cotton duck, poplin, sailcloth
Overalls	1 1/4	1 1/4–2	2 3/4	3 1/4–3 1/2	3 1/2–4	Denim, corduroy, cotton duck, poplin, sailcloth
Jacket	1	1 1/4-1 3/4	2 1/4	2 1/4-2 1/2	2 3/4-3	Denim, corduroy, wool, poplin
Coat	1	1 1/2-2 1/2	3	3 1/2-4	3-3 1/2	Wool, corduroy

* Based on 60-inch-wide fabric. Knit fabrics are usually 60 inches wide. Woven wool fabric is usually 54 inches wide. Other clothing fabric is usually 45 inches wide.

Storing Fabric

Focus on practical fabrics such as denim, cotton and wool flannel, broadcloth, corduroy, unbleached muslin, and various pant-weight cottons. Terry cloth is good for bath towels, while muslin is used for dish towels and bandages. A supply of leather can also be useful. Thru-Hiker makes kits for down jackets and other outdoor gear. Also, store fabric that can be traded for the services of a seamstress.

Sewing Machines

A typical sewing machine requires electricity, which may make it nonfunctional in a prolonged crisis if electricity is unavailable or limited. Although hand sewing can be just as strong and durable, it is a less efficient option.

Treadle Sewing Machines

Another alternative to an electric sewing machine is a foot-powered treadle sewing machine. Janome makes a nonelectric machine, but it does not come with a treadle. Treadles and cabinets may be purchased from Lehman's, or you may find a vintage treadle machine to refurbish.

It is also possible to adapt your electrically powered machine to a hand machine with a pulley drive. For heavy-duty sewing on canvas and leather, you might consider an electric commercial or upholstery sewing machine.

Sewing Tools and Notions

To be successful with home sewing, you'll need proper tools as well as sewing notions. Table 35.3 lists basic sewing tools, and table 35.4 lists basic sewing notions.

Table 35.3 Basic Sewing Tools		Table 35.4 Basic Sewing Notions	

· Basic patterns
· Bobbins
· Cutting mat
· Hand needles, sharps
· Pins, sharp, colored pinheads
· Pincushion
· Rotary cutter and blades
· Safety pins
· Scissors
    Embroidery
    Dressmaker's
    Pinking
	· Seam ripper
	· Sewing gauge
	· Sewing-machine needles
	· Sewing-machine oil and brush
	· Tailor's chalk
	· Tape measure
	· Thimbles
	· Thread, all-purpose
		· Bias tape
		· Buttons
		· Elastic
		· Grommets
		· Hem tape
		· Hooks and eyes
		· Interfacings
		· Iron-on patches
		· Linings
			· Nylon tape fasteners
			· Rivets
			· Seam tape
			· Snaps
			· Thread, assorted colors
			· T-shirt ribbing
			· Twill tape
			· Zippers

Knitting and Crocheting

Store yarn and the proper needles if you want to be able to knit mittens and sweaters. Hats, gloves, and scarves take about four ounces of a three-ply sport weight yarn, while socks take eight ounces. Sweaters require 1,500 to 2,000 yards of average worsted weight yarn, depending on the gauge of the stitch and the size of the sweater.

Production Skills

Although beyond the scope of this book, you may want to consider developing skills in furniture making, weaving, spinning, textile making, tanning, leatherworking, shoe repair and shoemaking, rug making, basket making, rope making, etc. If a production art appeals to you, look for instructions and advice online or from guilds or support organizations.

Bedding

You should have enough bedding to keep each person warm if there were no heat. That could be three to four good blankets per bed or more. See table 35.5.

Table 35.5 Bedding Recommendations	
· Bedspreads	· Rubber sheets
· Blankets	· Pillows
· Comforters	· Pillowcases
· Duvet covers	· Quilts
· Mattress pads	· Sheets
· Mattress protectors	

Sleeping Bags

Sleeping bags make excellent emergency bedding for unexpected guests. If you use liners, you can keep them cleaner, and they will last longer. Down is warm but expensive and, as mentioned, loses its warmth when wet. PolarGuard is a quality synthetic insulator. Quallofil and Hollofil were among the first synthetic insulators but are heavier, bulkier, and less-effective insulators. Synthetics dry quickly and still insulate when wet.

Homemade Quilts and Comforter

Quilts quilts and comforters are traditionally made from flannel, cotton, and muslin. Comforter covers can be made from bedsheets and home-decorating fabrics. They can also be made from cloth scraps, recycled clothing, and salvaged parts of worn-out blankets. See table 35.6.

Batting

Batting can be purchased in prepackaged sizes or by the yard at fabric stores or online. Store half-inch batting for quilts and one- to one-and-a-half-inch batting for comforters.

Table 35.6 Yardage Requirements for Standard-Sized Comforters				
Size of Bed	Comforter Dimensions	45-inch fabric	54-inch fabric	108-inch fabric
Receiving blanket	36" x 36"	2 yards		
Crib	28" x 52"	3 yards	3 yards	1 yard
Twin	68" x 86"	10 yards	7.5 yards	4.75 yards
Full	78" x 86"	10 yards	7.5 yards	4.75 yards
Queen	86" x 90"	10 yards	10 yards	5 yards
King	102" x 90"	12.5 yards	10 yards	5 yards

Quilting

A quilting frame makes quilting so much easier. You will also need #5 to #8 quilting and darning needles and thread for quilting. Use decorative threads, yarns, and strings for tying quilts and comforters.

Alternative Insulation

Spare blankets could be used for warm robes or to line jackets and coats. Although it's not ideal, newspaper can be used for emergency insulation if necessary.

36 Heating, Cooking, Lighting, and Refrigeration

Energy is a critical concern for survival during crises. It will be in short supply during an extended power outage or prolonged disruption of distribution channels.

If you live in a cold climate, you could be faced with trying to heat your home without the benefit of natural gas, fuel oil, or electricity. Wherever you live, it could be a challenge to cook food and light your home at night. The list of appliances requiring electricity is extensive. While some are conveniences, others, like refrigerators, hot-water heaters, and radios, are more essential.

You will need contingency plans for energy shortages or if energy sources are nonexistent. This area of preparedness can be very complex. Your plans will depend on your circumstances, desires, and perceived needs. Although a camp stove with a small supply of fuel and battery-powered lights may be sufficient for short-term emergencies, complete energy independence will require much more extensive planning and preparation.

This chapter covers possible alternative methods of heating, cooking, and lighting. You will have to decide how extensive your preparedness needs to be, but you should have at least one backup method for each energy need. Redundancy is always desirable, so two or three methods are even better.

Heating with Wood or Coal

First, assess what your situation would be if your normal heating system failed. Realize that most modern systems operate on things—like thermostats, fans, blowers, fuel injectors, and ignition switches—that require electricity. Also, in a home with little or no heat, the water pipes in the home may be at risk for freezing.

FIVE THINGS YOU CAN DO NOW

1 Stockpile fuel for an existing heat source. If possible, acquire enough for the winter season.

2 Decide on an alternate heat source for your home—at the minimum, a way to heat one room. Obtain adequate fuel.

3 Purchase a Dutch oven and try it out. Purchase a Dutch oven cookbook or gather recipes online. Store unused briquettes in a plastic tub.

4 Try cooking using solar energy. Experiment with different solar-cooker designs.

5 Research different options for emergency lighting, then decide which kind of nonelectric lighting you want to purchase. Become familiar with how to use it and purchase any batteries or solar panels needed.

Wood- and Coal-Burning Stoves

One potential solution for heating large areas in a home during a crisis is an airtight woodstove or coal stove/furnace. Wood and coal heaters have an advantage in a prolonged crisis because they do not rely on the power grid or gas utilities. Wood-burning stoves are common primary and secondary heat sources in parts of the United States where wood fuel is inexpensive and easy to obtain. In 2015, the EPA passed strict regulations on wood-burning heaters, and new stoves were required to meet these standards.

Coal is no longer a common fuel for heating homes and is used primarily in rural areas in states where anthracite coal is mined. Over half the homes that burn coal are in Pennsylvania. Despite its lack of popularity, a coal stove is a good heat source.

Both wood and coal are readily available in many areas of the United States and could be legitimate sources for heat. With forethought and planning, they could be stockpiled in every part of the country. Whether you choose an alternative heat source that burns wood or one that burns coal will depend on your preferences, the local restrictions, and which fuel you can more easily obtain.

Comparing Wood and Coal

Modern wood- or coal stoves burn with up to 85 percent efficiency. Historically, the Franklin Stove, designed by Benjamin Franklin and later improved upon, burned at only 35 to 45 percent efficiency. Coal burns about twice as hot as wood and requires a firebox made to withstand higher temperatures. A coal stove is built differently than a woodstove is, with a heavier bottom and a grate where the coal is burned.

If you are trying to decide between the two, you will find convincing arguments for both. Coal-stove supporters like that the stoves burn hotter and that the fuel is denser and takes up about half the space of wood for the same amount of energy.

Coal stoves, however, are dusty and sooty. They emit an unpleasant smell and contaminate the air with sulfur dioxide and other pollutants.

Woodstove advocates like the warming, radiant heat wood-burning stoves give off. They are economical and can be less expensive than fossil fuels. Today's woodstoves are more efficient and less polluting. Keep in mind that if you procure your own wood, you'll need tools for splitting and chopping.

PERSONALLY SPEAKING

When Jack and I built our first home, we decided we needed a backup heat source to our gas furnace. We did our research and ended up going with a Lopi fireplace insert installed on a concrete-and-brick hearth in our living room. It was amazing how much heat that little stove could put out. It heated our entire main floor and all the bedrooms on the top floor. In fact, we ended up closing the bedroom doors so the bedrooms would not get too warm for comfortable sleeping.

Combination Coal and Woodstoves

Wood- and coal stoves are not generally interchangeable. As mentioned, coal burns too hot for the typical woodstove. However, you can get hybrid stoves and furnaces designed to burn either fuel. Combination stoves for wood and coal are not as efficient at burning as either of the single-fuel stoves. A dual-fuel stove should have an interchangeable grate, a removable liner, an ashpan, and a larger draft opening than found on woodstoves.

Multifuel Stoves

Multifuel furnaces that use fossil fuels like gas, oil, or propane combined with a wood/coal capacity allow you to use regular fuel until it becomes necessary to switch over.

Another option is a wood/coal forced-air furnace that can be used as your regular heat source. Or you could consider a wood/coal furnace or stove that could be connected to the regular heat ducts in your home to serve as a backup. Or you could install a wood/coal boiler furnace for hot-water or steam heating.

Pellet Stoves

Pellet stoves are an especially efficient, clean, and good source of heat. Their downside is that they only burn pellets, which can be expensive if you must buy them from a retail store. If you can purchase them directly from a local manufacturer, this alternative source of heating may be worth considering. But you will need to store a sizeable number of bags of pellets, since they may be difficult to obtain in a prolonged crisis. They also require electricity to function, which makes them vulnerable if the power grid fails, so consider having a back-up generator.

PERSONALLY SPEAKING

We have a pellet stove, and we love it! It produces a warm, radiant, comforting heat, and my husband Craig chose it because it is much cleaner and much more convenient than burning wood in a woodstove. It is thermostat controlled and supplements our gas furnace and keeps the entire basement warm—and during winter in Utah, that's hard to do! The stove is located at the bottom of the stairs, so some of the heat rises to the next floor—we turn on the fan in our central heating system to distribute the heat. We go through about two tons of pellets a winter and figure it saves us about half on our heating bill. Of course, a major drawback from a short-term survival point of view is that it needs electricity to work, but a gas generator solves that problem. We know that for a long-term survival situation, gas for the generator and the pellets themselves will be hard to come by—but we are working to come up with a solution for that!

Fireplaces

Fireplaces are more aesthetic than functional, with typical efficiencies of less than 10 percent to about 20 percent. Heat exchangers help somewhat. If you want a fireplace with a higher efficiency, look for an EPA-certified high-efficiency fireplace. Using a fireplace insert can raise the efficiency to the 70 percent range, but it will look much like a stove.

Heating Precautions

Preventing Chimney Fires

It is important to maintain your chimney. Over time, creosote can build up in the interior of the chimney and eventually lead to a chimney fire. Burning unseasoned wood, restricted airflow, or cooler than normal chimneys can all contribute to creosote buildup. Burn wood that burns hot—hardwoods burn the hottest and make the best firewood. But if you have access to an evergreen forest, try to select Douglas fir, larch, and lodge-pole pine and avoid more pitchy woods. Burn long enough to get a good, hot airflow up the chimney to prevent condensation.

Stove Safety

Whichever stove you get, be sure to install it on a solid masonry foundation and with a wall shield. Take precautions to prevent children from accidentally burning themselves. Table 36.1 lists tools for taking care of your stove. You will also need proper tools for cleaning the chimney.

A stove thermometer can help you gauge how hot the fire is burning so you'll know when to adjust the damper.

Table 36.1 Stove Accessories	
· Ash pan	· Poker
· Extension rods and ropes	· Shovel
	· Stiff brush
· Hearth brush	· Stove thermometer
· Metal bucket	· Wire chimney brush

Other Heating Fuels

Oil Stove

An oil stove can be used to heat a home with radiant heat and without electricity. You can also store a large quantity of fuel oil for an oil stove, but in an extended crisis, once the fuel is used up, you may find it difficult to obtain oil.

Kerosene Heaters

If you live in a centrally-heated apartment building or only need moderate heating for the short term, consider a highly efficient pressurized kerosene heater. There are many models to choose from. You will need to store sufficient amounts of kerosene. Also, be sure to follow all safety precautions listed by the manufacturer.

Any type of fuel burning will use up oxygen and release poisonous carbon monoxide. Misusing a kerosene heater could lead to asphyxiation. To prevent a dangerous buildup of gases or a depletion of oxygen, heaters should not be used in a confined space but in well-ventilated areas with an outside air source, such as a partially opened window. Select a well-vented heater with a built-in oxygen sensor.

Make sure your heater is UL-approved and that you're using the proper grade of kerosene. Rinnai and Suburban make similar heaters that use propane or natural gas. White gas and propane can also be used in flameless catalytic heaters.

Catalytic Heaters

Catalytic heaters convert fuel, primarily propane, to warm, infrared heat. There are no open flames, and if in good working order, combustion is nearly 100 percent, eliminating the potential for carbon-monoxide poisoning. However, because they deplete oxygen, venting is necessary. They are effective at heating small spaces but require storing the appropriate fuel. Look for a catalytic heater that comes with a low-oxygen sensor and shuts off if knocked over.

Sheepherder Stove with Pipe

One last option is the time-honored, wood-burning sheepherder stove with pipe. Made from sheet metal, it is heavy but portable, gives off a good quantity of heat, and can be used to heat a tent, cabin, or similar structure. Cooking can be done on its top. The one pictured is the Alaska Jr. Deluxe stove by Kni-Co Manufacturing.

Survival Cooking

To begin with, if you have a gas range or oven, you need to know it still works when the power is off. You will need matches or a lighter to ignite the gas since the igniters are electric. That way, even if the power is out, you'll be able to cook a meal.

To be prepared for crises, you may want an alternative cooking source. Off-grid cookstoves use wood, kerosene, or propane. These stoves are quite sophisticated and used extensively by the Amish.

Wood-Burning Cookstove

A wood-burning stove with a flat top will usually get hot enough to allow you to do minimal cooking. A wood-burning cookstove provides a cooking surface and an oven. It may have an optional water reservoir or jacket, called a waterfront, that will heat water as it cooks. The water reservoir can be connected directly to your home's plumbing system. This type of stove will need a reinforced floor under it. A double-wall construction will help keep the kitchen cooler in summer. Lehman's in Dalton, Ohio, sells woodstoves and other useful off-grid appliances. Obadiah's Woodstoves in Kalispell, Montana, specializes in wood-burning stoves and stoves using other fuels.

The Challenges of Cooking on a Wood- or Coal Stove

Cooking and baking with wood or coal are very different from cooking with gas or electricity. It is much harder to regulate the temperature, and you'll need to learn the nuances of attending to the fuel. You will also need to learn how to use trivets to raise and adjust pans on a too-hot stovetop. Also, using and caring for cast-iron pots and pans takes some adjustment. Table 36.2 lists the tools you'll need for cooking on a woodstove.

Table 36.2 Tools for Cooking on Woodstoves	
· Trivets	· Heat-proof long-handled
· Cast-iron pots	wooden spoons and forks
· Cast-iron pans	· Tongs
· Lid lifter	· Oven cleaning rod

Kerosene or Propane Cookstove

While a kerosene stove is much like a wood-burning stove in design and function, propane stoves have a more modern appearance. Both types require a reserve of fuel for a long crisis. Unique Off-Grid Appliances in Ontario, Canada, specializes in propane appliances.

Cooking with Fireplaces

Although it's possible to cook in a fireplace, it's challenging. The best way is to have a swinging arm crane or fixed bar and chains to hang cast-iron pots from. Tongs and a long ladle also help. Use a Dutch oven for baking and cast-iron pans and griddles for stewing and frying. A fireplace grate will work if there is nothing else.

Rocket Stoves

Rocket stoves are outdoor stoves that use small pieces of wood, sticks, twigs, and other biomass for fuel. They burn efficiently, produce very little smoke, and can be used for cooking, boiling water, and heating space. They must be vented properly if they are used in an enclosed space. Rocket stoves come in a variety of designs and sizes, but all have the common features of a fuel chamber, a stovepipe or chimney, and an outer insulation chamber. Rocket stoves are commercially available and can also be made following DIY designs.

Camping Stoves

Portable camping stoves that use propane, butane, white gas, alcohol, or kerosene are less-permanent alternatives. Two or three burners are adequate for an average family. Coleman makes a small oven that fits on top of a burner for baking. The obvious disadvantage to these stoves is that they depend entirely on stored fuel.

Heavy-Duty Outdoor Cookers

Camp Chef makes a heavy-duty outdoor cooker, affection- ately known as a Cache Cooker (it was invented in Cache County, Utah). The one pictured is the Explorer Two-Burner. These stoves are high-BTU propane camp stoves that sit on sturdy legs, have two or three burners, and are capable of heating large pots and large grills or griddles. They can be folded up and come with a carrying case. People like to use them at tail-gate parties, hunting camps, and large group cookouts. In a crisis, they could be used to cook food for a big group. They need a source of propane fuel.

PERSONALLY SPEAKING

I learned about the Cache Cooker when I married Craig. I was amazed when he cooked corn for over 1,500 people at the annual Ag Day Barbecue at Utah State University using two cache cookers and large enamel canning pots. He's also used them to cook breakfast for hungry Boy Scouts and for church parties and family reunions. In disaster situations, Cache Cookers could be used to feed a neighborhood of hungry people. They would also be great for heating large quantities of water.

Short-term Cooking Sources

Stoves using canned heat (e.g., Sterno), heat tabs, or fuel bars are strictly for short-term emergencies and are useful for bringing water to a boil or for warming small quantities of food. They have very limited cooking ability. Charcoal burners or barbecue grills have much higher heat output and more capabilities but should only be used outdoors where there is adequate ventilation.

Backpacking Stoves

Backpacking stoves are designed for one or two people. They can be used for both recreation and preparedness and are a good option because they are small and easy to transport.

PERSONALLY SPEAKING

When it comes to personal stoves, the backpackers I know all have their favorites. Most of them care about how compact and lightweight a stove is, how easy it is to light and operate, and how efficiently and quickly it will heat up water. They also pay attention to stream-lined design and extra features like a built-in screen to disperse any wind and an integrated cooking pot. On the downside, backpacking stoves may require specialized fuel and aren't very useful for bigger groups.

Biofuel Stoves

Sierra Zip Stove

The portable Sierra Zip Stove burns about any solid fuel: e.g., twigs, bark, pine cones, scrap wood, and charcoal. It weighs less than a pound and uses a single AA battery for its blower

Solo Stove

The Solo Stove is an innovative, extremely efficient wood-burning outdoor stove, its design simple and elegant (see picture). It burns twigs, pine cones, sticks, etc. Models range from a small portable backpacking stove to a large backyard firepit. Cooking pots and other accessories for it are available.

Solar Cookers

Solar cookers, limited to midday cooking on sunny days, are a good option, especially if you live in a sunny climate. Two solar cookers to consider are the All American Sun Oven and the GoSun Solar Cooker.

Sun Oven

The Sun Oven (pictured) uses shiny reflective panels to direct sunlight into a cooking chamber. It will bake, boil, and steam food. It can also be used to purify water, dehydrate foods, and rehydrate freeze-dried and dehydrated foods.

The GoSun Solar Cooker works by placing a tray of food inside a vacuum tube surrounded by compound parabolic reflectors. This solar cooker will steam, bake, roast, and even fry. It comes in several versions that feed from two to fifty people.

PERSONALLY SPEAKING

I taught fifth- and sixth-grade science for twelve years. For a culminating activity in the unit on heat, I had my students design and build their own solar cookers. They worked in teams and were given aluminum foil, cardboard, masking tape, and Styrofoam plates. For an extra challenge, we often tried the cookers out on a sunny day in the middle of winter when there was snow on the ground. Most of them could successfully bake a few apple slices, melt chocolate, and roast a marshmallow. Since then, I have noticed that there are many plans online with directions for making sophisticated solar ovens. They all work on the same principle of reflecting and focusing sunlight to cook food. I encourage you to have some fun experimenting with them.

Dutch-oven Cooking

Dutch-oven cooking is fun and a good example of how you can combine family recreation with preparedness. You can find Dutch-oven cook-offs in many parts of the country—and the dishes can be quite tempting and delicious. There are also plenty of experts who share their knowledge in recipe books and online.

Cast-iron Dutch ovens come in several sizes. Ten-inch and twelve-inch are about right for a family, but there are larger sizes for bigger groups. They can be used on an open fire, but you'll get more consistent results if you use charcoal briquettes. You can increase these ovens' efficiency by using them along with a reflector oven. Dutch ovens can be stacked to cook several dishes at a time and to make the best use of the briquettes.

Outdoor or camp Dutch ovens should have three legs to allow for stacking and heat circulation. They should have a flat lid with a lip for holding coals and a firm wire handle for lifting. You may also want a charcoal chimney lighter, a lid lifter, and heavy leather work gloves. A tripod is nice for hanging the pot over a campfire. And, of course, you'll need to store plenty of charcoal briquettes if you plan to use them. Table 36.3 shows you how many briquettes you need. Store bags of briquettes in large plastic storage bins.

Table 36.3 Oven Temperature to # of Coals	
300° =	12–15 coals
325° =	13–17 coals
350° =	14–18 coals
375° =	15–19 coals
400° =	16–20 coals
425° =	17–22 coals
450° =	18–23 coals
475° =	19–24 coals

Cast-Aluminum Dutch Ovens

Dutch ovens made of cast aluminum don't rust, weigh and cost less than traditional cast-iron ovens, and work much the same as cast iron but heat up more quickly, retain heat a little less longer, and can be washed in the dishwasher. Although it isn't necessary to season cast-aluminum pots, like cast-iron, seasoning them will make food less likely to stick. (Don't wash seasoned pots in the dishwasher!)

Off-Grid Lighting

A light source can be indispensable, especially for emergency repairs, caring for the ill, and calming frightened children. It also extends the functional hours in the day, particularly during winter.

Battery-Powered Lamps

Flashlights and battery-powered lanterns are obvious choices for quick, always-ready portable lighting. Their obvious limitation is battery life. Solar-powered and rechargeable batteries help solve this problem. Store spare bulbs and batteries for flashlights and lanterns.

Rechargeable Batteries

NiMH (nickel-metal hydroxide) rechargeable batteries are newer technology and have replaced NiCd (nickel-cadmium) batteries. They have two to three times more energy capacity and are most often sold as AA or AAA batteries. They are less likely to develop memory effect, and they gradually lose their energy capacity. They are also more environmentally friendly since they are not made with heavy metals.

Lithium-ion or Li-ion are the standard in portable batteries and are most often used in consumer electronics. They have a high energy density, a low-memory effect, and weigh less than NiMH batteries. They are, however, significantly more expensive, and they can't be substituted for regular D-cell batteries. The efficiency of new Lithium batteries continues to improve.

Solar-Powered Battery Chargers

A multivolt solar array can be used to charge batteries or power small radios, flashlights, and cell phone chargers, etc. Lights, radios, and cell-phone chargers that come with their own solar cells and built-in chargers are also handy.

Kerosene Lamps

For situations without electricity, Aladdin kerosene lamps are unsurpassed when it comes to nonelectrical lamps. They have a round wick, compared to other lamps with a flat wick, they give off a white light equivalent to a sixty-watt light bulb, and they burn silently with minimal odor for forty-eight hours on a gallon of kerosene. They also give off a small amount of heat.

You can purchase them equipped for electric light bulbs with a conversion kit to adapt them to kerosene when needed. The Aladdin MAXbrite 500 is their most recent update. It's important to store extra mantles, wicks, chimneys, flame spreaders, and a selection of spare parts.

PERSONALLY SPEAKING

 The electric conversion Aladdin lamp we purchased over thirty years ago sits on a dresser in our guest room. It has a beautiful classic, white shade and brass base, its light soft and subtle. We have stored parts we need to convert it to a kerosene lamp.

Gas Lamps

Propane lamps and white-gas lanterns are brighter than kerosene lamps but hiss and flare when burning. They give off light equivalent to about 100 to 120 watts for the single mantle and 150 to 200 watts for the double. They're not recommended for use indoors because of their carbon-monoxide emissions.

A gallon of propane or white gas will burn about forty hours in a single-mantel lamp and about thirty in a double. White gas is more dangerous because it is liquid and can be spilled, while propane is pressurized and enclosed. Be sure to store extra mantles.

Cheaper wick and hurricane lamps are often used for nostalgic decorations. These lamps are not desirable for preparedness because they are much less efficient. They give off an amber light and are only about 10 percent as bright as an Aladdin lamp.

Candles

Candles are useful primarily for short-term emergencies. Although they do give off enough light for some activities and a small amount of heat, their open flame is dangerous. Still, they are safe and easy to store, so keep a supply of them, but do not rely on them as your main light source. Candles come in a variety of shapes and sizes. Some have two or three wicks to give off more light and make efficient use of the wax. The burn time of a candle depends on its size, the kind of wax used, the wick size, and the number of wicks.

It is easier to store actual candles, but if you plan to make them, you should store paraffin, wicking, and molds. You may also want a supply of candle holders or lanterns.

Other Nice-to-Have Backup Light Sources

Light Sticks

The light in glow sticks is made when two chemicals unite and form a temporary chemical glow lasting several hours. Often used to keep kids entertained, glow sticks add another layer of light to the environment. They can also be used to identify hazards, keep track of children, keep a group together, mark a trail, signal for help, or create a perimeter alarm. Select a quality grade rather than an inexpensive toy version. Several are described in chapter 33.

Kinetic Flashlights

Also known as Faraday or shake flashlights, these flashlights are activated by shaking or flipping them back and forth. Look for ones with a ratio of low shake time to a high light time. You should be able to find flashlights with a ratio of 30 seconds of motion to 15 minutes of light or more. You generally get what you pay for, so look for a quality brand. A carabiner clip and waterproofing are also nice features.

Small LED flashlights

Small LED flashlights come various styles and shapes. For easy access, keep these ready-to-go flashlights in bedside drawers, in kitchen utility drawers, near every outside door, in emergency evacuation kits, and in vehicle glove compartments. Hanging hooks and magnets are nice features.

Hand-cranked Lights

A minute of cranking can offer twenty minutes of light. There are several choices. The one pictured is made by Ivation.

Head Lamps

LED headlamps allow you to use both hands instead of just one. One creative use is to strap it onto a water-filled gallon milk carton with the light facing inward so the light diffuses and makes a glowing ball of light.

Off-Grid Hot Water and Refrigeration

Hot Water

Hot water is useful for bathing, cleaning clothes, and washing dishes. Water may be heated in large pots using your wood- or coal stove or even over an open fire, but there are better methods if you plan for them now.

Solar Water Heaters

A credible option for heating the water you use in your home, solar water heaters can cut hot water costs significantly as well as offer hot water when other power sources fail. These systems include solar collectors and storage tanks. With some effort, solar panels can be fitted to use with existing water heaters.

ENERGY STAR certifies efficient solar water-heating systems. Most of the time the systems are designed to be used with an electric or gas water heater as backup. There are both active and passive systems. Both systems require a backup system for cloudy days.

Portable Hot Water Devices

For portable hot water, you might use a camp shower that is simply a sturdy black bag or black PVC container that you fill with water and place in the sun. They hold five to ten gallons of water. Expect to pay twenty to fifty dollars, depending on size and features. The one pictured is the Advanced Elements Summer Solar Shower.

Portable Propane Water Heaters

Portable propane water heaters and showers consist of a propane-fueled water heater connected to a shower nozzle. They are recommended for outdoor use only. Camp Chef, Ditsli, EccoTemp, and Zodi are among the most recognized brands. The price for these water heaters starts around a hundred dollars. You will need to store propane to use them.

Wood-Fired Water Heater

Lehman's wood-burning stove can do double duty as a water heater if jacketed and equipped with a coil of metal tubing around and through the firebox. It is also possible to make a wood-fired water heater out of an old gas one by welding a firebox to the bottom. Be sure to install a temperature gauge and pressure-relief valve, available at hardware or building-supply stores.

Refrigeration

Most of us would miss the convenience of refrigeration in an emergency without electricity. However, refrigerators and freezers can be powered by sources other than electricity. Dometic offers the Servel Americana series dual-electric-LP gas or kerosene refrigerator-freezer. These can also be converted to butane or natural gas.

Lehman's Frostek freezer uses LP gas, kerosene, or propane. There are other refrigerator/freezers made for the RV and marine markets that operate on electricity or LP gas, including the Explorer, Novakool, and Norcold brands. SunFrost, Solarfridge, and Polartech refrigerators use one-fourth the usual electricity.

Spare Parts

While you are planning for alternate energy sources, be aware that the equipment can break down—often at the least opportune time. Be sure to keep your alternative-energy appliances in good working order. Assess the needs of your equipment and store a supply of spare parts so you can at least fix the most common repairs.

Managing Energy Sources in a Crisis

Some of the biggest decisions you must make as you prepare for a crisis is which backup fuel systems you'll use for heating, cooking, and lighting. This chapter will help you compare various types of fuels and energy sources so you can decide which ones best fit your needs. Again, it's always a good idea to use redundant systems so that if one fails, you'll alternatives.

Carbon-Monoxide Poisoning

Before we get into hydrocarbon fuels, I want to offer a warning about carbon-monoxide poisoning. Carbon-monoxide (CO) poisoning occurs when people incorrectly use an alternative energy source. All carbon-based fuels give off CO when burned. This dangerous gas is odorless, colorless—and deadly. It is absorbed into the bloodstream more efficiently than oxygen. Symptoms of carbon-monoxide poisoning include headache, nausea, vomiting, dizziness, weakness, confusion, and sleepiness.

To prevent carbon-monoxide poisoning it's critical to properly maintain and fully vent fuel-burning appliances and use them only as directed. It's also important to install one or more carbon-monoxide detectors.

Stockpiling Solid Fuels – Wood

In a prolonged emergency, the best fuels are those that presumably could always be replenished locally, like wood or coal. Wood is abundant in many locations. In locations where wood is scarce, be aware that it may become scarcer during a crisis. It's true that trees are considered a renewable resource wherever they grow naturally, but trees take years to accumulate enough biomass to be used for fuel.

FIVE THINGS YOU CAN DO NOW

1 Examine your heating bill and become familiar with how many BTUs you use to get an idea of your energy use.

2 Purchase a gasoline container(s) and gasoline stabilizer to have extra gas stored for your vehicles or generator. Store and rotate your gasoline storage.

3 Purchase a shut-off wrench for your home's natural-gas valve. Teach responsible people in your family how to use it.

4 Purchase a solar battery charger for electronics or solar-powered flashlight or emergency radio.

5 Find out the price of adding off-grid solar energy to your home.

Comparing the BTUs of Different Woods

All woods are chemically similar. The amount of energy or BTUs in a cord of wood depends on its density and moisture content. Hardwoods are denser than softwoods and offer more energy for the same volume of wood. Hardwoods also store longer. See table 37.1 for the BTUs of common firewood in different parts of the United States.

The most practical and economical woods will be those that grow in your area. For example, oak, hickory, and maple are common top-rated hardwoods. Most coniferous trees are considered inferior firewood because of their lower energy output and their tendency to spark and smoke. Douglas fir and juniper are higher rated among the softwoods. You may also want to consider that a knotty wood is more difficult to split.

If hardwood is at a premium where you live, consider using a softer wood, like pine, to start your fires and then add harder woods for slower burning and sustained heat. Also, use the softer firs and pines in the spring and fall, saving the hardwood for more extreme winter cold.

Table 37.1 Average BTUs for Common Woods*					
Fuel	Million BTUs per Cord	Pounds per Dry Cord	Fuel	Million BTUs per Cord	Pounds per Dry Cord
Western Species			Eastern Species		
Douglas fir	26.5	3,075	Shagbark hickory	27.7	4,327
Western juniper	26.4	3,050	Black locust	26.8	3,890
Western hemlock	24.4	2,830	Oak	24.0	3,757
Lodgepole pine	22.3	2,580	Maple	24.0	3,757
Ponderosa pine	21.7	1,520	White ash	23.6	3,689
Quaking aspen	18.0	2,400	Birch	20.2	3,179
Cottonwood	16.9	2,225			
*Compiled from World Forest Service Firewood BTU Ratings.					

Look for a highly recommended dealer and inspect the wood before you purchase it. It should be clean and free from dirt or sand and neatly stacked. Use a tape measure to verify the size of the pieces and stack. Pieces should be several inches shorter than your firebox. Purchase and stack wood in the spring so you can control the seasoning process.

Storing Wood

Store wood outdoors and shielded from the weather so it will stay dry. It is best to store it off the ground to avoid rotting. Firewood should last up to five years if it is quality to begin with and protected from the weather. Like all consumable commodities, it is best to use and replenish it.

It takes about twelve months for wood to season well. A good practice is to obtain wood for the following year so that it has time to become well-seasoned. Burning green wood can reduce efficiency by up to 25 percent. The moisture from it also contributes to creosote buildup in chimneys, which can lead to chimney fires.

A Cord of Wood

A standard cord is 8 x 4 x 4 feet of tightly stacked wood. The 128 cubic feet only contains about 80 to 90 cubic feet of actual wood because of space between logs. A cord of air-dried softwood weighs about 1¼ tons, while hardwood can weigh over two tons. A face cord is 8 x 4 times whatever length

Table 37.2
Firewood Tools

· Ax	· Sledgehammer
· Bucksaw	· Splitting maul
· Crosscut saw	· Splitting weights
· Chainsaw and fuel	· Wood-stacking crib

the wood is cut in, making it one-third to one-half a standard cord. Again, be sure the wood you purchase is a size that will fit your firebox. The estimated time needed to chop and split a cord of wood is eight hours. If you plan to use a chainsaw, you'll need to properly maintain it and store fuel for it.

Stockpiling Solid Fuels – Coal

The United States' extensive reserves of coal are mined in twenty-six states, and should be considered for use as a basic fuel during a crisis in those areas. You will need a stove designed to burn coal. It is not recommended you burn coal in a woodstove or fireplace. The construction of most woodstoves is not designed to withstand the heat generated by burning coal.

Anthracite, or hard coal, is a desirable heating fuel that burns hot and clean and is found almost exclusively in Pennsylvania. However, it is harder to start a fire made with anthracite coal. Softer bituminous coal is easier to start but less suitable because it produces more soot and smoke. Although coal burns dirtier than wood, it has several advantages. It does not need to be split, does not absorb rainwater, and does not carry unwanted insects into your home. Also, three to five times as much energy can be stored in the same space as wood.

Storing Coal

One ton of coal takes up about forty cubic feet—it varies depending on the type of coal and the size of pieces. For best results, use the recommended size of coal for your stove or furnace. Coal should be stored in a dark, dry place, such as a lined pit, coal bin, or shed. How much coal you need will depend on the size of the space you are heating, insulation factors, and temperature preferences. But generally, a two-thousand-square-foot home will use four to five tons a year.

Table 37.3	
Average BTUs for Wood Pellets and Coal	
Fuel	Million BTUs per Ton
Wood Pellets	16.5
Hard Coal (Anthracite)	26.0
Soft Coal (Bituminous)	24.0

Fire Starters

You will need fire starters or lighters to start coal fires—they would also be helpful for wood fires. Also, be sure you store disposable lighters and wood-friction matches. Waterproof the matches by coating them with paraffin wax.

Homemade Fire Starters

For homemade fire starters, use egg-carton candles, paraffin or oil-soaked cardboard, newspaper twists, gauze rolls, or cotton pads. Connecting shreds of steel wool to the ends of an ordinary D-cell battery also works.

Calculating How Much Wood and Coal to Store

The amount of fuel you store will depend on what crises you anticipate. It will also depend on the size of the space you are heating and the climate in which you live. As you decide on amounts, consider that in a crisis or survival situation you will likely be more conservative and use less fuel than you do right now. You also may choose to heat only a portion of your home.

Calculating Based on Current Usage

Use your current utility power statements as a starting point for how much fuel you should store. In all cases, calculate the total number of BTUs you use in a year and then divide it by the number of BTUs the desired alternative fuel source provides (see tables 37.1 and 37.3).

Remember to account for stove inefficiencies. Since wood- and coal-burning stoves are less than 100 percent efficient, you'll need to add fuel to make up the difference. The most efficient woodstoves burn at around 85 percent efficiency.

If you are already using wood, coal, oil, or propane to heat your home, you probably have a good idea of how much fuel you use in a heating season. In that case, depending on the type of crisis you anticipate, stockpile the amount you would need for a short-term emergency or for an entire season.

If you plan to use an alternative wood or coal heating source for emergency heating that you currently use infrequently, begin by noting how much fuel you use in a day or week. That number should help you estimate how much fuel you need to store for an entire heating season.

Calculating Using Your Energy Utility Statements

If you are currently relying on outside power sources, begin by examining your energy utility statements. If you heat with natural gas, that bill should tell you how many DTH (dekatherms) you use. A dekatherm is equal to one million BTUs (British Thermal Units). Multiply the number of DTHs by one million to determine the number of BTUs you use in a given period. In tables 37.1 and 37.3, you'll find the BTUs listed for common fuels. With that information, you can calculate and then estimate the total amount of fuel you would need to heat your home for any determined length of time.

> **QUICK FACTS**
>
> ### Useful Energy Facts
>
> - Artificial light is needed about 1,400 hours a year.
> - Freezers must operate a minimum of four hours per day to keep food frozen.
> - A family of four uses about six million BTU's of gas or nineteen million BTU's of wood or coal to cook meals for a year.

If you heat or cool your home with electricity, begin by comparing your highest energy-use months with your lowest to get an approximate idea of how much of the electricity is being used for heat, air conditioning, and other electricity usage. Convert kilowatt hours (kWh) to BTUs by multiplying the number of kWh by 3412.14. (1 kWh equals 3412.14 BTUs).

Calculating Using Heat-Day Degrees

You can also estimate heat needs if you know the number of heating-day degrees (HDD) for your location. HDD are determined by comparing average outside temperature with desired inside temperature. So, if the average outside temperature for a day is 35 degrees and the desired inside temperature is 65 degrees, the difference is 30 degrees. And if that were the case for the entire month, then the HDD would be 900 (30 degrees x 30 days) for that month and 5400 (30 degrees x 180 days) for six months. As examples, the average heating-day degrees is 8200 in Minneapolis, 7383 in Concord, New Hampshire, 6052 in Salt Lake City, 4711 in Kansas City, and 2773 in Sacramento. You can find calculators for your location on sites like http://www.degreedays.net/.

To give you an idea, the average 1,500 square-foot, two-story home with R-30 ceiling, R-20 wall, and R-5 foundation insulation and using weather stripping, caulking, and storm windows and doors would use 13,300 times the heating-day degrees in BTUs per year. A less insulated home could require more than double that amount, while a home using passive solar methods could require a lot less.

Stockpiling Liquid Fuels

Some energy choices require liquid hydrocarbon fuels. Of course, such fuels will eventually run out. The only solution for a long-term crisis is to stockpile a large quantity and hope for renewed availability or an alternative source by the time you deplete it.

Stockpiling liquid fuel can be dangerous if not done correctly. Furthermore, in the case of gasoline, the amount you can legally store varies by location. Many urban locations limit you to five or fewer gallons. Storing gasoline can also void fire insurance. Nevertheless, gasoline can be stored safely. Do not store volatile fuels in your home or near heat, flame, or motors. Ideally, they should be stored in a separate, detached structure.

Volatile fuels should be stored in a separate, detached structure.

Underground Storage

The best way to store a large amount of fuel is underground in properly vented tanks covered with at least four feet of earth. This not only prolongs the life of the fuel, it also reduces evaporation and helps protect it from, fires, accidents, theft, and gunshots.

Before burying and filling a tank, check for leaks using five pounds of air pressure or water, and install and test all fittings. Leaking stored fuel could be disastrous, and if water gets in it, that will also be a problem. Unless you use gravity fed, have at least one hand pump dedicated to pumping fuel.

Steel Drums

You can use fuel containers that range from fifty-five-gallon steel drums to large tanks that hold thousands of gallons. Steel tanks will last more than ten years underground when protected from corrosion.

Fiberglass Tanks

Fiberglass tanks last at least thirty years. Metal detectors cannot locate them, if that is a concern to you. Be aware that EPA regulations on underground tanks are extremely strict.

Aboveground Storage

Fuels stored above ground are much more subject to temperature variations, causing alternating expansion and contraction. This must be considered for container size and venting. If you have no alternative but to store containers directly in the sun, take precautions to keep temperature variations to a minimum. Storage containers can be wrapped with insulation or painted with reflective silver paints.

Preventing Fuel Degradation

Fuel Deterioration

All liquid hydrocarbon fuels deteriorate over time. Kerosene is by far the most stable liquid fuel. The components of gasoline oxidize to form gums, lacquers, and peroxides that reduce performance, cause stalling, clog fuel filters, foul carburetors and spark plugs, and even damage pistons and valves. Antiknock compounds in high grades of gasoline will also decompose.

Diesel fuel separates into its various components. Little streamers of paraffin form in it and bacteria and fungi can pollute it. The resulting sludge and slime plugs filters and strainers, fouls fuel probes and lines, gums up injectors and combustion chambers, and breaks down tank sealants and coatings.

Deterioration Rates

Heat is the main enemy of stored fuels, greatly increasing the oxidation rate. Gasoline is presently formulated to be used within six months under normal conditions and may deteriorate in less than two months in a hot garage. Diesel fuel and heating oil deteriorate within a year or so. Kerosene may last for two to five years.

Fuel Stabilizer Additives

Fortunately, for a minimal cost, there are stabilizer additives that will retard chemical breakdown and preserve the fuels longer.

Stabilizing Gasoline

Gasoline has a very short shelf life and destabilizes quickly. It should only be stored for three to six months unless a stabilizer is added. Treated gasoline will keep about six months in direct sunlight and up to fifteen months protected from sunlight in an airtight gas can. It will last two to five years underground.

There are several brands of effective stabilizers that extend the life of gasoline. One gallon of stabilizer will treat between 1,200 and 2,000 gallons of fuel. The shelf-life for stabilizers is two years if tightly capped and kept cool and dry. Older products will not cause any harm, but they may not be as effective. Repeat the stabilizer treatment annually.

For small two-cycle engines, which require a gasoline-and-oil mix, the two-cycle oil may have a stabilizer already added. Check before adding more stabilizer.

Run any fuel that has been stored for a long time through a filtered funnel or other fine filtering agent, such as chamois leather, to strain out additional contaminants. If gasoline smells like varnish, the gums have already formed. It may still be usable when diluted with two to three times as much fresh fuel. Fuel revitalizers can also be added to dissolve gum deposits.

Stabilizing Diesel

Diesel fuel can be stored for six months to a year without any degradation if kept cool, clean, and dry. Diesel fuel and heating oil keep about five to ten years when treated. Larger amounts extend the storage life further, and you can re-treat.

Diesel-fuel stabilizers should be specifically made for diesel fuel. Check the product directions, but plan on one gallon per six hundred gallons of diesel fuel. Look for stabilizers that contain an antigel for winter use and a biocide to prevent algae and bacteria contamination.

Storing Gasoline

Besides being used in automobiles, gasoline is also important for generators, small engines, chainsaws, etc. Gasoline is highly volatile and burns fast and hot. Spilled fuel evaporates quickly. Gasoline will be scarce during a severe or prolonged crisis. On top of that, gasoline pumps require electricity, which will prevent access unless the gas station is equipped to transfer to a generator. Store gasoline in several locations away from living areas for safety and security.

Storage Containers for Gasoline

For aboveground gasoline storage, one container that increases storage safety is ExploSafe jerricans. A honeycomb aluminum mesh is packed into the jerrican. It takes up less than 2 percent of the space and disperses heat buildup so quickly it prevents explosions. Pictured is a cross-section of the ExploSafe jerrican. Jerricans and other portable fuel cans should have adequate spouts.

Other UL-approved gas cans, five-gallon jerricans, and drums may be used but are more hazardous. Any fuel container that cannot be easily lifted when full should have a hand pump or manual siphon. Of course, never siphon using your mouth because ingested gasoline can cause fatal chemical pneumonia.

You can also store gasoline in larger fourteen- to thirty-five-gallon containers, which are often equipped with wheels and a siphon or rotary pump.

Gasoline vapors can be extremely explosive. To minimize vapor formation, keep containers nearly full, with only a few inches for expansion. Volatile fuels like gasoline, propane, and butane have heavy vapors that settle in lower areas and can puddle, possibly causing suffocation or an explosion.

How Much Gasoline to Store

If you will be using gasoline for a generator, you need to know how much gasoline your generator uses in one hour. Begin by calculating your needs for three days. Next, determine how many hours a day you think you might need to use your generator. Remember, you can conserve energy by only using the generator for several hours a day for critical electricity needs. With this information, you should be able to calculate your gasoline needs.

If you have a destination where you plan to go for an emergency crisis, store enough gasoline to reach that destination. Calculate the gasoline needed by dividing the distance by your vehicle's average miles per gallon. Also, consider that highways will likely be congested in an evacuation and you'll use more gasoline driving at slow speeds. Add 25 percent for good measure.

Finally, as mentioned, it's good practice to keep your vehicles at least half full. Refill them when half empty to increase your gasoline storage.

Rotating Gasoline

Perhaps the best solution is to conscientiously rotate stored gasoline. Ideally, rotate it every six months. Simply pour stored gasoline in your vehicles and replenish your gasoline cans.

Storing Other Fuels

Fuel Oil

In some parts of the country, fuel oil is plentiful and a practical option for heating your home. Although you are dependent on being serviced by a fuel-oil company, you are not dependent on an elaborate grid system. In an extended emergency, you have more control over your options. But you run the risk that it may be in short supply during a prolonged crisis. Fuel oil will deteriorate over time unless a stabilizer is added.

Diesel

Like gasoline, diesel will also be hard to obtain during a prolonged crisis. If you own diesel vehicles or appliances, plan to store this fuel. It is less volatile than gasoline, safer to store, and deteriorates more slowly. Stabilizers should be added to extend normal storage life. And, of course, continually use and replenish this fuel.

Kerosene

Kerosene is used in lamps, portable stoves, and space heaters. Although readily available now, like all liquid hydrocarbon fuels, it will likely be hard to find during a prolonged crisis.

Kerosene is the most stable liquid hydrocarbon and stores the longest without deterioration. It is also reliable and burns hot, clean, without odor, and silently in modern, efficiently designed burners. Its flashpoint is much higher than gasoline, which makes it safer to store. It produces carbon monoxide when burned and should be well ventilated. Spilled kerosene does not evaporate quickly.

It is most economical when purchased in five-gallon or larger quantities. Buy only pure, certified "water-clear," 1-K-grade white kerosene for home use. To avoid contamination, store kerosene only in clean, sealed containers. Use blue fuel containers intended specifically for kerosene to distinguish it from other fuels.

Once opened, kerosene should only be stored from one to three months. However, if it is stored in sealed containers in cool, dark conditions, kerosene can be stored two to five years. A fuel stabilizer will prolong its storage life.

White Gas

White gas is also known as camp fuel and is used for small stoves and lanterns. The Coleman and Crown brands are the most common. It is clean-burning and has a high-energy output. For safety, it should only be used outdoors. Fuel can be stored in unopened containers at room temperature for five to seven years. It is best to use opened containers within a year.

Propane

Propane is produced from crude-oil refining and natural-gas processing. Naturally odorless, propane manufacturers add a distinctive odor to make leaking more detectable. It is used for generators, heaters, grills, stoves, lights, and refrigerators. To prevent carbon-monoxide poisoning, do not use outdoor lanterns, stoves, grills, or heaters indoors. Follow appliance manufacturers' directions for use indoors or outdoors.

Storing Propane

Propane can be stored in cylinders, portable tanks, or stationary tanks. One-pound cylinders, typically used for camping, may be refillable or prefilled and disposable. A common size tank is the twenty-pound refillable tank typically used for grilling. There are also a variety of less-common sizes that may suit your needs. Large, stationary propane tanks are used to power home heating systems.

Propane itself has an indefinite shelf life. The only limitation with storage is the quality of container it is stored in and the valves, regulators, and connector hoses. It is important to maintain the storage tanks to prevent rust or deterioration. Propane storage containers should not be stored in the house or in an attached garage or shed. Store them outside in a well-ventilated space where the temperatures will not exceed 120 F (48 C).

Liquid Petroleum Gas (LPG)

LPG mixture of hydrocarbon gases is another liquid fuel stored under pressure and vaporized to burn. It is reliable, clean, odorless, easy to use, inexpensive, and can be used for heating, cooking, and refrigeration. It's good for backup generators and as a source for supplementing solar energy. It comes in smaller portable cylinders and in large stationary tanks. LPG stores for many years without deterioration.

Every responsible person in the household should know when and how to shut off the natural gas in case of a break in the line.

Natural Gas

Many homes are equipped with natural gas, which is used to run furnaces, stoves, and dryers. Its distribution may be interrupted during a crisis, and you may not be able to rely on it.

Shut Off Natural Gas

Like propane, natural gas is odorless, and gas companies add odor to make leaks more detectable. It is important to know how to shut off your natural gas if pipelines are broken and you smell gas or hear hissing. Typically, there is a shut-off valve on the gas meter running parallel with the pipe. Use an emergency shut-off wrench, a tool specifically designed for this purpose, or a regular twelve-inch or larger adjustable wrench. Simply give the valve a quarter turn in either direction. This will put the valve crosswise in the pipe, cutting off the gas. Every responsible person in the household should know how to do this before an emergency. Gas companies recommend you not try to turn the gas back on yourself.

> **QUICK LOOK**
>
> ### Improvised Fuel
>
> - Plain newspapers that are folded, soaked in water containing detergent, and then rolled and then dried burn as clean and hot as normal wood.
> - Make homemade charcoal by "cooking" pieces of hardwood, black walnuts, and fruit pits.
> - Fire starters and fuel can be made by pouring a paraffin and sawdust mixture into paper egg cartons, cupcake papers, or tuna cans filled with rolled cardboard.
> - Other things that burn include peat, corncobs, straw, grains, manure, and old motor oil.

Canned Cooking and Chafing Fuel

This category of fuel will augment your other fuel preparations, especially for short-term emergencies. Canned fuels are typically jellied and contain alcohol, ethanol, or methanol. They are used to heat and cook a small amount of liquid or food and are safe to use indoors. Common brands include BioFuel, Blaze, Safe Heat, and Sterno. These products have low heat value, especially when cost is considered. They should be stored in a cool, dark place.

Charcoal

Charcoal offers more heat per cubic foot than wood and is a good option for cooking outdoors. But it must *not* be used indoors because it produces carbon monoxide. A charcoal fire can be started with lighter fluid or by concentrating it in a chimney starter.

Charcoal can be purchased and stored in several sizes of bags. Store them in closed metal or plastic bins. Charcoal can also be made at home. You will need an enclosed metal container, such as a metal barrel with a metal lid. One method for making charcoal involves tightly loading small cuts of hardwood in an enclosed metal container and building a bonfire around it to "cook" the wood.

Producing Electricity with Generators

Electricity is an indispensable convenience, and in a crisis, you may not want to do without it completely. There are several ways you can produce electricity.

Human-Generated Power

The most basic method of energy creation is your own power. One of the simplest is a bicycle with a small generator attached to the wheel for powering small lights and recharging batteries. How much electricity can be generated will depend on the fitness level of the cyclist, but an efficient bike generator and a reasonably fit cyclist can create about one hundred watts of electricity in an hour—enough to power a twenty-watt bulb for about five hours. Bicycle generators are not efficient considering the small amount of energy they generate versus cost. If this energy source interests you, consider building your own.

Fuel-Powered Generators

Fuel-Powered generators are commonly used for immediate backup power. They can be powered by fuels such as gasoline, diesel, LP, and natural gas. They will provide large amounts of power but are not really designed for continuous operation. It is best to run your generator frequently for a short time each week to help prolong its life, reduce maintenance, and increase efficiency and reliability.

In an emergency, a generator can provide power, but its demand for fuel will make it impractical to run around the clock. You will need to make decisions about how to maximize fuel usage and prioritize which appliances to run. High priority should be given to heat, refrigeration, communication, and safety.

Fuel Requirements

Expect to use from eight to sixteen gallons of gasoline per hundred kilowatt-hours of power. The exact amount will depend on the generator model you select. Diesel-powered generators will use about 25 percent less than their gasoline-powered counterparts. For planning purposes, the average 1,500-square-foot home normally uses about twenty kilowatt-hours per day. Under crisis conditions, you could probably get by on half that or less.

PERSONALLY SPEAKING

 If your home is fitted with natural gas, I recommend you consider a generator powered with natural gas. When my husband built our home, he had a natural-gas generator installed. It automatically kicks on when the power goes out. It doesn't power everything in our home, but it takes care of the essential things like heating, refrigeration, cooking, washing, and lights. It is programmed to cycle on once a week for a few minutes to keep it in good working order.

Generator Size

The correct generator size is determined by first adding up the total wattage of all the appliances, tools, and lights that will be used at any one time. Next, add in the wattage needed to start any motors, compressors, and blowers. Finally, add an extra 25 percent margin. Smaller generators tend to use less fuel, but it depends on the model and the operating load.

For planning purposes, the average 1,500-square-foot home normally uses about twenty kilowatt-hours per day. That is the equivalent of 174 amp-hours at 115 volts. Under survival conditions, you could probably get by on half that amount or less.

All generators located in enclosed areas should have their exhaust vented to the outside. Store motor oil to keep the generator lubricated.

Producing Electricity with Solar Energy

The use of solar energy is nothing short of an enormous energy revolution. It has great potential for providing an alternative energy source without relying on hydrocarbon fuels or a tenuous grid system. Advances in solar-energy options are being made every year, and you should consider how you can take advantage of this energy source in preparing for a crisis.

In a solar-energy system, photovoltaic solar cells convert sunlight into direct current (DC) electricity. Power inverters change DC electricity into the alternating current (AC), which is used for appliances in homes. A solar array is the total number of photovoltaic solar cells, modules, and panels. A photovoltaic system includes other components, like wiring, switches, inverters, mounting parts, etc. Power generated may be used immediately or, if you are off-grid, stored in batteries for use during nighttime and other times when sunlight is unavailable. If you are attached to the power grid, the excess power is usually net-metered and credited to you for later use.

Is Solar Energy Right for You?

Solar energy for crises can be classified into permanent and portable applications. For either, you will want to first consider the solar potential for your location. You can do this by knowing the average number of peak sun hours or the intensity of the sunlight in your location. A peak sun hour is an hour of sunlight with at least one thousand watts of photovoltaic power per square meter.

Sun hours are determined by how directly the sun shines and are maximized when the sun is highest in the sky—during the middle of the day, during the months closest to the summer solstice, and in locations closer to the equator—and, of course, when there is no cloud cover.

How much sunshine do you get? To make solar energy viable, you must average at least four peak sun hours daily. You can also figure four to six hours of direct sunlight daily, or at least most days. Many parts of the United States have good solar potential. Whether going with solar is an economical choice for you will depend on the cost of your present electrical power.

Ideally, for crisis preparation, the best option is a combination system that can be on the electrical grid in normal circumstances but switches over to batteries when the grid goes down. Knowing how much energy you need in grid-down situations will help determine how much battery capacity you need.

Single Solar Cells

The simplest form of solar energy is a single photovoltaic solar cell. Small solar cells are often attached to small battery-operated items like calculators, flashlights, or radios. Larger single solar cells or several solar cells may also be attached to a battery charger that can be used to charge cell phones and other small electronics.

Portable Solar Arrays and Solar Chargers

Solar arrays collect energy from the sun and either directly charge devices and appliances, or the energy is stored in a battery that holds it until it's needed to charge devices. As there are many portable solar-energy options to choose from, you'll have to determine which system best fits your needs. Portable solar chargers can be flexible or foldable, mounted on rigid frames, or enclosed in suitcases. For simplicity, there are two basic types of portable solar arrays—small and large.

Small Portable Solar Energy Options

QUICK CHECK
What to Look for in a Portable Solar-Energy System
✓ Lightweight
✓ Sturdy
✓ Portable
✓ Waterproof
✓ Charges quickly
✓ Plenty of battery charge cycles
✓ Capable of charging devices directly
✓ USB ports and power outlets
✓ Can be tilted for most efficient sun direction
✓ Has enough wattage for your needs
✓ Efficient monocrystalline solar cells
✓ Can also be charged with AC current

Small arrays, generally with a twenty-five-watt or less capacity, can power small electronics and are good for charging devices like cell phones, flashlights, emergency radios, and possibly laptops. If the sunlight is consistent and bright, it may be possible to charge directly from the solar array, but if not, it is better to let the battery charge and then use it for charging devices.

There are many models to choose from, and manufacturers are continually improving them. Evaluate the solar capacity and quality of the array and battery capacity, the amount of time required to charge, the number of charge cycles, and the expected longevity of the charger. (See Quick Check.) Companies, such as Sunjack, for example, specialize in smaller arrays for high-speed charging of personal devices. The Sunjack 25W portable charger is pictured. It comes with three power-banks and costs about $190.

Large Portable Solar Energy Options

Larger solar-cell arrays have more cells and a greater capacity for collecting energy. They may be foldable and very portable or rigid and less portable. Both types are paired with a lithium battery that may be an integrated part of the system or a separate power pack. Larger arrays may have monocrystalline or polycrystalline solar cells. Monocrystalline cells are more expensive but more efficient. The larger-capacity arrays can capture and store enough energy to selectively power medical devices, power tools, small and large household appliances, and other critical devices.

Portable Solar Generators

The biggest limitation with portable solar generators is the capability of the battery. The technology continues to improve, however, and there are batteries on the market that will fully charge with as little as three and a half hours of sunshine. They also can provide over 1,200 watts of energy.

Solar generators work in tandem with solar panels, the solar panels capturing the photons from the sun with photovoltaic cells and charging the generator battery. The best batteries are lithium-ion. They hold a charge longer, can be recharged multiple times, and are superior to acid-lead batteries. If AC power is desired, an inverter is needed to change DC power into AC power, but inverters are usually included as part of the system.

GoalZero Solar makes complete portable systems that include solar panels, power packs, and connecting hardware and has options for compact, midsize and heavy-duty power requirements. (The GoalZero Yeti 3000 Lithium Power Pak is pictured.) Solar energy is a rapidly changing industry .

Grid-Tied Rooftop Solar with Battery Back-Up

Solar rooftop panels can be net-meter grid-tied, completely off-grid, or a hybrid of the two. Point Zero Energy, specializing in hybrid and off-grid solar energy, has the technology to provide backup power to your grid-tied system or to secure independent portable power with a solar generator. When the power goes out, this technology allows you to either manually or automatically switch from your grid-tied solar system to battery back-up.

Off-Grid Solar Panels

Given the right circumstances, you can power a small home or cabin with off-grid solar panels and a backup storage battery. This is practical where solar energy is highly available and where energy demands are relatively low, and especially where other energy options are unavailable, though it requires an investment in equipment and batteries.

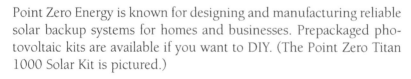

Point Zero Energy is known for designing and manufacturing reliable solar backup systems for homes and businesses. Prepackaged photovoltaic kits are available if you want to DIY. (The Point Zero Titan 1000 Solar Kit is pictured.)

Twelve-Volt Systems

You may also consider a twelve-volt system that runs off storage batteries charged by solar energy or any electricity-generating source. A twelve-volt system does not work for regular household appliances but can be used with the wide variety of equipment and accessories made for use in campers, motor homes, and boats. It can also be powered temporarily with your car battery.

Producing Electricity with Hydroelectric Power

If you have an independent, continuing source of water with enough flow and vertical drop (head), you can use a water turbine and generator to produce electricity. The amount of potential power in kilowatt-hours per month can be estimated by multiplying the average flow in cubic feet per minute (CFM) times the head in feet and then multiplying by the system efficiency factor (0.5 to 0.8 is about the range).

For example, a four CFM flow with a fifty-foot head flowing twenty-four hours per day through a 60 percent efficiency system could provide about 120 kilowatt-hours each month. You probably would not be interested in producing a whole lot less than that. Hydroelectric power can be the best and cheapest form of energy if you have the water source.

Producing Electricity with Wind-Generated Power

Wind generators can be used only in those relatively few locations where the average monthly wind speed is at least ten miles per hour or greater. Contact your local airport, weather station, or National Weather Service to make sure your location has enough wind prior to purchasing any wind-based generator equipment. You can also determine this by renting an anemometer and placing it on a temporary tower thirty to fifty feet high. Record the wind velocities for a calendar month, then find a local weather station that records the wind velocities daily and compare your wind speed with theirs for that month. That will help you extrapolate to find the wind-speed probability for an entire year. Wind power can be used to complement solar power.

Off-Grid Power Storage

Not only is generating electrical energy important in a grid-down situation, but also equally important is storing the energy produced. When power is generated, it is not produced or used in a steady flow, so it is necessary to have off-grid power storage to provide power at night and on sunless or windless days and to even out variations in power usage.

> **QUICK CHECK**
>
> ### What to Look for in an Off-Grid Storage Battery
>
> ✓ What is its capacity? How much electricity can it store?
> ✓ Are the batteries "stackable" to increase capacity?
> ✓ What is the battery's power rating? How much electricity can it deliver at one time?
> ✓ What is its depth of discharge (DoD)? You want one that is high—around 90 percent.
> ✓ What is its round-trip efficiency? What percent of the energy that went in can be used?

Deep-Cycle Storage Batteries

Storage batteries, often called deep-cycle batteries, can be lead acid or lithium and are designed to store energy and slowly discharge it as needed.

Unsealed Lead-Acid Batteries

Deep cycle, unsealed lead-acid batteries must be used regularly, require ventilation, and need their water level maintained. They are a good choice for full-time applications but not intermittent use. They should be protected from heat, sunlight, and cold.

AGM and Gel Lead-Acid Batteries

AGM (absorbed glass mat) batteries and gel lead-acid batteries are sealed and require no maintenance. They are designed for intermittent use and work well as backup batteries. Gel batteries have a longer life cycle than AGM batteries but are limited in the amount of power they provide.

Lithium Batteries

Lithium deep-cycle storage batteries are an important developing technology. They began as power sources for small electronics, but continual innovations have made them viable in solar-battery applications. Although they are more expensive than lead-acid batteries they offer some big advantages. See the Quick Look for a list of the advantages of lithium batteries.

> **QUICK LOOK**
>
> ### Advantages of Lithium Batteries
>
> - High-energy density—more storage capacity in less space
> - High efficiency—more energy is stored
> - High charge/discharge rate
> - Long life—10,000 cycles or more
> - High depth of charge—more available energy
> - No maintenance needed
> - No memory problems
> - Sealed, no emissions
> - Initial higher cost, but potentially less cost per kWh

How Many Storage Batteries Do You Need?

Figuring out how many storage batteries you need is an important step in putting together an off-grid system. The amount of battery storage you need is based on how much energy you generate and your energy usage. Energy usage is measured in kilowatt-hours (kWh). If an average house uses thirty kWh per day, and if a battery has a ten-kWh capacity, you'll need three batteries to store energy for one day. If you're going to be totally off-grid, you'll need more batteries for times when no energy is generated and stored.

Wholesale Solar has a nice off-grid load-evaluation calculator that can help you figure out your energy usage (see page 432). You can also use your electricity power statement to calculate your average daily kWh usage.

Battery Lifespan

Battery storage life ranges from five to fifteen years and depends on the quality of the battery, the number of charge cycles it has, how often it goes through these cycles, and how well it is maintained. Study the specifications of batteries so you know how to best care for them. You might consider buying spare batteries for backup.

Car Batteries

With an inverter to convert DC to AC electricity, your car battery can be adapted for limited backup energy and used to charge cell phones, computers, and small electrical appliances. A car battery is not intended for this kind of use and is not a practical alternate-energy option, but in an emergency, it could have limited use.

Inverters for Car Batteries

An inverter converts the twelve-volt direct current (DC) in a car battery to a 110- to 120- volt alternating current (AC). This makes it possible to power many ordinary plug-in devices for a short time. Look for an inverter that has both battery clamps and a socket plug so that it can be used directly on a stand-alone battery or inserted into the accessory outlet on a vehicle. Be sure it can convert enough power for the wattage needs of the appliances you want to use as the power requirements of the appliance will drain and limit the use of the battery.

Making Your Own Fuel

Stored fuel will eventually run out, but it is possible to make your own fuel if you have the raw materials and proper equipment. These biofuels are made using digesters and stills.

Methane

Methane digesters that capture methane from decomposing organic waste can be used in place of propane and natural gas to power lanterns, heaters, water heaters, ranges, refrigerators, and furnaces. Most methane digesters are for large-scale agriculture, but you can find sources for smaller home versions. For example, Home Biogas makes a simple bio digester that converts organic waste into enough cooking gas to power a simple cooking stove. (See page 428.)

Alcohol and Other Biofuels

Alcohol or ethanol stills are capable of distilling alcohol fuel from wood products, excess grains, and agricultural waste. Biofuels are used to fuel appliances such as ethanol fireplaces and cooking stoves.

Conserve What You Have

Consider how you can conserve the fuel you have. An earth-sheltered home that utilizes both passive and active solar techniques might be the ultimate in conservation, but there are many other things nearly anyone can do to reduce energy needs. See the Quick Check on increasing the thermal efficiency of your home to determine your home-heating efficiency.

> **QUICK CHECK**
>
> ### Increasing the Thermal Efficiency of Your Home
>
> ✓ Add insulation to ceilings and walls.
> ✓ Insulate heating ducts and electrical outlets.
> ✓ Insulate water tanks and pipes.
> ✓ Use storm windows and doors.
> ✓ Install double-pane windows.
> ✓ Heat only a portion of your home.
> ✓ Take advantage of passive solar energy.

Sanitation and Personal Care

Inadequate sanitation and hygiene during a major disaster can be deadlier than the disaster itself when disease and virulent epidemics follow. It requires considerable forethought and preparation to provide proper sanitation and good hygiene when you are without running water, a functioning sewer system, and regular garbage service.

Sanitation in a Crisis

Sanitation lines can become blocked, broken, or infiltrated with excess stormwater. Pumps can malfunction in an extended power outage. In these circumstances, you'll be confronted with three challenges. First, you need to be able to properly dispose of human waste. Second, you'll need to take care of liquid waste and garbage. Third, it is important to keep people and their living accommodations clean.

The Sanitation Challenge

Without the two to six gallons of water required for each flush and a sewer system to accept it, disposing of human waste can quickly become a significant problem. A satisfactory solution should remove the waste from sight, eliminate odor, and prevent water contamination and the spread of disease by insects and rodents.

Short-Term Human Waste Disposal

A simple portable toilet using plastic bags or one using chemicals and a built-in holding tank can be good short-term solutions, but they are dependent on how long the chemicals last, the number of plastic waste bags you have, and the storage accommodations for those bags.

Improvised Toilet

First, consider using your existing toilet by lining it with a heavy-duty disposable plastic bag. Disconnect the flush lever or otherwise restrain it to avoid inadvertent flushing. Although you will not be able to flush the toilet, it will be reasonably comfortable to sit on. Use a chemical toilet deodorant after each use. Reliance makes both Double Duty

---

FIVE THINGS YOU CAN DO NOW

1 Stock up on toilet paper, paper towels, and hand sanitizer for several months.

2 Purchase a flyswatter and flypaper.

3 Purchase and store a bag of lime or other disinfectant.

4 Purchase heavy-duty waste bags.

5 Purchase a toilet seat to fit a five-gallon bucket.

Sanitation and Personal Care

Toilet Waste Bags and Bio-Blue Toilet Deodorant Chemicals. They also make Bio-Gel Waste Gelation, which solidifies liquid waste and makes it easier to move. (See page 428 in the resource section.)

Alternatively, a five-gallon plastic pail, bucket, or paint can with a tight-fitting lid could be fitted with a toilet seat. Several commercial emergency toilets, such as the Luggable Loo from Reliance, are made from a five-gallon bucket and an attached toilet seat.

Waste Removal

The next challenge is what to do with the waste. You will need larger containers to empty the smaller toilets into but not so large you can't move them. A ten-gallon container is about the right size. It should be watertight, lined with heavy-duty liners, and have attached, tight-fitting lids.

You could also place a grocery sack inside another one the same size and fill the space between sacks with shredded newspaper or a plastic bag liner. One person will need five to ten gallons of space per two-week period for waste storage.

Portable Toilets

Portable toilets that use water usually have two holding tanks—one for fresh water and one for waste. Of course, when this toilet is full, you still have the problem of where and how to properly dispose of the waste.

Disinfecting Waste

Bleach is often recommended as a disinfectant, but it can interfere with natural composting, so check with the guidelines for your portable toilet. A small quantity of bleach or disinfectant should be added after each use to control germs and odors. Also store at least a quart of fly-and-insect spray to control insect problems. Table 38.1 recommends the sanitation supplies you should store.

Waste Disposal

Without better facilities or options, you'll eventually be forced to bury accumulated waste. Dig a hole deep enough that you can cover it with at least one foot of earth to prevent insects from breeding and to discourage animals from digging it up. The hole should be located at least fifty feet downhill from any water source and where it will be unaffected by flooding or surface runoff.

A QUICK LOOK

When the Sewer Fails

- If the sewer system fails, absolutely do not flush your toilets. It will cause your sewage to comingle with the sewage in surrounding pipes and backflow into your drains—tubs and sinks included.
- Ensure that all household members, including children, understand the seriousness of the situation.
- Do everything possible to prevent flies, rodents, and pets from encountering sewage. Flies especially will fly from fecal material to your food, spreading fecal-borne pathogens.

Table 38.1
Sanitation Supplies to Store

· Bowl cleaner	· Lime, several bags
· Chemical disinfectants	· Plastic liners, heavy-duty hospital grade
· Deodorizer	· Portable toilet disinfectants
· Disintegrating toilet paper	· Quart of fly and insect spray
· Duct tape or large heavy-duty rubber bands	· Sawdust, straw, or cedar shavings, peat moss
· Fly paper	· Storage bags
· Gallon of bleach	· Wood ash
· Kitty litter	

Lime

Lime is important for decontaminating human waste and for controlling odors in latrines and outhouses. It effectively kills the pathogens in solid wastes by controlling the environment needed for their growth by both altering the pH and increasing the temperature. It also reduces odors and discourages flies and other insects.

Store lime in bags in airtight and watertight containers. Though not as effective as lime, other absorbents that aid in decomposition include sawdust, cedar shavings, wood ash, peat moss, and shredded straw.

Human Waste Management for an Extended Crisis

Straddle-Trench Latrine

For more extended use, dig a straddle trench latrine in soil, one foot wide, two and a half to four feet deep, and two to four feet long. Leave the soil in a pile near the trench to be used to cover excrement after each use. As the latrine becomes full it should be covered over and a new one dug.

For more comfort, a makeshift toilet seat or a regular wooden chair with a hole cut in it may be placed over the trench. Reasonable privacy can be provided for by screening the area from view.

Urine Soakage Pits

Urine-soakage pits for men and boys can reduce the use of the trenches. These pits are lined with rock or other nonporous material. Its diameter will depend on how many will use it. It should be about four feet deep.

Deep-Pit Latrine or Outdoor Privy

A deep pit latrine or outdoor privy is more permanent. Begin by selecting a site downwind and 150 feet away from any water sources and away from areas that may flood.

It consists of a hole that is two feet wide, three and a half feet long, and ideally six to eight feet deep. This width will accommodate two toilet seats. More shallow holes will need to be moved more frequently.

Toilet seat lids should be fly-proof and, ideally, self-closing. Dirt should be packed around the toilet box to prevent flies from entering. This latrine will provide the average-size family with facilities for a couple of months.

> **QUICK CHECK**
>
> ## Outdoor Privy Safety
>
> ✓ Ventilate the pit itself with a vent and screened pipe.
> ✓ Ventilate the top of the outhouse with cross-ventilating screened openings.
> ✓ Do not use chlorine bleach; it forms toxic chloramine vapors when mixed with ammonia.
> ✓ Do not let small children use it unaccompanied to prevent them from falling in.
> ✓ Avoid using cells phones or valuables in an outhouse to prevent loss. Do not try to retrieve valuables.
> ✓ Do not use matches or cigarettes in an outhouse.

Point-of-Use Individual Packet Latrines

Another option is a self-sanitizing, single-use, biodegradable bag designed specifically for personal sanitation. These are lightweight and easy to transport, may be used along with a reusable bucket, and have a two-year shelf life. You will need to store one per person per day. It's vital they are handled hygienically and are safely disposed of—not in open areas. Two brand-names are Peepoople and Go Anywhere Toilet Kit (Wagbag).

Outhouses

Traditionally, outhouses have been built of wood, but materials that are not porous, such as concrete or stone, make them easier to clean. The outhouse itself should have ventilation and natural light, but not direct sunlight. Openings should be covered with screens to cut down on flies.

Waterless Composting Toilet

Composting toilets are considered safe, clean, and efficient. They have an aerobic chamber where, over time, the waste is turned to safe humus. However, they are expensive. A composting toilet may be self-contained, where the composting takes place in a chamber within the toilet. Or the composting unit may be separate from the toilet. The composting mechanism may be slow- or active-composting.

The best models are completely hands-off and come with fans and heaters to speed up the composting process and alleviate odor. This means they require a power source, which may be supplied with 12v or 110v systems. Some come with a solar panel for power. Also look for one with a urine separator and DIY plans that are effective and reduce cost.

Septic Tank with Leach Field

Perhaps the best long-term solution is to have a septic tank with a leach field to back up the regular sewer system. A properly designed system, given minimal care, will serve for decades. A valve that can easily switch from the sewer system to the septic system is ideal. Another option is a holding tank, but it would require regular emptying.

Liquid-Waste Disposal

Though not as critical as human-waste disposal, thought must be given to how to dispose of wastewater. This includes liquid kitchen wastewater, bathwater, and water used for cleaning laundry.

Kitchen Wastewater

Kitchen wastewater has bits of food and dish soap in it. It can be disposed of in a soakage pit or trench or in an evaporation field. All three methods should use a grease trap to separate grease from the water.

Grease Trap

A grease trap is a strainer that separates and collects the grease from kitchen wastewater. Use it to prevent grease buildup from attracting insects and rodents. To minimize grease in your wastewater, remove as much grease as possible from dishes and pans before washing.

For small amounts of kitchen wastewater, you can make a strainer from material you can burn later. Strain wastewater through crisscrossed sticks layered with matted grass. Or make a strainer by punching holes in a cardboard box and layering it with straw, hay, sawdust, burlap, fine bark chips, or any natural thing that burns. You may also use a layer of ashes to filter grease. Burn the trap when it is filled.

Gray Water Use and Disposal

Gray water is the water used for bathing and laundry. It can be captured and used in many ways, just not for drinking or food preparation.

Garbage and Rubbish Disposal

Garbage may be defined as food waste and rubbish as nonfood waste. As piles of garbage and rubbish will attract insects and rodents, be serious about minimizing garbage and rubbish in the first place. Take seriously the advice to "reduce, reuse, recycle," separating garbage and rubbish into categories based on how you'll dispose of it.

How to Dispose of Garbage and Rubbish

Food waste can be fed to animals, especially pigs. Dogs enjoy leftover lean meat scraps and small amounts of animal fat drizzled over dry dog food. Compost unusable vegetative garbage in compost piles.

Garbage that cannot be composted should be buried under at least one foot of earth. Garbage that must be stored indoors for some time (such as in a fallout shelter) will keep better if first drained and then wrapped with newspaper and grocery sacks. Garbage cans should also be lined with paper. Do not store garbage in airtight plastic bags because the decomposing garbage produces gas that could make the bag explode.

Rubbish also includes paper, plastic, glass, and metal. Zoning laws may regulate open fires, but in a crisis, you may need to burn unusable paper trash. It is best to burn in an enclosed device, such as a burn barrel or chimenea. Any remaining refuse, if not reusable, must be stored and then periodically hauled away or buried in a deep pit. Keep it as clean and compact as possible. Flatten cans to reduce bulk. Have one or more twenty- to thirty-gallon cans with tight-fitting lids for this storage.

Garbage Pests

Store rat traps and poisons to keep the rodent population under control. Owning a cat also helps.

Keeping Things Clean

Personal Hygiene

Personal cleanliness is important in preventing disease and fungal infections. It also contributes to mental and emotional well-being. Obvious health practices should be scrupulously observed. Wash hands well before preparing and eating food and after going to the toilet, assisting an ill person, or handling any material that may be contaminated. Hand sanitizers are useful as well.

When running water is not available, baths may be taken by pouring a small amount of heated water into the tub, assuming you have sewer capacity. A sponge bath can be taken with as little as a cup of water. Showers can be taken with a Stearns SunShower (pictured) or Zodi propane shower. The SunShower costs about thirty dollars. Zodi propane showers begin at about $200. (See page 339 in chapter 36.)

Stockpiling

Many of the items you might need for sanitation in a crisis can be homemade, but storing them is easier and less stressful. One way is to determine needs by tracking current usage. Or you can estimate. Based on averages, store twenty-six rolls of toilet paper and twelve to eighteen bars of soap per person per year.

PERSONALLY SPEAKING

During the COVID-19 Pandemic of 2020, the toilet paper quickly disappeared from the shelves. Though there was no actual shortage, people hoarded it in a panic-buying frenzy. No one wants to be without toilet paper. So just how many rolls of toilet paper you should have in reserve? In the United States, on average, people use 100 rolls of toilet paper per year. For a family of four, that adds up to 400 rolls of toilet paper or about twelve giant 32-roll packages!

For me, toilet paper falls under "Things I Can't Live Without" — my list of essential things that make life comfortable and more pleasant. I try to keep about a six-months' supply. My list also includes Kleenex, acetaminophen and ibuprofen, sunscreen, hydrocortisone cream, dishwashing liquid, and dishwasher and laundry detergents. And there are personal things, like hand cream and body lotion, toothpaste, my favorite hair products, basic makeup and makeup remover, and a stash of dark chocolate. My husband's list would include Chapstick, Q-tips, aftershave, and contact solution. Give some thought to what might be on your list of things you can't live without.

Feminine Hygiene

Feminine hygiene products are typically disposable pads and tampons. Stockpile enough for regular use, but also consider using or at least storing reusable pads or menstrual cups.

Babies

Babies need diapers. Store both disposable and cloth types as well as baby wipes for cleanup. If you store cloth diapers, make sure to store plastic pants and diaper pins.

Home Cleaning

Make sure you have a supply of your favorite cleaning products. Table 38.2 lists common household cleaning supplies and equipment. Store all cleaning supplies in a dry location away from food. Liquids should be tightly capped to prevent evaporation.

Carpeting

Carpeting can be a tremendous breeding ground for insects if left uncleaned, so at least have a manual carpet sweeper. Consider that wooden floors with areas rugs are easier to keep clean.

Personal Care

Table 38.3 lists basic personal-hygiene supplies

Table 38.2 Cleaning Supplies and Equipment	
· All-purpose liquid cleaner · Ammonia · Brooms · Carpet shampoo · Detergent cleaners · Drain cleaner · Dustpan · Flexible cable clean-out · Floor mops · Furniture wax	· Household bleach · Mop buckets · Oven cleaners · Paper towels · Rags · Scouring powders · Scrub brushes · Sponges · Toilet-bowl cleaner · Toilet plunger

A straight razor with strop, stone, brush, and shaving soap could replace razor blades. An average can of shaving cream lasts for about sixty shaves.

Table 38.3 Personal Hygiene Supplies		
· After shave · Combs · Deodorant · Emery boards · Facial moisturizer · Facial tissue · Hairbrushes · Hair-cutting tools	· Hand cream · Head-lice treatment · Insect repellant · Lip balm, lipstick · Nail clippers · Nail files · Personal wipes	· Razor blades · Shampoo and conditioner · Shaving cream · Sunscreen · Towels · Washbasin · Washcloths

NOTES

Medical and Dental Emergencies

Being ill or injured is bad enough under normal circumstances, but it could be catastrophic should medical assistance be severely limited or not available at all. And any crisis may compound the challenges related to a chronic health condition such as diabetes, heart disease, or COPD.

A crisis that results in many injuries and physical trauma or a widespread epidemic or pandemic will strain all parts of the medical system. You must be prepared to help yourself as much as possible.

This chapter is not intended to offer medical or dental advice but to give you a starting point for making your own decisions about medical and dental preparedness.

Disease

It is important to be aware of the potential for disease so that you can act to prevent it. Depending on the crisis, there are several ways disease can impact you.

During a major extended crisis caused by a natural disaster, water sources will likely be compromised and sanitation inadequate. Diseases such as typhoid fever, cholera, and dysentery proliferate in contaminated water. Insects and animals spread diseases like malaria, plague, typhus, giardia, and tularemia.

Without adequate refrigeration, sanitation, and clean water sources, food-borne diseases caused by bacteria pathogens, such as salmonella, listeria, *and E. coli*, and those from viruses, such as norovirus and hepatitis A, will be more likely.

Poor nutrition will weaken the body's defenses against diseases, such as influenza, hepatitis, meningitis, and pneumonia, that are spread by personal contact. In a prolonged crisis, malnutrition can lead to diseases like scurvy and rickets.

FIVE THINGS YOU CAN DO NOW

1 Update all immunizations.

2 Take care of preventative checkups for all family members.

3 Make a list of items you want to put in your first-aid kit.

4 Purchase a backup supply of the over-the-counter medications your family regularly uses.

5 Purchase a medical handbook or manual.

Preventative Medical Care

Preventative Immunizations

The best way to prepare for many diseases is to keep immunizations current. Health experts advise children be routinely vaccinated for diphtheria, pertussis, tetanus, polio, measles, rubella, and mumps, and tested for tuberculosis. Likewise, adults who were fully immunized in childhood should have a tetanus/diphtheria booster every five to ten years. Older adults should get pneumonia and varicella immunizations to prevent shingles.

Annual influenza vaccines are recommended, and vaccines may be advised for type-B hepatitis and types A and C meningococcal meningitis. Vaccines are also available for typhoid and paratyphoid, typhus, cholera, yellow fever, and plague. Vaccines are no longer given for smallpox since it has been eradicated, but be aware that a bioterrorism threat exists. Consult your physician on the timing for vaccinations and booster shots.

Preventative Check-Ups

Have regular medical checkups and get preventative or corrective surgeries done. Regular eye exams are also important. If needed, store spare eyeglasses in practical frames with a record of the prescription, and have an eyeglass-repair kit. Keep old eyeglasses in case you need them someday. Hearing should also be checked and the necessary items stored.

Medical Skills and Training

Improve your medical skills by taking courses in basic first aid, advanced first aid, and CPR. Consider home nursing, EMT (emergency medical technician), and paramedic training. The Red Cross and/or technical schools and colleges may offer training. The tools and supplies you stockpile will be more valuable if you have the skills and training to use them.

Nontraditional Medicine

Ignoring modern medicine is not recommended. However, it can be helpful to learn what plants and herbs may be used to alleviate common ailments in the absence of modern medicines.

Emergency Medical Kit

During a crisis where medical help is scarce, you may want to acquire a more extensive medical kit, one that goes beyond your typical first-aid kit. Whether you make up your own or buy a commercial kit, it needs to be adequate for the situations you anticipate. Augment commercial kits to meet your needs.

Table 39.1 suggests basic medical equipment and supplies, most of which can be purchased locally. Table 39.2 recommends more specialized, advanced medical equipment and supplies, most of which can be purchased online or through a medical supply store. Some of the items are intended for much more than simple first aid. Consider acquiring the suggested items even if you are not sure how to use them. In a crisis, there may be a skilled medical person available but a lack of the necessary equipment and supplies.

A plastic fishing tackle box is an excellent container for your medical kit. It is noncorrosive and portable, allows instant access, and keeps contents organized.

Table 39.1
Basic Medical Equipment and Supplies

Basic Medical Equipment	Basic Household Medical Supplies	
· Angular bandage scissors · Bulb aspirator · Bedpan · Enema kit · Fingernail clippers · Graduated plastic cylinder · Heating pad · Ice pack · Medicine dropper · Penlight · Pocket magnifier · Razor blades (4 single-edged) · Rubber sheeting · Stretcher and blanket · Snake bite kit · Thermometers, (oral/rectal/forehead) · Tweezers	· Aloe lotion · Baby oil · Baking soda · Calamine or Caladryl lotion · Chapstick · Cold compresses/ instant ice pack · Cornstarch · Dental floss · Desitin · Epsom salts · Fels-Naphtha bar soap · Gatorade (electrolyte replacement powder) · Hand lotion · Heating pad · Hydrogen peroxide · Iodine shampoo	· Isopropyl alcohol (70%) · K-Y Jelly · Mentholatum ointment · Moleskin · Olive oil · Petroleum jelly · Q-tips · Safety pins · Sanitary napkins · Sanitizer or germicide powder · Spenco 2nd Skin · Sunscreen · Table salt · Talcum powder · Zinc oxide · Ziploc bags, assorted sizes

Table 39.2
Advanced Medical Equipment and Supplies

Advanced Medical Equipment	Specialized Medical Supplies	
Surgical Tools · Operating scissors · Splinter forceps (tweezers) · Scalpel and surgical blades (#10, #11) · Suture scissors · Tooth forceps · Hemostats (1 straight, 1 curved) · Needle driver **Syringes and Needles** · Disposable syringes, assorted sizes (3-cc, 5-cc, 10-cc with drawing needles (#22 x 1½" needles) · Normal saline bottle (250 mL-500 mL) with tubing and #18 needle · Airway intubation kit · Nasogastric tubes (#12 and #16) · Stethoscope · Sphygmomanometer (blood-pressure cuff) · Baby oil	**Surgical Supplies** · N-95 respirator masks (2 per person) · Surgical scrubs (4-6) · Nitrile gloves · Betadine or Povidone disinfectant prep pads or solution (20) · Hibiclens antiseptic skin cleanser · Sutures, nylon with needles, assorted (10) · Sutures, absorbable with needles, assorted (2) · Steri-Strips, assorted · Surgical sponges (4) **Dressings** · Surgi-pads, assorted sizes (10) · Trauma dressing 18" x 22" (2) · Gauze pads, assorted sizes (100) · Telfa nonstick gauze bandages, assorted (10) · Butterfly bandages (10) · Adhesive eye pads (4) · Specialized coverlet bandages for fingers knuckles, knees (30)	**Tapes and Wraps** · Kerlix gauze bandage rolls (2) · Gauze rolls, assorted · Vaseline gauze, 3" x 36" (12) · Coban self-adhesive wrap · Adhesive tape rolls, 1" and 2", 2 rolls each **Broken Bones** · Fiberglass splint rolls (2) · Resin plaster bandage roll (2) · Cast liner (2 3" rolls) · Triangular bandages (2) **Miscellaneous** · Ipecac · Activated charcoal (100 g) · Burn sheet (sterile) · Tongue depressors (10) · Tourniquet, rubberized

Emergency Childbirth

To prepare for emergency childbirth, in addition to the listed items, you should store obstetrical pads, sterile cord ties, cord clamps, sanitary napkins, an eight-ounce peri squeeze bottle, diapers, and a receiving blanket.

Medications for a Crisis

Just as other life-sustaining necessities will be in short supply during a crisis, so will over-the-counter (OTC) medicines and prescription drugs, so be sure to store a good selection to meet your anticipated needs and to see you through a crisis.

Over-the-Counter Medications

Most of us take over-the-counter medications for granted because they are so easy to get. But in the panic that comes with a crisis, expect the shelves in your local Walgreens to be stripped of OTC medications. The solution is to rationally collect a supply of OTCs for your family's needs before a crisis occurs. Begin by listing all the OTC drugs your family regularly uses or might use in a crisis. Table 39.3 lists common OTC medications, but you may have favorite brands and other medications. The most important medications to include are pain relievers, such as ibuprofen and acetaminophen, as well as antibiotic ointments.

Table 39.3 Medications	
Pain Relievers · Aspirin (adults only) · Excedrin (acetaminophen, aspirin and caffeine) · Tylenol (acetaminophen) **Anti-inflammatory (NASID)** · Advil, Motrin (ibuprofen) · Aleve (naproxen) **Antihistamines (diphenhydramine)** · Benadryl (diphenhydramine) · Claritin (loratadine) **Antiseptics** · Alcohol · Hand sanitizer · Hydrogen peroxide	**Ointments** · A&D Ointment (skin protectant) · Aloe gel (soothes burns) · Hydrocortisone cream (anti-itch) · Neosporin ointment (antibiotic) · Tinactin Cream (antifungal) · Silver sulfadiazine (antibiotic) · Skin antiseptics **Antacids** · Maalox, and Mylanta (acid neutralizers) · Omeprazole (acid blocker) · Zantac (ranitidine) **Antidiarrheal Medications** · Kaopectate · Imodium (loperamide) · Pedialyte (rehydration electrolyte)

*Usually requires prescription

Prescription Medications

Prescription Medications for Chronic Conditions

If you depend on a prescription medication for a chronic condition, it should be a top priority for you to acquire an emergency supply in case the supply chain is interrupted during a crisis. Develop a trusting relationship with your doctor, who can help you acquire needed medications. One strategy is to conscientiously fill your maintenance prescriptions on the first day they become eligible to refill. Gradually, you can build up a reserve.

For type-1 diabetes, store at least a two- to three-month supply of insulin along with instant glucose, syringes, and other supplies.

Antibacterial and Antiviral Prescription Medication

It can be difficult to treat many life-threatening diseases without prescription medicines. In a survival situation, bacteria-fighting drugs, primarily antibiotics, may be the most important drugs to have in your medical kit.

There are a few precautions that go along with storing antibiotics:

- Get proper medical advice; do not rely on self-diagnosis.
- Get complete directions for their use, including dosage and duration.
- Be aware that using antibiotics when they are not needed can contribute to bacteria building up resistance to antibiotics.
- Be sure to complete any course of antibiotics prescribed so that bacteria are completely gone.

Also, consider antiviral drugs such as Tamiflu (oseltamivir) and Relenza (zanamivir).

Pain Relievers

Pain relievers can also be important drugs in an emergency. Consult with your caregiver about how to use OTC pain relievers for maximum benefit. Prescription pain relievers may be difficult to obtain, but don't expect your doctor to prescribe controlled substances, like opioids.

Natural Antibiotics

Table 39.4 lists natural substances that have antibiotic properties. You will need to research how best to use them.

Alternative Acquisition of Medications

Be responsible in how you acquire and use prescription drugs. My recommendation is that you consider the possibilities and decide what makes sense for you.

Internet Pharmacies

A legitimate internet pharmacy will require a prescription, will be licensed in the state where it is located, and will offer the opportunity to speak to a pharmacist. Be wary of internet sources that sell drugs without a prescription. There are too many unknowns about what you are getting—drugs may be counterfeit, contaminated, or past their expiration date.

Table 39.4 Natural Antibiotics
· Apple-cider vinegar
· Cloves
· Echinacea
· Garlic
· Ginger
· Grapefruit-seed extract
· Horseradish
· Manuka honey
· Onions
· Oregano oil
· Thyme
· Vitamin C

Veterinary Medicines

Sometimes, animal and human drugs are identical. In the United States, the FDA tightly regulates pharmaceuticals, both for humans and animals. Animal pharmaceuticals approved by the USP (United States Pharmacopeia) are often made in the same manufacturing plants as human pharmaceuticals and contain the exact same ingredients. By law, every pill, tablet, or capsule approved by the FDA must look unique from all others. Each has its own distinct color, pattern, and shape, helping you identify identical drugs.

However, medical experts advise against using drugs meant for animals. Unless they are identical (as described above), you cannot be guaranteed they are safe for humans or be confident of the correct dosage. In addition, self-diagnosing is not a wise practice and could lead to the misuse of antibiotics or other drugs.

Drugs from Foreign Countries

You may want to investigate prescriptions acquired abroad at often considerably lower prices. Be aware that there are rules about how to legally import foreign drugs. Also, the FDA does not oversee the genuineness and safety of foreign drugs.

Storing Medications

For maximum shelf life, store medications in a cool, dry, dark location. Also, it's good practice to store medications under lock and key, especially controlled substances.

Expiration Dates for Medications

It's important to use medication before the expiration date stamped on them, which is required by law. Check that your pharmacist includes this information.

> **QUICK CHECK**
>
> ### Tips for Storing Medications
>
> ✓ Organize medications systematically.
> ✓ Make a list of your required prescriptions.
> ✓ Record the proper use and dosage.
> ✓ Label all medications.
> ✓ Store in original containers.
> ✓ Keep an electronic copy of prescriptions.
> ✓ Rotate medications, using oldest first.
> ✓ Store in a cool, dry, dark location.

The expiration date is important because drugs lose their potency. However, it has been shown that most medications remain at least 90 percent effective long past their expiration date. Be aware that some drugs, like tetracycline, become toxic with age. A few drugs let you know when they are no longer good (aspirin smells like vinegar, for instance), but others do not. Ideally, rotate all drugs before their expiration dates to assure effectiveness. A pharmacist will be able to help you with shelf life and any unique storage conditions needed for medications. For other tips for storing medications refer to the Quick Check table.

Preparing for a Pandemic

A pandemic is a widespread infectious disease transmitted from person to person. Whether a disease is classified as a pandemic is determined by the World Health Organization (WHO). An epidemic is more regional in scope. Begin preparing by following common-sense preventative health practices, such as washing hands, sanitizing surfaces, avoiding contact, getting enough sleep, etc. Also, be sure to get recommended immunizations, including those for influenza and secondary infections, like pneumonia.

Pandemic Emergency Measures

- Use reliable news sources, such as the CDC and local health departments, to keep up-to-date on diseases and their potential to cause problems, and respond accordingly.
- Wash hands properly and frequently.
- Use hand sanitizers and wipes; keep them at work, in your car, and in your purse.
- Avoid touching surfaces in public places (door handles, elevator keys, escalator handrails, restrooms, etc.).
- Use disinfectant wipes on shopping carts, door handles, work spaces, etc.
- Wipe down the airplane tray and armrests with sanitizing wipes.
- Teach children good health practices, especially proper handwashing.
- Where possible, work from home.
- In an extensive pandemic, practice social isolation by staying home and preparing to shelter in place.
- Use masks when traveling or in public places.

PERSONALLY SPEAKING

It was a lot of fun teaching fifth- and sixth-grade science. Especially anything that was "hands on." In the microorganisms unit, I wanted my students to understand how germs are spread. I learned about an activity from one of my teacher friends. It began by "infecting" one of the students by rubbing an "invisible" GloGerm powder on his or her hands to represent germs (this powder is visible under a black light). The student was then instructed to shake hands with three other students, who then shook hands with three more students and so on until every student had shaken hands. We turned out the lights, and I shined a black light on the students' hands. Some part of the GloGerm powder, or "germs," was evident on all students' hands. This activity impressed upon the students (and me, as well) how easily germs are spread. When the activity was finished, I told the students they could go wash their hands. When everyone was finished washing, I surprised them by turning out the lights again and once more checking their hands with the black light. Not surprisingly, quite a few students still had GloGerm on their hands. This part of the lesson showed them that a thorough washing is needed to get rid of germs.

How to Wash Hands in a Public Place

1. Look to see what hard surfaces you must touch in the process.
2. If you must touch a handle on a paper-towel dispenser or faucets handles, get your paper towels ready first and set them aside.
3. Wet hands with water, hot or cold.
4. Apply soap and lather hands for twenty seconds, or the length of time it takes to sing the "Alphabet Song."
5. Wash between fingers, on the backs of your hands, and under fingernails.
6. Rinse your hands.
7. If you must touch the faucet handle, use a paper towel to turn off the water.
8. Dry your hands.
9. Use the paper towel to open a door if you must touch the door handle.
10. Dispose of the paper towel.

Pandemic Emergency Kit

It is highly likely that during a pandemic, hospitals and care facilities will be inundated with sick people. Plan now to take care of yourself and your family to the extent possible. Make simple preparations for a pandemic now. When the threat is high, these items will be hard to get. Some of them are listed in previous tables, but for convenience are again listed in table 39.5.

Because many viruses and bacteria enter through eyes, nose, and mouth, be sure to use masks, respirators, and goggles. Hands spread infection to these entry points, so keep hands meticulously clean. Have a good supply of antiseptics, like soap, hand sanitizers, and hand wipes. Stockpile disinfectants, like Clorox wipes and Steramine disinfectant tablets, for cleaning hard surfaces. Store a supply of disposable nitrile gloves to use when caring for the sick.

Table 39.5
Pandemic Supplies
· Antiviral mask
· Particulate respirator
· Goggles
· Nitrile exam gloves
· Soap
· Hand sanitizers
· Antiseptic wipes
· Disinfectants
· Sanitizing tablets to mix with water
· Pain relievers
· Cough and cold medicine
· Stomach medicine
· Antidiarrhea medicine
· Antiviral medication, like Tamiflu or Relenza

You may need to quarantine and seal off part of your living spaces and dedicate a room as a sick room. Store duct tape and plastic sheeting for this purpose. As explained in other parts of this manual, be sure to have a supply of food and a way to purify water in case you will be isolated.

Medical Library

Information in hardcopy format will be important in a prolonged emergency without modern communication systems, such as the internet or video channels. Medical references should be part of your survival library. They will be critical in an emergency and for getting the most from your stockpiled medical supplies.

There are many volumes containing quality information, and it's easy to end up with a long list of books costing hundreds of dollars. The problem comes in recommending a smaller number of books without too much duplication.

Best Survival Medical Reference Book

The Survival Medicine Handbook: A Guide for When Help Is Not on the Way by Joseph Alton MD and Amy Alton ARNP (updated 2013). Not only are the authors experts in the medical field, they have thoroughly researched effective historical and nineteenth-century practices that would be useful when modern medical help is unavailable. The authors are dedicated to helping people survive in a long-term survival situation. The book includes an exhaustive list of instructional videos as well as recommendations for a concise medical library. Refer to pages 432–433 in the resource section for more recommendations for your medical library.

Dental Emergencies

Preventative Dental Preparation

To prevent dental problems, practice good oral hygiene and regularly visit your dentist to correct any problems that may develop. Have all preventative and corrective work done as needed.

Store soft-bristle toothbrushes, dental floss, toothpaste, and mouthwash. For false teeth, keep an extra set, if possible, and any necessary supplies.

PERSONALLY SPEAKING

I checked with my dentist – and you should check with yours – to find out what he or she recommends for dental emergencies. Mine emphasized the importance of preventative dental care, the idea being that a small fix now will prevent a big problem down the road. For pain relief, he recommended ibuprofen augmented with Tylenol, as well as Orajel. In a worst-case scenario, use Super Glue to reattach a loosened crown, but be sure to reattach the crown right away to prevent other teeth from moving into the space.

Emergency Dental Kit

For dental emergencies that may arise during a crisis, purchase an emergency dental kit that will alleviate pain and provide a temporary cavity-filling compound.

DentalMedic by Adventure Medical Kits and Dr. Stahl's Emergency Dental Kit-Deluxe include oral anesthetic and a temporary cavity-filling compound. These products can be purchased individually as well.

Besides the dental supplies listed in Table 39.6, you will need detailed instructions on how to use them. A good reference is *Where There Is No Dentist* by Murray Dickson (2012).

Table 39.6 Dental Supplies
· Temporary filling mix of a half ounce of zinc oxide powder USP
· ¼-ounce tube of Eugenol oil of cloves
· 5- to 15-g tube of Orabase with benzocaine
· Cotton pellets
· Bottle of benzocaine 0.5%
· Benzalkonium chloride toothache gel or drops
· Bottle of dental wax
· Dental cement
· Cotton rolls
· A mixing block with spatula
· A dental mirror

Communications, Transportation, and Protection for a Crisis

40 Communications

The availability of communications before, during, and after a crisis is vital. Communication prior to a crisis keeps you up to date and warns of dangerous situations as they develop. Good communication keeps family and friends informed and allows for more personal control over a situation.

With good interpersonal communication, you can learn about events that may be ignored or suppressed by the media or government officials. During a crisis, it may be the only way to stay in contact, coordinate activities, know what the present situation is, and call for help. After a crisis has passed, it allows for quick assessment and reorientation.

The main types of communication are radio receivers, including AM/FM and shortwave radios, two-way radios, mobile devices like cell phones, and the internet.

Radio Receivers in a Crisis

AM/FM Emergency Radios

An emergency radio should be durable, compact, and lightweight. Select a portable radio that can be charged with several power sources—a combination of regular AC, regular or rechargeable batteries, solar, or hand-crank. Many radios are capable of recharging small electronics, like your cell phone, and may come with a built-in flashlight. A USB port will help you connect to electronic devices.

QUICK CHECK

What to Look for in a Basic Emergency Radio

- ✓ Good sound quality
- ✓ Wide range of AM and FM frequencies
- ✓ Durable
- ✓ Compact and lightweight
- ✓ Multiple power sources
- ✓ USB port
- ✓ Solar-power capability
- ✓ Hand-crank power capability
- ✓ Digital tuner
- ✓ Headphone jack
- ✓ Extendable antenna
- ✓ Built-in flashlight
- ✓ Emergency Alert System frequency
- ✓ Weather-band capability

FIVE THINGS YOU CAN DO NOW

1 Purchase a basic emergency radio that has one or more alternative power sources.

2 Learn more about shortwave radio and consider it a survival skill to develop.

3 Learn more about ham radios and the requirements for a license.

4 Read about and investigate the American Radio Relay League. Evaluate whether ham radio would be a good survival skill to develop.

5 Make sure you have two or more ways to charge your cell phone.

Look for a radio receiver that has clear reception and a wide range of AM and FM frequencies. Also look for one with the NOAA weather band (WB) frequencies and Emergency Alert System messaging.

Satellite Radio

Satellite radio is subscription-based and uses satellites to broadcast its content. The value in a crisis is that the radio signals are broadcast over a large geographic area and the programming is the same coast to coast. You may still get news and information even if local stations are disrupted. Satellite radios are more expensive than conventional radios and require monthly service contracts. New cars are often equipped with them.

Scanners

You can use a multichannel, programmable AC/DC scanner to listen to local police, fire, ambulance, public safety, aviation, marine, and local, state, and federal government transmissions. However, to listen in on the many agencies that use trunk tracking (frequency hopping), you'll need a scanner with that capability.

You can also listen to scanners online or using smartphone apps. Broadcastify is a popular radio communications platform that streams public safety live audio over the internet.

Shortwave Radio

Shortwave broadcasts use the radio band lengths above the AM radio band. which are those higher (shorter) than 1,500 kHz. In the past, general-coverage receivers capable of picking up the basic shortwave (World Band) spectrum have been useful if you wanted to know what was going on outside the country or if you were looking for perspectives and interpretations of current events from a broader viewpoint.

But the popularity of shortwave has diminished significantly in recent years because of internet availability. There are fewer government and commercial international broadcasts in North America, Europe, and Australia than in the past. Those that remain offer fewer programming hours, and the program content is primarily religious or political. Shortwave broadcasts are more common in Africa, the Middle East, Asia, and South America, where internet service is not as widely available.

Shortwave Community

Despite its loss in popularity, shortwave radio is still used by amateur stations worldwide, and shortwave and amateur radio enthusiasts could be a powerful voice in a time of crisis and a legitimate source for broadcasting up-to-date news and information. The SWLing Post blog (www.swling.com) is dedicated to teaching about shortwave-radio communication and offers a forum for

QUICK LOOK
Reasons to Consider a Shortwave Radio
• A basic, simple technology
• Uncomplicated to learn
• Works everywhere on the planet
• Ultimate free speech medium
• Cannot be easily monitored or censored by government but could be jammed unless encoded
• No buffering or speed variations
• Inexpensive to the listener
• No apps or subscription fees required

the shortwave community. In the SWLing blog post "Does Shortwave Radio Have a Future?", Thomas Witherspoon argues for the benefits of shortwave radio in the internet age. The reasons are summarized in the chart "Reasons to Consider a Shortwave Radio."

Selecting a Shortwave Radio

If a shortwave radio is an important part of your preparedness objectives, begin by choosing a radio that has the reception range for the broadcasts you desire to listen to. The farther you are from the places you want to listen to, the better your radio and antenna should be.

You'll also need to decide whether you want a small radio for portability or a larger, permanent radio with better sound and receiving capabilities. In-dash car radios that receive shortwave are also available, as well as radios that can be adapted to run off a car battery.

The receiver should have digital tuning, a good signal strength, and an illuminated display, unless you want to avoid detection. It should include jacks for headphones and external antennas and allow for both AC and DC power sources. Make sure to store fresh batteries and the chargers you need to power whatever equipment you choose.

Multiple or double conversion is a necessity, and synchronous selectable sideband increases adjacent-channel rejection while reducing fading distortion. Direct-access tuning via keypad and preset memories is a convenient feature. Single sideband (SSB) capabilities will also expand your listening options.

As models change and specific recommendations soon become dated, refer to the SWLing Post blog for reviews and recommendations for the latest on shortwave equipment. No longer published, the Passport to World Band Radio was the primary reference for shortwave communication. Past issues are still in print and may be found online.

Reception

Long-distance reception is best at night and in the winter, when radio waves travel better. Higher frequencies work better than lower ones during the day and in the summer, but they are also most affected by sunspot activity.

Radio Antennas

The antenna may be even more important than the radio. Since the built-in antennas in portable radios may not be adequate, you may want to add a separate one. An outdoor antenna works better than an indoor one. Effective outdoor antennas can be made from fifty to a hundred feet of sixteen- to eighteen-gauge insulated copper wire. String the wire between the highest point possible, with insulators at each end. Directional dipole antennas and active antennas are also possibilities.

Radio Storage

Radios should not be stored with their batteries in place and should be protected from moisture, sunlight, and freezing temperatures.

Protecting against Electromagnetic Pulse Charges

Although there is some controversy about the extent of damage that can be caused from an EMP, it is good practice to take precautions. Radios can be protected by leaving them disconnected from electrical outlets and removing antennas over ten inches long. If possible, shield them by placing them inside metal, aluminum foil, or in boxes that have been electrically grounded and are covered with metal window screening. Small battery-operated radios that use only built-in short-loop antennas will not be affected by EMP.

Two-Way Radios

An effective way to communicate, there are several options of two-way radios with varying power and transmission capabilities. See table 40.1 for a comparison of the range for two-way radios.

Table 40.1 Average Direct Radio Range for Two-Way Radios				
Type of Radio	Power	Base to Base	Vehicle to Vehicle	Person to Person
Ham HF SSB	100 watts	30+ miles	15+ miles	xxx
Ham HF SSB Pack	20 watts	xxx	xxx	10+ miles
Ham VHF FM	20 watts	17 miles	7 miles	6+ miles
	5 watts	15 miles	5 miles	2 miles
CB SSB	100 watts	20 miles	11 miles	5 mile
	12 watts	15 miles	8 miles	3 miles
CB AM	5 watts	10 miles	4 miles	1 mile
MURS VHF FM	2 watts	12 miles	4 miles	1 mile
GMRS UHF FM	5 watts	9 miles	2 miles	½ mile
FRS HT	½ watts	2 miles	1 mile	¼ mile

Based on information from HFLINK at hflink.com/hfpack/radiorange

GMRS, FRS, and MURS Radio

Family Radio Service (FRS) and General Mobile Radio Service (GMRS) are private, short-distance, two-way voice and data transmission services. FRS has 22 channels and uses frequencies around 462 MHz. GMRS has 30 channels and uses frequencies around 462MHz, which it shares with FRS and 467MHz. GMRS can expand its capability by using repeater channels that allow it to transmit up to 100 miles. MURS (Multi-Use Radio Service) radios have 5 dedicated frequencies and channels around 151MHz and154 MHz. The radios are typically functional and rugged and can be useful in a crisis.

GMRS requires a license from the FCC and can be operated using up to fifty watts of power, but they usually function in the one-to-five-watt range. FRS and GMRS radios are allowed specifically for personal and business communication, operate with fewer than 2.0

watts, and do not require a license. Note that communication over these dedicated radio frequencies is not private and the lines may become overloaded during a crisis. In an urban setting, GMRS and FRS as well as CBs, would become swamped and practically useless.

CB Radios

The most common two-way radios are citizen band (CB). They operate on forty different channels within the 27MHz (11m) band. Channel 9 is the emergency channel. They have a legal maximum four-watt output and a normal range limit of a few miles. They are good for communicating at close range or in a caravan. They are another option to the phone system and can be used to report road and traffic conditions, pileups, roadblocks, washouts, and other problems.

Cobra, Uniden, and Midland make excellent models for less than $100. Small 100-milliwatt, handheld CB units with up to a one-mile line-of-sight range are less expensive. Mobile CB units are the most versatile, but base stations, which are normally not portable, have a wider range. Single sideband (SSB) units are more static free, with less interference, and have two to three times the power and range of the normal four-watt output.

Although modification is currently illegal, in an extreme crisis, modifications could be useful. With proper knowledge, the power of CB radios can be boosted up to four hundred times using linear amplifiers. Illegal frequency "sliders" also allow operating outside normal CB frequencies.

Amateur (Ham) Radio

The best long-distance, two-way communication method is the amateur, or ham radio. These radios have the longest range and, with proper backup power, are not dependent on the electrical grid. The argument can be made that they will be the only viable communication in a worst-case crisis.

Amateur Radio Licensing

To operate ham radios legally, you'll need an Amateur license from the FCC. You will also need knowledge about basic electronics and radio communications to pass the thirty-five-question entry-level test for the Technician license. Proficiency in Morse code is no longer required to get advanced radio licenses, but you'll need additional knowledge for advanced General or Extra Class licenses.

These advanced licenses will not only give you credibility with other ham operators, you'll be able to practice and develop a useful survival skill. You may want to join a club and learn the skills of operating and maintaining a ham radio.

American Radio Relay League

The American Radio Relay League (ARRL) is the United States' amateur radio organization. They publish a reference handbook each year with current information in radio technology, principles, and practices. They have an online presence at www.arrl.org and offer great information about preparing for the licensing tests.

Ham Equipment

Ham equipment is available in hand-held, mobile, and base units. As there are many variables in selecting amateur radio equipment, be sure to educate yourself to determine what is best for you. You can often get used equipment through ham radio organizations such as ARRL. It is also important to have a reliable power-backup plan. This could include a battery bank, solar energy, or a portable generator. You will also need a proper antenna, replacement parts, and a repair manual.

Mobile Devices and the Internet

Cell Phones

Cell phones and smartphones are essentially two-way radios that use local repeaters or cell towers to transmit their signals. Cell phones are convenient and can be used almost anywhere and with anyone in the world. Other devices, such as tablets, may use WIFI or a direct internet connection and allow you to download and store data.

Benefits of a Smartphone During a Disaster
· Store emergency phone numbers
· Easy access to contact information
· Sharing and receive up-to-date information
· Practical and emergency apps
· Recruit help through social media

However, this interconnected system can easily be disrupted, and some locations have limited access. During disasters, cell phone towers may be damaged or without power and unable to transmit. They are also susceptible to network gridlock when everyone is trying to use their cell phone during a crisis.

Regardless of their shortcomings, mobile phones are helpful in many crises. They can also dramatically increase your safety in situations as you can call for immediate help. Smartphones offer many useful features and applications. A few of the apps specific to emergencies and survival include GPS, Smart Compass, Emergency Radio Application, Boot Print-Pocket Survival, US Topo Maps, Weather Bug, Home Food Storage Tracker, SAS Survival Guide, American Red Cross Emergency, and American Red Cross First Aid.

PERSONALLY SPEAKING

One of my sons was hiking with a friend in a remote area in the Badlands of South Dakota. They became disoriented and were not sure where they were or how to get back to where they started. They did not have a map or a GPS device, and they could not figure out their location. Luckily, he had cell-phone coverage, so he called his brother. His brother found a GPS track in their vicinity on a hiking forum. The lost son was able to describe some of the geographic features nearby. Using Google Earth, his brother helped him pinpoint where they were. They then figured out which way to go to get back to an area they recognized and found a good route to take them back to the highway. There are several lessons to be learned from this story, but the one I would like to emphasize is that technology can be very useful in many different types of emergencies.

Cell Phone Chargers

Another big downside to cell phones is that they need to be charged frequently. A backup charger for emergencies is essential when the power grid is down. One option is a conventional portable charger that will recharge your device four to six times before it needs a new charge. Also, make sure you have all the connecting cords you need.

You might consider an emergency charger that runs off solar power or hand-cranking. Solar chargers come in an array of sizes and shapes and charge reasonably well if sunlight is available. Look for hand-crank models that have a low ratio of time spent cranking to charging ability. Do not expect either alternative energy to charge as well as a wall plug, but you get enough of a charge to make emergency calls.

The Internet

The internet is the communication medium of choice for sending information and getting the latest news from just about anywhere in the world. It was designed to work in times of major disaster by routing traffic around malfunctioning sections. You can access it with a variety of devices and connections, but it can become quickly overloaded during emergencies, and, of course, devices require some form of power to operate. It is also subject to hackers and cyber terrorists and can be critically damaged by destroying only a few key locations.

Security

If secure text and phone messages are important to you, consider mobile apps that encrypt your phone and text messages. If you desire secure or protected computer files, use encryption software for your programs. Signal and Silent Phone are examples of mobile apps that will encrypt phone and text messages. To encrypt emails and other files, use an encryption program such as PGP (Pretty Good Privacy). The open-source version is GPG (Gnu Privacy Guard).

NOTES

Transportation

Be Prepared on the Road

As Americans spend an average of one hour in their vehicles each day, it's important to consider what you'll do if your vehicle breaks down and you are stranded. Your vehicle will be your temporary shelter, and you want to be prepared if that happens. First, you'll need to store tools to make minor adjustments and, if possible, help you get back on the road. Second, you'll need protection from the elements.

Precautions are especially important if you are traveling long distances in open country or during the winter, when the weather can quickly change and make driving conditions treacherous. Excessive heat can also be dangerous.

Basic Maintenance

First, take care of basic maintenance, servicing your vehicles regularly. Keep your gas tank at least half full, especially in the winter. Routinely top off other fluids, such as antifreeze and windshield-wiper fluid. Make sure windshield wipers, lights, and tires are in good condition.

Preparing for Road Emergencies

The three road emergencies most likely to cause vehicle problems are running out of gas, a flat tire, and sliding off the road or getting stuck. For all three emergencies, you'll need flares, flashing lights, or emergency reflective triangles and at least one good flashlight, preferably one that clamps on or mounts magnetically, or a headlamp. Insulated overalls and gloves are also nice to have. See table 41.1 for a list of items to keep in your vehicle to manage emergencies.

Table 41.1 Things You Need in Your Vehicle for Roadside Emergencies	
· Air-compressor pump	· Jack and lug wrench
· Auto emergency shovel	· Jumper cables
· Bag of course sand	· Reflective triangles
· Chains	· Spare tire
· Deicer pellets	· Tire gauge
· Emergency flares	· Tire-plug kit
· Flashlight, clamp-on	· Tow strap
· Headlamp	· Water
· Insulated overalls	· Work gloves

FIVE THINGS YOU CAN DO NOW

1 Develop the habit of keeping the fuel tanks in your vehicles at least half full.

2 Regularly maintain your vehicle.

3 Keep a supply of maintenance fluids.

4 Purchase several tools that will help you maintain your vehicle.

5 Evaluate your existing vehicles to see how well they will serve you in an emergency

---

Running Out of Gas

The solution, of course, is not letting your tank run too close to empty. Know how many miles you can go on a tank of fuel. For long trips, plan your stops for refueling, or when you refuel, look ahead to where you'll refuel next. But no one is perfect, and running out of gas happens. Consider keeping fuel on board in an approved container, such as the ExploSafe jerrican pictured in chapter 37 on page 347.

Flat Tire

Be sure to keep a working, inflated, spare tire on hand. You will need a jack and a lug wrench to change a flat. Make sure you are familiar with how to use them. It is also useful to have a tire gauge. As an extra precaution, carry an air-compressor pump and good-quality tire-plug kit. Liquid tire repair sealants are not recommended—they may cause more problems than they fix.

Sliding Off the Road or Getting Stuck

If you get stuck in mud or on icy/snowy roads, several items will come in handy. You can use a shovel to clear snow or mud from around the tires. A bag or two of coarse sand, pea gravel, or even kitty litter in your car can improve traction. Deicer pellets help in icy conditions. And it's smart to keep a tow strap and chains in your vehicle.

Dead Battery

Frigid winter temperatures and the high heat of summer can wreak havoc with batteries, especially older or weaker batteries, so as a precaution, have a set of jumper cables. Be sure to replace batteries as they weaken.

Stranded–Keeping Safe

Being prepared requires that you respect the forces of nature. In chapter 5 (see page 35), I shared the tragic story of the family who perished on an innocent little trip to return their grandmother to her home after Thanksgiving. What can you do to prepare for and prevent such a catastrophe? See table 41.2 for a list of items for your vehicle to keep you safe if you are stranded.

Table 41.2 Vehicle Survival Kit	
All-Weather Emergency Supplies	
· Bottles of water · Emergency radio · Emergency snacks · First-aid kit · Map · Paper and pen	· Paper towels, wipes · Tissues or toilet paper · USB mobile device charger · Walking shoes
Cold Weather	**Warm Weather**
· Hand warmers · Ice scraper, snow brush · Matches, fire starter · Sleeping bags, blankets · Space blankets	· Water (extra), two quarts per person · Windshield heat shade · Tarp for shade

Cold-Weather Survival Kit

Especially during cold weather, keep warm clothes, hand warmers, and blankets in your vehicle. Sleeping bags tightly stuffed in bags make efficient use of space. Also, be sure you have bottles of water and a few snacks.

PERSONALLY SPEAKING

A few days after Christmas many years ago, Jack and I, and our two-month-old baby were traveling from Western Oregon, where my family lived at the time to our home in Utah. It was a fourteen-hour trip, and we were driving it in one day. Our car started having serious engine problems halfway into the trip, and by the time we were in the middle of cold, wind-swept Idaho, we were barely limping along. Our gas mileage was really suffering, and about five miles out of Burley, Idaho, the worst happened—our car sputtered and we coasted to a stop on the shoulder of the freeway. Now what? I felt panicky—no roadside assistance, no way to call for help, below-freezing temperatures, and nothing but to rely on the compassion of strangers. Luckily, in just a short time, a nice guy stopped to see if we needed help. He was kind enough to take Jack to the next town to get a can of gas. While I waited, I was grateful for warm blankets for my baby and a warm coat for myself. It ended well for us, but I've always remembered how vulnerable I felt.

Warm-Weather Survival Kit

The most important warm-weather survival item is water. You should have at least two quarts per person, store them out of the sunlight, and rotate them frequently. The ideal place is in a dry cooler inside the trunk. Also use a windshield heat reflector to keep heat from radiating into the car.

Roadside Service

Consider joining a roadside service, such as AAA, especially if you travel extensively or in remote areas. Their assistance can ease minor inconveniences and make a difference in an emergency. A satellite assistance such as Onstar is also helpful in an emergency. Plans vary and cost between $150 and $350 a year.

Your Transportation Needs

Your transportation needs will vary depending on the type of crisis you face. You may simply need alternative transportation for routine activities during a crisis. Or you may need to temporarily evacuate your home and quickly move your family and a few possessions. Or perhaps you foresee the need to get away to a distant location or retreat during a prolonged crisis. This may require that you be prepared to travel a long distance.

Various options for transportation are discussed in this chapter. As you consider them, try to imagine what your transportation needs will likely be during a crisis. No single mode of transportation will be ideal for every circumstance, and it is likely that several transportation options could benefit you.

General-Use Survival Vehicles

General Use Vehicles

It makes sense that any vehicle you choose for preparedness should also fill multiple purposes in your everyday life. Consider a general-use vehicle that is capable of hauling loads and towing. Heavy-duty pickup trucks, sport utility vehicles (SUV), and heavy-duty vans are best for the job. Pick the one that best accommodates the combination of passenger and cargo space you require.

PERSONALLY SPEAKING

In 1978, Jack and I bought a brand-new Chevy Blazer, but really, it was his vehicle. He very purposefully considered all the options he would get in this SUV. He saw this as a survival vehicle even back then and was deliberate about his choices. He somehow convinced me that we should not get air conditioning in it—too much could go wrong with it, and he wanted to keep it as simple and fixable as possible. He chose options to make it a tough machine—at least in 1978. It came with four-wheel drive, heavy-duty front springs, heavy-duty shock absorbers, engine-oil cooler, an extra-capacity fuel tank, a cold-weather package, a trailering package, and heavy-duty tires. He didn't choose options that made it look "bad" but internal options that made it functional and durable. Now, over forty years later, my son has the Blazer and is updating all the systems. The pretty, metallic-red exterior is sun-faded, and the buckskin interior is worn, but this vintage vehicle is still a solid little SUV.

Traveling to a Safe Place

If you're concerned about being able to get to a safe place or retreat, you'll want a vehicle that is dependable and able to safely carry the expected load of passengers and gear. Ideally, it should have good ground clearance and be able to travel off-road and navigate ice and snow and other perilous road conditions. It must also have adequate fuel capacity to travel the distance required.

These criteria eliminate nearly all passenger cars and, again, suggest heavy-duty pickups, SUVs, and heavy-duty vans. A three-quarter- or one-ton pickup truck is best if you plan to pull a big load. See the Quick Look "Options in a Survival Vehicle" for the

Table 41.3 Options in a Survival Vehicle		
• Cold-weather package • Electric or PTO winch • Front brush guard • Fuel tank, oversized • Heavy-duty alternator	• Heavy-duty cooling system • Heavy-duty dual battery • Heavy-duty rear bumper	• Skid plates • Towing suspension package • Towing gearing

things to look for in a "survival" vehicle. See the Table 41.3 for the things to look for in a "survival" vehicle.

Older Models

Although modern vehicles have the advantage of newer parts and come loaded with many options, they also have computers and an increased complexity that might make it hard to diagnose and fix problems. One solution is to acquire an older model with fewer electronics and simpler mechanical features, though the engine, transmission, differential, driveline, wiring, suspension, braking, and cooling systems may need to be repaired or refurbished so the vehicle is in proper working condition. If you have the time and want to completely refurbish an old model yourself, you can guarantee your vehicle is in top condition and learn the skills necessary for upkeep and repair.

Vehicle Options

Big Engine versus Small Engine

Up to this point, we have discussed the value of full-size SUVs, and heavy-duty trucks and vans. If you are more concerned about cost and fuel economy and really do not need one of these big vehicles, look for a smaller work vehicle with some of the same capabilities.

Gasoline versus Diesel

Diesel engines may be preferred because of their better fuel economy, longer engine life, and especially engine torque and pulling power. However, they cost more and require special tools and training for maintenance and repair. And the biggest downside in a crisis is that because they burn diesel, it might be difficult to get fuel. Not only are there fewer diesel pumps, but in a major crisis, there is a good chance diesel will be rationed and available only for long-haul trucks that transport necessities and disaster-relief supplies.

Four-Wheel Drive versus Two-Wheel Drive

What about four-wheel drive (4WD)? It will depend on your situation and what your needs are. The gap between four-wheel drive and two-wheel drive is narrowing as two-wheel-drive trucks now come with the same suspension as four-wheel. Two-wheel-drive vehicles can be modified for higher ground clearance, improved traction, and improved suspension. They can also be equipped with locking differentials and other off-road equipment. Two-wheel-drive vehicles with a locking differential can be almost as effective as four-wheel drive in many situations. Careful driving and skill can also minimize the need for a four-wheel drive.

A four-wheel drive is helpful for towing loads if there is poor traction or for slippery surfaces. They definitely have the edge in extreme weather conditions and for off-road driving where traction is limited. However, they cost more and use more fuel. They are also more complicated and may require more specialized maintenance. If you do decide to get a four-wheel drive, consider getting one with solid front and rear axles. Solid axles are simpler, stronger, and easier to maintain and service.

Locking Differentials

If you decide to go with a four-wheel drive, consider differential lockers. They can mitigate the effects of low or no traction. By locking the differential, the wheels on the same axle spin together. This can be helpful in mud, snow, ice, sand, or other low-traction terrain.

Some four-wheel drives come equipped with rear differential lockers, but they can be added aftermarket as well. Generally, automatic lockers are more reliable, but manual lockers give the driver more control. The mechanisms used to engage the locker (air compressors or electrical switches and wire) add a level of complexity that increases potential failure of the system.

Standard versus Automatic Transmissions

Most vehicles today come with automatic transmissions—fewer than 2 percent have standard (manual) transmissions. However, you might consider a standard transmission if you're looking for simplicity because they are simpler to fix and can be maintained with basic tools. They are also less expensive to begin with.

Automatic transmissions, on the other hand, require specialized tools and training to properly maintain or repair. It is beyond the skill of most mechanics to repair automatic transmissions, let alone the average casual mechanic. However, if cared for properly, the automatic transmissions in today's vehicles are very reliable and seldom need repairs. It used to be that standard transmissions got better gas mileage, but that is not the case with today's automatic transmissions, which get as good, or better, gas mileage.

Fuel Economy and Simplicity

When fuel economy is important in selecting a vehicle, it may be less critical than other objectives. If you may have to service and maintain the vehicle yourself, choose simplicity. Be sure to have plenty of spare parts and select a common model so those parts are easier to acquire.

Towing and Winches

A towing package will come with heavy-duty suspension, a receiver/hitch for towing, and heavy-duty brakes. A winch is worth considering to help with rescue or getting unstuck. A power-takeoff (PTO) winch is preferred, but an electric one is a viable option. A manually operated come-along provides some of the same benefit as a winch at much less cost.

Under the Hood

Get the highest-rated battery available. A dual battery is preferred if available to provide backup and to power electronics and communications without endangering restarting. Get the heaviest-duty alternator and cooling system (radiator, water pump, fan). If you live in a cold climate, get a cold-weather package.

Other Options

You may want a heavy-duty rear bumper, skid plates, and possibly a front brush guard and roll bar. Also, you may want to consider buying an oversized fuel tank or having an auxiliary fuel tank installed. Tires should be load-bearing, all-terrain type and mounted on steel rims. Finally, you may want an inconspicuous color for the exterior so as not to draw attention.

Vehicle Maintenance

If you expect to maintain your vehicle during an extended crisis, you'll need adequate tools, supplies, and spare parts. The lists in tables 41.4, 41.5, and 41.6 are quite comprehensive, and you'll need to decide which items you should acquire and store for vehicle maintenance. At the minimum, store the tools and supplies necessary for changing your oil and oil filter and for repairing, changing, and rotating your tires.

Basic Tools

Your decision about which tools to store will depend on your skill level and to what extent you are likely to maintain and repair your vehicles during a crisis. It will also depend on how self-reliant you anticipate being. If you are not inclined to repair your own vehicles, just store those tools you feel are necessary for simple maintenance and safety. (See Table 41.4.)

Table 41.4 Basic Tools for Vehicle Maintenance, Repair, Safety, and Comfort			
Maintenance and Repair		Safety and Comfort	
· Bolt cutters · Chisel, ¾" cold · Continuity tester · Drive-socket set with T handle and 6" extension · Hammer, ball peen · Hammer, 2 lb. · Jack, hydraulic, 5-ton · Jack, high lift · Key set, hex · Key set, torx	· Mini hacksaw with extra blades · Pliers, channel-lock pliers · Pliers, linesman's · Screwdrivers, regular and Phillips · Spark plug socket · Wrench, adjustable · Wrenches, combination ⅜"–1"	· Air-pressure gauge · Inflation pump · Lug wrench · Plywood, foot-square ¾" · Tire chains · Tire-repair kit · Collapsible water container, 2-gallon · Jumper cables, 4-gauge · Siphon · Flashlight, spare batteries · Hand spotlight · Headlamp	· Ax · Come-along hand winch · Crowbar, 60" · Nylon tow strap, cable, or rope · Shovel · Blanket · Bottled water · First-aid kit · Fire extinguisher · Ice scraper/snow brush · Safety flares

Basic Supplies

You may also need basic supplies to repair and maintain your vehicles. Even if you are not skilled in maintaining your vehicles, you may want to keep these supplies handy so that a person with skills can help you. (See table 41.5.)

Table 41.5 Basic Supplies for Vehicle Maintenance and Repair			
· Baling wire · Brake fluid, 1 gallon · Coolant, 4–6 gallons · Denatured alcohol, 1 gallon	· Duct tape · Electrician's tape · Motor oil, 60 quarts	· Power-steering fluid · Radiator stop-leak · Starting fluid · Tire-patching supplies	· Transmission fluid · WD-40 or LPS25 · Wheel-bearing grease

Spare Parts

Some spare parts are necessary for regular vehicle maintenance. Others you may want in case a major part fails. Become an expert on your vehicle and store the spare parts that commonly break down, along with any special tools. Consult a good mechanic and check the evaluations in off-road magazines.

Use online videos and join internet auto forums for your vehicle to learn how to maintain and repair it. During a crisis, you may not be able to access internet sources, so get a detailed factory repair manual for each vehicle. The repair manuals by Chilton and Haynes are also good. Table 41.6 on page 394 lists common spare parts that may need to be replaced.

Table 41.6 Basic Spare Parts for Vehicle Maintenance and Repair			
· Air filters · Batteries, 1-2 · Battery cables Belts, replacement set · Brake master cylinder · Clutch plate, pressure plate and release bearings (manual) · Coil	· Friction plates, steel plate, rings, front and rear seals, complete gasket and rubber set (automatic transmission) · Fuel filters · Fuel pump · Fuses · Generator or alternator with spare set of brushes	· Headlights · Lights, bulbs · Oil filters · Radiator cap · Radiator and heater hoses · Shock absorbers · Solid ignition cables · Spark plugs, 1-2 sets · Starter brushes	· Thermostat · Tires, 2 spares · Tires, extra set · Water pump · Wheel bearings, seals · Wheel-cylinder rebuilding kits · Windshield wipers, 2 sets

Welding

If you acquire a heavy-duty off-road vehicle and use it as it was intended, you may occasionally need to make structural repairs on the mounts and frame. Explore the possibility of welding your own repairs. Since most welding is electric, you'll need a reliable power source, such as a generator. You might also consider braze-welding with torches, although this method requires oxygen and acetylene tanks, which may not be refillable in a crisis. Portable MIG welding kits that use rechargeable batteries are also an option.

Storing Basic Tools, Supplies, and Spare Parts

Begin by organizing tools and supplies into categories, such as hand tools, tools for tire repair, tools for getting unstuck, etc. Use toolboxes or totes for the various types of tools. You will likely want to store some of the tools in your vehicle. Others should be available for easy access in case of evacuation.

Batteries

Batteries are responsible for more "no-go" situations than any other vehicle component. Since nearly all batteries are almost impossible to repair or rebuild, store one or two spares. Optimally, they should be stored in a maintenance charger, which is a device that monitors your battery's charge and keeps the charge from falling below 12.4 volts. Extra batteries should be stored in dry, cool locations.

Tires

Consider storing extra tires. How many will depend on your perceived needs. The spare tire in most vehicles is intended as a temporary solution, and so you cannot rely on it for long-term use. It's a good idea to have an extra set of all-season tires designed for your type of vehicle. Ideally, they should be stored on a second set of rims because mounting and balancing tires can be challenging. If you live in a winter climate, your spare set could be a winter-/snow-tire set. Select the largest size suggested by the factory for your vehicle with the highest mileage range possible. You may want high-performance or all-terrain tires. Avoid oversized tires, which can cause additional stress on the steering components.

Storing Tires

Tires should be stored in a clean, dark, cool, dry location, preferably in plastic bags. Remove as much air from the bags as possible and secure with tape. This will help prevent the loss of lubricating oil in the tires. Store upright if possible, but stacking is also all right if they are not stacked too tall. Ideally, prior to storage, tires should be mounted, inflated to operating pressure, spun balanced, and deflated to about ten psi. If tires are mounted, they can be hung from hooks, but hanging unmounted tires will distort them.

Tires should not be stored near petroleum products or electrical equipment. Electrical equipment produces ozone, which damages the rubber.

Storing Vehicles

If you're going to store a vehicle for a long time, use a dark, dry, and cool location. Siphon out as much fuel as possible and run until completely dry. Remove the battery. Drain all the coolant from the radiator, flush the cooling system, and refill it with fresh coolant. Change the oil and oil filter, loosen all belts, and place the vehicle on blocks to prevent the tires from flat spotting.

If you want to store fuel for the vehicles, see chapter 37.

Trailers

Trailers can be used to increase the load capacity of vehicles. Standard trailers are an option for basic transporting, but they can be difficult to maneuver and turn, and their ground clearance is usually poor if you need to travel over rough terrain. However, there are several heavy-duty utility trailers made to handle off-road hauling.

Morris Mule Trailer Company offers several field- and trail-grade trailers. Located in Anniston, Alabama, their trailers are simple, strong, and flexible. They also sell do-it-yourself frames so you can build a custom trailer.

The Xventure line of off-road trailers made by Schutt industries are "severe" off-road trailers and military inspired. One of their models is pictured.

Other Motorized Land Vehicles

Mopeds and Scooters

A moped is a small motorcycle that uses pedals for starting. They usually use a one to two horsepower 50cc engine and will drive up to forty miles per hour. They weigh about a hundred pounds and can get eighty to two hundred miles per gallon.

Although the name implies motor and pedal, they are much harder to pedal than a regular bicycle. Get one with full suspension, a multi- or variable-speed transmission, and some way to carry cargo.

Scooters are also small motorcycles but without pedals. They weigh more and go faster than mopeds.

Off-Road Dirt Bikes

Like mopeds and scooters, off-road dirt bikes get excellent gas mileage. They are quick and maneuverable and excellent for getting around when roads get clogged and pathways are narrow. On the negative side, they are noisy, can carry only one other person and very little gear, and offer little protection from the elements or an attack. They should be under 250cc so that they will not be too heavy or hard to control in off-road use.

All-Terrain Vehicles

ATVs weigh 160 to three hundred pounds, have 50cc to 700cc engines, get around fifty miles per gallon, and can hit speeds of over seventy miles per hour. They go in sand, mud, ice, or snow and can haul small loads over rough terrain.

Snowmobiles and tracked all-terrain vehicles can be used in winter conditions.

Electric Trike

The Rungu Electric Juggernaut is a battery-powered, pedal-assisted, bicycle-like trike designed for off-road travel. It navigates sand, mud, and snow with ease. It is quiet, which is important if concealment is important to you. It is also more maneuverable and stable than a four-wheeler. The battery range is 15 miles, and the upper speed is 35 mph. It will carry 375 pounds (including the rider), and there are several versions ranging in price from $2,500 to $6,000.

Nonmotorized Transportation

Transportation modes independent of stored fuel supplies could become valuable if stored fuel supplies are all used. They are also valuable as backup systems. There are several options to consider.

Bicycles

Bicycles are the most common and efficient mode of self-powered transportation and would be excellent during a prolonged crisis if chosen wisely. The best choice is an all-terrain mountain bike. Most are made with an extremely strong, lightweight material and have multi-gears, oversized brakes, and suspension systems.

Table 41.7 Spare Bike Parts to Store	
· Air pump	· Patching kit
· Bicycle-repair manual	· Rear gear sprockets
· Bicycle tool kit	· Spokes
· Brake pads	· Tire rims
· Chains	· Tires
	· Tire tubes

Be sure to add a luggage-carrier rack, folding side baskets, or saddlebags for carrying loads. Also, mount a pedal generator and light. Be sure to store spare parts (see table 41.7).

Consider buying or building a two-wheeled bike trailer for hauling loads up to one hundred pounds.

Electric Assist Bikes

A pedal-assist bicycle, sometimes called an e-bike or throttle-assisted bicycle, uses electric power to help provide propulsion. A battery-powered electric motor helps the cyclist pedal the bike. The motor assist may be on either the back or front hub, where it either pushes or pulls the bicycle, or it may be in the center and integrated with the crank and gears, which makes it more efficient and better for climbing hills. Most electric bicycles use lithium batteries, which are lighter in weight and have a longer charge and lifespan than lead-acid batteries. Prices begin around $1,000, but expect prices to continue to go down.

Horses and Mules

Horses and mules can be used for riding, packing loads, or doing work around a homestead, like plowing and pulling wagons. A good horse can cover forty miles a day over rough terrain and more than twice that in an emergency. A packhorse or mule can carry one-fifth of its weight in gear or supplies, and a team will allow you to farm fifty to one hundred acres. Stallions are more difficult to control than geldings but are more desirable to breed with mares.

A mule has more endurance than a horse, makes a better pack animal, and is a good choice for draft work. Along with burros, they are more adaptable than horses to drier, less-vegetated areas.

Caring for Horses

Stalls are usually twelve feet square. Horses also need at least eight hundred square feet of corral and prefer two to four acres for grazing. Depending on the quality of grazing available, you may need to supplement that with up to two tons of grains, three tons of hay, and fifty pounds of salt for each horse per year. For a list of horse tack to store, see table 41.8.

Horseshoeing and Medications

If you expect to shoe your own animals, you'll need horseshoes and nails plus a forge, hammers, tongs, and files. If you'll want to haul loads, you'll need a wagon or cart with harnesses and hitches.

Store worming supplies and other medications to keep your animals healthy, and have them properly vaccinated. Daily exercise is important to keep them in good condition too.

Table 41.8 Horse Tack to Store	
· Bridle	· Nail trimmer
· Grooming equipment	· Ropes
· Halter	· Saddle blanket
· Hobbles	· Western rop-ing saddle
· Hoof pick	
· Horseshoes	

Sled Dogs

In areas with lots of snow, sled dogs can be a good form of transportation. A good dog can pull about two to three times his own weight at ten to twelve miles per hour for extended periods. You will need a freight sled, tow lines, and harnesses. They can also be used to pull carts when there is no snow and can carry approximately one-third of their weight. Sled dogs eat about 750 pounds of dog food per year!

Skis and Snowshoes

Skis and snowshoes can be effective for short-distance winter transportation. For cross-country travel, backcountry or alpine touring skis are best. They are lightweight and designed for varying terrain. Although regular cross-country skis can be used, the heavier mountain skis allow the use of heavy boots and are preferred. Ski skins give additional traction for mountaineering. Snowshoes are better in hilly terrain and not as dependent on the condition of the snow. Store snow goggles or glasses to protect eyes and prevent snow blindness.

Handcarts

Carts can be useful around the homestead for hauling heavy loads of dirt, compost and firewood. They can also be used as a last resort for carrying a sizeable number of belongings over long distances. Large tires help in rough terrain and a suitable cart can be homemade using twenty-six-inch bicycle tires or plastic motocross bike wheels. A waterproof tarp or other top is also beneficial.

The Polymule utility cart (pictured) is a high-performance utility handcart that is designed with uphill assist. It will carry 400 pounds over rough terrain. It also has a cargo carrier option for attaching to a vehicle.

Survival Firearms

General survival weapons include firearms, bows and crossbows, knives, machetes, and hatchets. They have multiple uses and can serve a variety of survival purposes. This chapter discusses firearms, while the next discusses other survival weapons.

The Second Amendment

We, as individuals, have the right and responsibility to defend our lives, our families, and our property. The Second Amendment of the U.S. Constitution recognizes that right. If a government does not trust its citizens with firearms and tries to disarm them, its motives should be questioned. Of course, with the right to own guns comes the obligation to be responsible for their safe use, handling, and storing.

If conditions such as civil breakdown, failure of law enforcement, or marauding gangs develop, firearms will be a vital survival tool. Even now, in many instances, you are the only one who can guarantee your safety. Having a defensive firearm and knowing how to use it to protect your family may be of even greater importance in the future.

> **"A well-regulated Militia, being necessary to the security of a free State, the right of the people to keep and bear Arms, shall not be infringed."**
>
> Second Amendment, The Constitution of the United States of America

PERSONALLY SPEAKING

Jack was knowledgeable about firearms, skilled in handling them, and a strong supporter of the Second Amendment. As you make preparedness decisions, you'll need to evaluate and decide what part firearms will play in your survival preparations. I recommend you try to approach your decision about gun ownership rationally and not rule out the possibility because of emotional arguments. Gain knowledge and the requisite skills to be comfortable with gun ownership if you do decide to have them as part of your plan.

Firearm Regulations

Firearm laws vary from state to state. In some states, it is legal to own and shoot fully automatic weapons. In others, most semiautomatic rifles are illegal. Check your local municipal and county laws as well. Some states have preemption of firearm laws, meaning that cities and counties cannot pass their own firearm laws. Some states do not, and you may need to follow more stringent city laws concerning firearms, ammunition, or ammunition magazines. For up-to-date laws, refer to the following two websites: www.opencarry.org/ and www.handgunlaw.us/.

Selecting Firearms for a Crisis

Firearms can be divided into two general categories for survival use—defensive guns and working guns. Defensive guns are primarily for self-defense. Working guns are used to provide food, to control predators and pests, and to protect you from dangerous animals, like wild dogs and snakes. Some guns overlap the two areas. Firearms can be further divided into handguns and shoulder-fired.

The Ideal Survival Gun

The ideal survival gun simply does not exist. The gun you need will depend on the specific survival situation and the tasks you require it for. Even the best gun in the hands of an untrained person will not be effective. On the other hand, a person skilled in the use of firearms can be effective with a mediocre firearm. Nevertheless, some guns are better than others for survival situations.

Selecting Firearms by Cartridge Size

Begin by choosing the cartridge sizes you want for your firearms. It will be easier to obtain ammunition for the most popular cartridges and those used by the military and police. Following those guidelines generally limits you to the cartridges listed in table 42.1.

Table 42.1 Recommended Common Calibers			
Semiauto Handguns	Revolvers	Rifles	Shotguns
· 9 mm Luger (Parabellum) · .45 auto · .40 Smith & Weston	· .22 long rifle · 357 Magnum · .38 Special · .44 Magnum	· .22 long rifle · .223 Remington (5.56 NATO) · .270 Winchester · .308 Winchester (7.62 NATO) · .30-06 Springfield	· 12-gauge shotshell

Take into consideration as you make your selection that some cartridges are interchangeable. For example, a .357 Magnum also shoots .38 Special, while the .44 Magnum also shoots .44 Special. Both are a good choice if you plan to reload because of their straight-wall shells. The fewer calibers you choose, the easier it will be to stock and use ammunition without confusion.

If you limit the number of firearm models, it will also make it easier to stockpile spare parts, and nonworking guns can be scavenged for parts, if necessary. For these reasons, redundancy is recommended.

Recommended Firearms for Preparedness and Survival

Recommendations for firearms will vary with the person making the them. Also, the number and type of firearms to stock in your survival arsenal will depend on your budget, philosophy, and the circumstances you anticipate.

Mel Tappan was an early proponent of preparedness and survival. His general firearm recommendations are timeless. (See Quick Look.)

Table 42.2 on page 403 lists good choices for survival firearms, both defensive and working, and includes rifles, shotguns, pistols, and revolvers. Of course, you may find others that suit your needs.

Handguns for a Crisis

Pistols and revolvers are the two main types of handguns. Their primary use is self-defense. They have the advantage that they can be concealed. Their range, however, is limited to fifty yards or less. Handguns come in single-action (SA) and double-action (DA, DA/SA), and striker-fired. Handguns, particularly semiautomatics, come with either metal or polymer frames. Metal frames are more traditional, but polymer guns (like Glocks and their clones) have gained immense popularity.

There are advantages and disadvantages to these options. Study their various features to decide which gun or guns are right for you, ultimately choosing the weapon you are most comfortable with. The internet article "Best Handguns for Beginners & Home Defense" at www.pewpewtactical.com has a good discussion of factors to consider when purchasing your first handgun. Some of you might appreciate this article "The Ten Most Popular Guns for Women" found at www.thewellarmedwoman.com.

QUICK LOOK

Minimum Gun Recommendations

Mel Tappan (1933-1980) was an early expert in the survival movement of the 1970s and was editor of *Personal Survival Newsletter.* He also authored the quintessential survival book *Survival Guns.* His minimum gun recommendations for each contributing adult are as follows:

- One hunting rifle plus an extra for every two adults
- One shotgun for every two adults
- One fighting handgun, plus an extra for every two adults
- One .22 rifle
- One working handgun
- One concealed handgun for every two adults

PERSONALLY SPEAKING

If you are buying your first handgun, there are a couple of things I recommend. First, do your homework. Talk to "gun people" you trust—people like my stepson, who loves and owns guns and is knowledgeable about them. Also, do your own research—it's easy to find informative discussions in gun magazines and online gun sites. Ask yourself why you want a gun. That will help you decide what type of gun you are most interested in. Next, narrow down the possibilities to come up with a couple of options. Now, choose a gun store or a sporting-goods store with a well-stocked gun counter. I recommend going in the middle of the day when the salespeople aren't rushed and can give you the attention you need. If you can, find a store that lets you test the guns. One of the best ways to experiment with several different handguns is at a gun range that rents handguns. Lastly, make sure to take the time to hold and handle the handgun and be sure it feels comfortable in your hands. Does it feel substantial but not too heavy? Is it smooth and ergonomically compatible with your hand? One gun I inherited from Jack felt fine except for an engraved cross-hatch pattern that irritatingly rubbed the crotch of my thumb when holding it. Little things like that can make a difference. Being a woman, I also especially care about the recoil. Once you have purchased a gun, learn how to shoot it safely and properly. Look for a gun range or a gun club that can help you.

Rifles and Shotguns for a Crisis

Rifles and shotguns are the two main types of shoulder-fired guns that can be used as defensive or working firearms. The two purposes, of course, may overlap. A .223 rifle is a popular defensive/working rifle and is most accurate under 250 yards. Shotguns are generally hunting guns but can be an effective defensive at ranges under forty yards.

The AR-7

The AR-7 is a popular choice because it is lightweight, corrosive resistant, breaks down to only sixteen inches, and uses .22 LR ammunition. It makes an excellent weapon for an emergency evacuation kit or cache.

The AR-15

The popular and adaptable AR-15 has a modular design that lets you build it to suit your preferences. (By the way, AR stands for Armalite Rifle, not assault rifle. Armalite is its original manufacturer.) Building one begins with the lower receiver, which the Bureau of Alcohol, Tobacco, Firearms, and Explosives (ATF) considers to be the actual firearm. The rest of the parts are "accessories" and can be purchased individually. The AR-15 shoots a common round (.223) that is easy to reload. It was designed to be field serviceable, lightweight, and reliable. There is almost no recoil, and so it's easy for anyone to shoot.

The M4

The civilian M4 is the carbine version of the M16—the military version of the AR-15. It features a 14-inch barrel and a collapsible stock and is versatile and effective in close quarters and urban and indoor environments. Though many consider it the defensive weapon of choice, you can't buy it legally. But a similar style firearm can be built using the AR-15 platform with a 14-inch barrel and collapsible stock. Legally purchasing a 14-inch barrel requires a background check.

The Mini-14

The popular Ruger Mini-14 semiautomatic rifle is often compared to the AR-15. A scaled-down civilian version of the military M-14, it's sometimes referred to as a ranch rifle. The Ruger Ranch Rifle has a wood or synthetic stock and a steel barrel. The wood stock makes the Mini-14 look like "just a hunting rifle" and gives it a lower profile than the metal/synthetic AR-15. The Mini-14 is known for its excellent trigger, maneuverability, and accuracy. Like the AR-15, it uses .223 cartridges.

Combination Rifle/Shotgun

The Savage 42 is a combination gun with rifle barrel sitting on top of a single-shot shotgun and is handy for taking care of different kinds of varmints. The earlier Savage 24 is no longer made but may be found at gun shows.

Table 42.2 Recommended Survival Firearms			
Defensive Firearms		**Working Firearms**	
Rifles	· AR-15 (.233) · .223 Ruger Mini-14 · .308 Winchester	Rifles	· .22 LR Ruger 10/22 · .22 Winchester 9422 · .22 Marlin · .22 Remington · .308 Winchester
Shotguns	· 12-gauge Remington 870 · Winchester Defender · Ithaca 37 · Savage 42 (combination)		
Pistols	· 9mm SIG, Glock, Beretta. Ruger, H&K · .45 Browning	Revolvers	· .44 Mag Ruger, Colt · .357/.38 Smith and Wesson · .22LR Ruger Mark II, Colt

Firearm Accessories

While there are many accessories you can add to a firearm, it's best to keep it simple. Begin by getting two to six extra magazines for any defensive weapons and good holsters for all handguns.

Shooting Aids

Adding a scope to rifles increases accuracy. A shooting sling provides additional stability to make the area between the rifle and the shoulder more rigid.

Suppressors

Suppressors, also known as silencers, may be a consideration if you want to avoid unwanted attention. Many states now allow their use while hunting. They are legal in forty-two states but require an application with the ATF. You can make your own suppressor, but legally, you'll need to register it and get preapproval from the ATF.

Modifying a Shotgun for Hunting

Shotguns designed for defense typically have shorter barrels with less choke. This makes them ideal for the closer ranges typically encountered in defensive situations. Shotguns used for bird hunting have longer barrels and tighter chokes to keep the shot closer together at longer distances. A good compromise would be a defensive shotgun with an extra ventilated-rib barrel and variable choke or interchangeable choke tubes. This would modify a defensive shotgun for hunting.

Storing and Caring for Firearms

Locking and Storing Guns

Any guns in your home should be stored safely away from children and to prevent unauthorized use. Besides locking the gun in a closet or gun safe, consider whether trigger or cable locks would work for your situation. Gun cases can physically protect guns but cannot prevent theft. Gun storage boxes attached to a closet wall or even inside a piece of furniture and equipped with a keyed entry pad offer quick access to defensive guns.

Gun Cabinets and Gun Safes

Guns can also be locked in steel gun cabinets or safes. Steel cabinets help prevent theft, are less expensive, and weigh less than a gun safe, but they also offer less protection. A gun safe is expensive, but it offers protection from not only theft but fires and floods. Be sure to purchase a gun safe large enough for your present guns and any guns you intend to purchase. Gun safes are also a good place to store valuable documents, jewelry, gold, etc.

Caching Firearms

Guns can be hidden either on or off-site in lengths of PVC pipe. Optionally, look for large commercial gun-vault burial tubes. Use an auger to dig a deep hole, place the pipe vertically in the hole, fill with the guns, keeping all parts together, and cover with at least one foot of soil. Use a high-quality preservative lubricant or rust preventative on all metal parts prior to placing them in the pipe. Be sure to have a method for locating your cache when needed.

Additional information on caching of weapons can be found in *Modern Weapons Caching* by Ragnar Benson.

Care and Repair of Firearms

If you want your guns to last, you'll need to learn how to care for them. Religiously clean and care for your guns after each use. Any amount of moisture can cause corrosion. Store cleaning kits with rods, tips, and brushes appropriate to your choice of caliber. Also, store plenty of cleaning patches, bore cleaner (solvent), and gun oil. Silicone or Teflon lubricants work best in cold weather.

> **You must take care of your guns if you want them to last. Get in the habit of cleaning your guns after every use.**

Spare Parts

It's always a good idea to have on hand spare parts for each gun to keep it operating. These normally include a spare firing pin, extractor, and assorted springs, but get the advice of a competent gunsmith if in doubt. Store the parts wrapped in an oily cloth.

Ammunition

Store at least two hundred rounds per firearm, more if budget allows. One thousand rounds per firearm would be better, although one thousand rounds of some large-caliber guns would be very expensive. In a crisis where law and order are marginal, all firearms and ammunition will be good barter items with a .22 perhaps being the best. Military-surplus ball ammunition should be boxer-primed so it can be reloaded. Buy high-base #4 buckshot and rifled slugs for defense and #6 shotgun shells for hunting.

Storing Ammunition

Military ammo cans are great for storing ammunition. For additional protection, seal the rounds in plastic before putting them in the can. Cartridges will be useable for decades if they are stored in a cool, dry location away from ammonia and oil-based products. Shotshells are more susceptible to moisture and may not last as long. Wrap them in plastic and store in airtight containers with desiccants.

Reloading

Another alternative to storing ammunition is reloading—a survival skill worth developing. It can also save on expenses. Hand-loads generally have a shorter shelf life than factory-loads and are not as reliable. Cartridges have a longer shelf life than individual primers and powders. You'll find a full discussion of reloading equipment and supplies at CrisisPreparedness.com.

Black-Powder Guns and Muzzleloaders

In many ways, muzzle-loading guns compare poorly with modern guns for accuracy, efficiency, and reliability. Muzzleloaders also are extremely slow to load and are single shot. There are some great ways to use a muzzleloader, and they can be quite safe with the use of Pyrodex instead of black powder. They also have massive stopping power. The guns themselves and their ammunition are not very expensive when compared to other modern guns, and muzzleloaders can be purchased without registration in all but a few states. Finally, they are a good backup to conventional firearms because you can cast your own bullets and make your own black powder.

One traditional muzzleloader, the flintlock long rifle, uses a flint-striking ignition system. The beauty of this rifle is its simplicity. It requires just knapped rock and black powder for its ignition. If all other ammunition sources are unavailable, you can make your own black powder from saltpeter, sulfur, and charcoal. If the muzzleloader is not a flintlock, you will need primers.

Air Guns

Air guns have a definite place in your survival arsenal. Effective air-rifle range is mostly in the twenty to fifty-yard range but can be up to about sixty-five yards, while pistols are limited to the fifteen to thirty-five-yard range.

The adult spring-piston type is the only type to consider. It is capable of propelling pellets at over seven hundred feet per second in rifles and over four hundred feet per second in pistols. Although the cost for these types of firearms can exceed $200, they are accurate and powerful enough to kill small game with little noise.

They also can be ordered unrestricted by mail. They are recoilless and provide inexpensive practice. The ammunition is lighter and takes up less space than other ammunitions. For reliability, be sure to buy only high-quality pellets.

Air guns will easily last a lifetime. Stock an oiling needle with some silicone chamber and spring oil. Also, have a spare-parts kit containing replacement mainsprings, O-rings, piston seals, and breech seals if your model uses them.

43 Other Survival Weapons and Tools

In addition to firearms there are other powerful survival weapons and tools you should consider. Many have the advantage of being silent, but they require practice to develop the skills necessary to be effective. Defensive weapons should also be part of your preparation. These also require skill so that they do not become weapons used against you.

Nonlethal Defensive Weapons

Most people prefer to avoid using lethal force. Firearms are expensive and in some jurisdictions are either illegal or require a lengthy application process. Of course, that doesn't prevent illegal weapons from being used against you.

One of the solutions is to acquire and learn how to use nonlethal defensive weapons. Plan and anticipate how you'll use them. Like all weapons, effective use requires practice. The purpose of many defensive weapons is to give you time to escape.

Chemical Deterrents

The family of chemicals known as lachrymators works by irritating the mucus membranes of the eyes, nose, and respiratory system. Pepper spray, mace, and tear gas are the main chemicals used for self-defense.

If you intend to use these products, be sure you know how to use them without making yourself an unintended victim.

Pepper Spray

Pepper spray is made from a chemical in chili peppers called oleoresin capsicum, or OC. It is most often pressurized in an aerosol container and causes inflammation of mucous membranes in the eyes, nose, and respiratory tract. Dog spray and bear spray are milder forms of pepper spray. Be aware that the blowback can cause problems for the user.

Pepper spray comes in different sizes and strengths. Leading brands include Sabre, Police Magnum, and Mace, Inc.

FIVE THINGS YOU CAN DO NOW

1 Research and list your top five survival tools.

2 Choose a nonlethal defensive weapon that is best for you.

3 Purchase an all-purpose utility tool.

4 Visit a knife specialty store and look at various options.

5 Sharpen the knives or tools you already have.

Mace

Chemical Mace is the brand name of proprietary defensive sprays. Because it was the original and is one of the most common brands, many refer to all pepper sprays as Mace. The original Mace was CN (*phenacyl chloride*), but today, Mace Brand spray is usually OC (pepper spray). Mace Brand also makes a combination spray containing OC, CN, and ultraviolet-marker dye.

Tear Gas

Though tear gas can be made of different combinations of chemicals, it's most commonly made from CS (2-chlorobenzalmalnonitrile), but it may also be made from CN (*phenacyl chloride*).

Tear gas is usually delivered in grenades. CS and CN tear-gas canisters can legally be carried in some locations but are strictly regulated by government entities, particularly in metropolitan areas, where they may be restricted to permit holders or banned altogether.

Tasers

Tasers can stop an attacker for at least five seconds. Their main purpose is to give you time to get away. They are illegal in a few states and regulated in others, with strict laws against improper use.

Personal Protection

Body Armor

There are other ways you can protect yourself besides weapons. Modern concealable body armor, made from Kevlar, Zylon, and other hi-tech materials is lighter, thinner, and more flexible than ever. It can be worn comfortably for long periods in almost any climate. It comes in different thicknesses for different levels of protection, with the heaviest capable of stopping bullets from .357 Magnum and 9 mm pistols.

Body armor is legal in most places in the United States, with some regulations and exceptions. Check your local laws to see if you are legally permitted to purchase and use body armor. Some companies also may only sell to law-enforcement agencies.

Self-Defense Skills

Finally, consider learning skills, like martial arts. Look for an instructor or school that focuses on self-defense. Gun training is another skill worth investigating. Gunsite Academy in Paulden, Arizona, and BSR offer tactical training.

Survival Training

Several survival schools teach outdoor survival skills and offer experiences in a wilderness setting. They are presented in different parts of the country and are usually led by an expert, often military trained. Each school has its unique niche and method. Besides physical survival skills, the best of these schools will teach mental and emotional strength, self-confidence, and environmental awareness.

Defensive Vehicles and Evasive Driving Tactics

If your situation warrants an armored car, you have a wide variety to choose from if you have the resources. Most are custom-built and have the "big, bad" look, while others, like the BMW X5 Security model, are not so obvious. Several companies will harden the car of your choice.

Evasive and precision driving tactics can be learned at Bill Scott's BSR driver schools in Summit Point, West Virginia, or Bob Bondurant's School of High-Performance Driving near Phoenix, Arizona. You can purchase armored cars, but they are expensive.

Crossbows, Bows, and Slingshots

Crossbows

Some consider bows the ultimate survival weapon because they are quiet and powerful. There are four main types of bows. The traditional bow has the least amount of power and is a challenge to shoot accurately. The recurve bow has curved ends that give it more power, but it is used mainly for archery and sport. The compound bow is an enhanced version of a traditional bow that uses cables and pulleys to add leverage. It is more powerful and more accurate than a traditional bow or a recurve bow and requires more strength to draw. The crossbow is light, quieter than other bows, and preferred by most bowhunters. It is easier to master than other bows and provides accuracy and power at ranges of fifty-plus yards. It requires considerable strength to draw it.

Bows

Bows can be purchased through the mail and require no registration. Like most weapons mentioned in this chapter, effective use requires regular practice. Store extra strings and quality arrows or bolts.

Slingshots

High-powered modern slingshots are engineered to be serious weapons for both hunting and self-defense. A good slingshot is strong, accurate, and silent. Like any weapon, it will take practice to perfect your skill.

Look for slingshots with an attached sight and a contoured handle with a comfortable grip. A slingshot may use rubber tubing or flat bands. Rubber tubing tends to be more durable and last longer. Flat bands spring back faster and allow for greater accuracy. Either way, be sure to have extra bands.

> **QUICK CHECK**
>
> ### What to Look for in a Good Slingshot
>
> ✓ Frame made of steel, aircraft aluminum, or strong modern plastic
> ✓ Tubing for durability
> ✓ Flat bands for accuracy
> ✓ Comfortable grip
> ✓ An attached sight
> ✓ Hollow handle

A hollow handle will allow you to carry extra ammo balls along with some basic survival equipment, like waterproof matches, fishing gear, a compass, etc. Some survival slingshots are specifically designed to shoot arrows as well as regular slingshot ammunition.

Knives

A good knife may be your most important piece of survival gear. It has many uses and can often substitute for a tool you may not have.

As with all gear, think through how you might use it, then match the knife to the uses. Most likely, you'll want one or more general hunting or survival knives. A fixed-blade sheath knife is preferred, but a locking-blade folding knife can be handy too.

Survival Knife

A good all-purpose survival and hunting knife begins with the blade. First, it should have a full tang that extends from the blade all the way through the handle. This will add strength to the knife and keep the blade from wobbling.

Look for a blade that is strong and holds its edge reasonably well. High-quality stainless steel is strong and holds an edge well. Carbon steel also holds an edge but is susceptible to rust.

The knife should be big enough to do the job but easy enough to maneuver. For most uses, look for a knife with a blade between four and seven inches long. Longer knives will be harder to wear and handle. The blade should be sharp on only one side and pointed at the end. The nonsharp side should be flat.

QUICK CHECK
What to Look for in a Good Survival Knife
✓ Full tang
✓ Strong, hard blade that holds its edge
✓ Comfortable grip
✓ Balanced
✓ Sharp, pointed blade
✓ Durable handle
✓ Well-fitting sheath

Handles come in many materials, shapes, and sizes. Choose one that is durable and feels comfortable in your hand. Be sure there is a guard between the blade and the handle to protect your hand from coming in contact with the blade during use.

Specialty Knives

The Wyoming Knife field-dressing knife, pictured, is unusual in design. It is small but efficient and has replaceable blades.

Pocketknives

A Swiss army knife is like a miniature toolbox in your pocket. There are also many other types you might want to consider, including boot and belt-buckle knives.

While there are many brands to choose from, be sure to look for quality. Swiss knives made by Victorinox or Wenger (the companies merged in 2013) are known for their craftsmanship.

Another classic pocketknife is the Schrade Old Timer knife. Despite a change in ownership in 2004, Schrade knives are probably the best buy for the money. The model illustrated is the Old Timer 960T Bearhead 2-Blade Trapper Knife. You can even get a paracord bracelet with a knife blade inside its buckle. The one pictured is the Para-Claw by Outdoor Edge.

Knives for Barter

Finally, because of their inherent value, knives are excellent for barter.

Crisis Preparedness Handbook

PERSONALLY SPEAKING

 I personally like the Victorinox Swiss Army knife because it's versatile and feels good in my hand. It's also compact and fits nicely in a pocket or purse. Two features I really love about it are the scissors and the tweezers. There seems to always be a reason to use them. Make sure you remove it before you try to go through airport security!

Machetes, Axes, and Saws

Machetes

A machete is another versatile tool with characteristics somewhere between a knife and a hatchet. It has a longer blade than a knife and is primarily used for slashing, cutting, and chopping. It can also be adapted for use as a weapon.

There are several style variations depending on the place of origin. Look for one that is durable and feels right in your hands. A wooden handle is warmer and less likely to be slippery than a plastic handle. Perhaps the best choice is a Micarta handle made of resin-impregnated laminates and natural materials.

> **QUICK CHECK**
>
> ### What to Look for in a Good Machete
>
> ✓ Appropriate blade style and length for your use
> ✓ Durable blade that holds an edge
> ✓ A full tang riveted in place
> ✓ Comfortable grip and balance
> ✓ Durable handle

The blade is usually between twelve and eighteen inches long but may be as long as twenty-eight inches. Shorter blades are easier to carry, but longer blades can do more work. Blades may be made of stainless steel, carbon steel, or high-carbon stainless steel. Carbon steel is tougher and less expensive than stainless steel but will rust and corrode. High-carbon stainless steel combines both rust resistance and durability.

Axes

Axes and hatchets are basic survival tools. A long-handled ax with a five-pound head, either single- or double-bit, is big enough for felling trees. Axes with lighter heads and shorter handles, such as the single-bit Hudson Bay or double-bit Cruiser styles, are excellent choices for homestead chores and general survival. A belt ax or hatchet is also handy.

Saws

Some for saws include felling and pruning trees, cutting lumber, and butchering animals. A good bone saw is helpful in butchering a large animal. Store extra blades. Quality folding saws are lightweight and will fit in a small space. One-man or two-man crosscut, bow, and bucksaws are useful for cutting logs.

Multi-Tools

Multi-tools are a basic tool kit in a pocket-sized package. They usually offer at least one knife blade, pliers, screwdriver blades, a can opener, files, and other useful tools.

There are many brands and models to choose from, each with its own strengths and weaknesses.

Decide on the size you want and the tools you desire, then compare models that fit your criteria. Locking blades are an important feature, as are wire cutters. Brands to consider are Leatherman, SOG, Gerber, and Victorinox. Avoid cheap imitations.

Honing and Sharpening Tools

Sharpening Knives

Knives, axes, and saws must be kept sharp if they are to serve you well. A sharpening steel hones (straightens) the edge of a knife to maintain the edge. It will not sharpen a dull knife blade. For actual sharpening, most experts recommend a manual rather than an electric sharpener.

A good manual stone-based sharpening system, such as the EdgePro Apex Sharpening System, Lansky Controlled Angle Sharpening System, or Gatco Five-Stone Sharpening System, is relatively easy to use and offers the flexibility of sharpening at any angle. Another option is a sharpening stone or whetstone. These may be made of diamond, ceramic, or natural sharpening stone. Electric sharpeners may be convenient and fast, but they can grind away the blade and damage a knife's tempering.

Sharpening Edged Tools

Sharpening tools is a skill worth honing. Blades on hoes, shovels, pruners, and garden shears will perform better if they are maintained.

Besides sharpening stones, you'll need files for tools like axes and hatchets that take a lot of abuse. Edges need to be filed straight before replacing them.

Table 43.1 Honing and Sharpening Tools to Store	
· Buck Honemaster · Ceramic sharpening sticks · DMT duo-fold stone · File · Honing oil · Pocket sharpeners · Rod-guided sharpening system	· Saw gauge · Saw-set tool · Sharpening steels · Strop straps · Whetstones, soft and hard

Other Tools

This chapter is not meant to be an exhaustive list of all the tools that might be useful for survival but to simply point out some of the more common and necessary ones. Table 43.2 lists other tools worth considering.

Table 43.2 Other Tools and Weapons to Store			
· Adze hoe · Backpack · Binoculars	· Compass · Crowbar · Mattock · Pickaxe	· Rope, cord, and twine · Sledgehammer · Splitting wedges	· Pry-bar · Tactical tomahawk · Woodchopper maul

Surviving Terrorism and War

What Is Terrorism?

Terrorism is a method of waging war used by the militarily weak or those who wish to hide their real agenda. The goal of terrorism is to make everyone feel vulnerable every moment and, therefore, to capitulate to the terrorists to have "peace" or safety.

Because there are so many ways terrorism can be carried out, it is extremely hard to totally prevent it. Excellent intelligence is the key to stopping it before it happens, though this may or may not be possible given the limitations of an open society. Terrorism has been growing worldwide, and you can expect increasing terrorist acts in the years ahead. This chapter will help you prepare for that possibility.

Surviving a Terrorist Attack

A QUICK LOOK
How to Increase Your Chances of Surviving Terrorism
1. You can decrease your risk by not living by or going near potential terrorist sites.
2. Prepare ahead of time. Store items that would be useful in responding to terrorist situations and their aftermath.
3. Know how to react to terrorist incidences properly. If you know how to respond to a situation, you can minimize its effects on you.
4. Good advice, in order of effectiveness, is to run, hide, or fight. The worst thing you can do in a terrorist attack is freeze.

Most people do not want to live in isolation and avoid worthwhile life experiences. Nor do they want fear to rule their lives. Most of the time, a terrorist act will not take place at even the most likely targets. It is, however, smart to consider the risks so you can make wise choices.

Assess the Risks

On January 29, 2002, in his State of the Union address, President George W. Bush stated, "We have found diagrams of American

In this order, the best things to do in a terrorist attack are run, hide, or fight.

FIVE THINGS YOU CAN DO NOW

1 Know how to respond to a terrorist attack and develop a family emergency plan.

2 Take steps to ensure your electronic access and computers are secure.

3 Purchase disposable N95 or similar gas masks and add them to your first-aid kit.

4 Obtain a copy of *Nuclear War Survival Skills* by Cresson H Kearney, either the manual or PDF version.

5 Purchase health supplements to help with radiation sickness, and potassium iodide to protect your thyroid from radiation.

nuclear-power plants and public water facilities, detailed instructions for making chemical weapons, surveillance maps of American cities, and thorough descriptions of landmarks in America and throughout the world." America and the rest of world are painfully aware that the threat of terrorism is real.

High Profile Public Places

Terrorists aim to wreak as much havoc as possible, seeking out high-profile public gatherings and venues, such as sporting events, concerts, nightclubs, airports, bridges, subways, churches, and schools.

Infrastructure

The destruction and disruption of infrastructure is a major terrorist objective. Military installations, transportation centers, dams, chemical and power plants, and biological labs are potential terrorist sites.

Types of Terrorism

Cyberterrorism

The prime targets for cyberattacks are often critical infrastructures, such as utilities and air-traffic control. The electric and internet grids are particularly susceptible. When one power grid fails, it can result in a domino effect, causing other grids to fail. Financial institutions are also a primary target.

Terrorism Using Hazardous Materials

Over two and a half million trucks have licenses to carry hazardous materials, and every one of them is a potential disaster. Explosions of LNG (liquefied natural gas) tanks and other fuels pose a risk. Even something as simple as a small crop duster can become an instrument of widespread disease and death.

Chemical Terrorism

Primary chemical agents include cyanides, mustard agents, nerve agents, and toxic industrial chemicals, and it's possible that new chemicals are being developed, which will increase the need for protective measures.

Cyanides

Hydrogen cyanide (HCN) and cyanogen chloride (CICN) are liquids that vaporize at room temperature. HCN smells like burnt almonds, and CICN has an acrid odor that causes choking. They are likely to be used in a confined area.

Mustard Agents

Mustard is a blister agent that attacks the skin and lungs. It may be clear to dark brown and has a garlic-like odor.

Nerve Agents

Sarin, tabun, and VX are nerve agents that disrupt the nervous system by blocking the transmission of nerve signals. They are difficult to produce without sophisticated chemical skill.

Toxic Industrial Chemicals

The trucks and trains that transport chemicals such as chlorine, phosgene, and organophosphate pesticides are potential targets for terrorist hijacking.

Biological Terrorism

Category-A Agents

There list of potential germ warfare agents is long (see table 44.1). The Center for Disease Control categorizes the most dangerous as category-A agents. This means they can be easily disseminated or transmitted from person to person, have high mortality rates, and can cause public panic and social disruption.

Table 44.1 Potential Biological-Terrorism Agents		
· Anthrax* · Botulism* · Brucellosis · Bubonic plague* · Foot-and-mouth · Listeria	· Melioidosis · Q fever · Salmonella · SEB toxin · Smallpox* · Tularemia*	· VEE virus · Viral hemorrhagic fevers* (Ebola) · Yellow fever
*Category A Agents		

Category-A agents include anthrax, botulism, plague, smallpox, tularemia, and viral hemorrhagic fever (Ebola). Each differs in how it is contracted, how it affects the body, what its symptoms are, and how it can be treated. Some of these agents have a mortality rate over 90 percent if left untreated.

Category-B Agents

Class B agents are also dangerous but have a low mortality rate. These include E. coli, hepatitis A, ricin toxin, salmonella, and West Nile virus.

Yet unknown biological agents may evolve and become a threat.

Radiological Terrorism

These weapons combine radioactive material with conventional explosives. Sometimes called "dirty bombs," they are not as catastrophic as nuclear weapons but can wreak havoc when spread over a small area. (See "Dirty Bombs," page 419)

Nuclear Warfare

Nuclear weapons have proliferated around the world, and nuclear contraband has been smuggled across borders and into unsuspecting countries for years. Not only do major powers like Russia and China have thermonuclear weapons, but many rogue nations and some terrorist groups potentially have access to radioactive materials and even suitcase nuclear bombs.

Terrorists do not need to be smart enough to design and build a nuclear bomb. All they need do is pack high-explosives with enough spent nuclear waste and blow it all over a population center. The resulting radiation could quickly kill many and contaminate an entire geographical area.

Other Forms of Terrorism

Truck and bus bombs may become a regular occurrence and not just in the Middle East. Recently, the trucks themselves have become the terrorist weapon, mowing down people in crowded places. Pressure-cooker bombs are another type of threat.

Making a Terrorism Survival Plan

There are several things you can do to prepare and lessen the potential threat of terrorism.

First, keep a low profile. Avoid high-risk targets and areas, especially during holidays. If you live in such an area, consider relocating.

Second, make a family crisis plan detailing what you would do if forced to evacuate, how you would stay in touch with each other, and where you could meet. Select both a local gathering site a few miles from home and one farther away that would most likely be unaffected by the crisis. If you have children, be proactive in planning and knowing what their schools' emergency plans are.

QUICK CHECK
Family Terrorism Plan
✓ Where will you go if you are forced to evacuate?
✓ What will you take with you?
✓ How will you communicate with each other?
✓ Where will you meet up?
✓ What is the emergency plan at your child's school?

Decide under what conditions it would be a good idea to evacuate if asked to and when it might not be best. Consider how well prepared and knowledgeable local authorities really are. Are you willing to trust them in a crisis?

Third, gather any supplies or equipment that would prepare you to survive a terrorist attack. This could be particularly important if you work in a city and live in an outlying area. Chapter 33 discusses emergency evacuation kits in detail.

Responding to a Terrorist Attack

The first thing to do is to limit your exposure to the problem. Get uphill, downwind, or as far away as possible or seek shelter immediately. Close off air entryways for biological and chemical attacks. Put as much mass as possible between you and a nuclear attack.

Decontaminate yourself and your surroundings as much and as soon as possible. People and animals can be wiped with bleach solutions and then washed and showered with antibacterial soaps and cleaners. Clothes can be sanitized by washing in detergent with bleach or placed in plastic bags and safely discarded. Radioactive dust (fallout) should be washed off and moved as far away as possible.

If you absolutely must leave the safe room or shelter to go into a still-contaminated environment, wear whatever protective clothing is available. When you return, decontaminate yourself and your clothing again.

Seek medical attention immediately for any person injured by an attack. If this is not possible, use whatever supplies and knowledge you have on hand to treat the injuries. Focus on treating symptoms and trying to make the victim as comfortable as you can. And, of course, seek professional help as soon as it is practical.

Terrorist Survival Kit

Although some of the items will likely already be part of your other crisis preparation, they will be listed here again. Many of the items may be part of your emergency evacuation kit discussed in chapter 33.

Water

Store at least a two-week supply of water, a filter or purifier, and another method of water purification, such as purification tablets or liquid bleach. A 0.2-micron filter will remove bacteria. A 0.01-micron filter will remove viruses and some chemical agents. Charcoal filters will also remove some chemicals.

Food

Store enough nonperishable food to last at least three days. Two weeks would be better if there is a possibility you could be quarantined in your house.

Portable Radio

You need to know what the authorities are saying about the current dangers and what to do about them. At the very least, get a radio with a hand crank and solar-power source. The best survival option is a ham radio. If you want to know what is going on outside the country, you'll need a shortwave radio.

Communications

Do not count on cell phones to work in a crisis. At the very least, the lines will be jammed. Nevertheless, have an off-grid method for keeping your cell phone charged in case you can use it. Landline access may also be unavailable. Walkie-talkies and CB radios may offer some communication.

First-Aid Kit

Every home should have one. Make sure to keep it stocked and the items fresh and useable.

Flashlights and Batteries

As attacks may not happen during daylight hours, keep a couple of flashlights handy. Either store fresh batteries or rechargeable flashlights.

Gas Masks

Adequate gas masks can protect you from chemical and nuclear attacks but are only effective for biological agents if you know you are being attacked. Most biological agents are odorless and colorless, so you would not likely know.

The first and best option for most people is an air-purifying mask. These are the most affordable masks and have replacement filter canisters. Ideally, they should cover your eyes as well as your nose and mouth and maintain an airtight seal. Be aware that facial hair makes them difficult to seal. The one pictured is the Honeywell North Full Face Respirator, 7600 Series.

The second option is a self-contained breathing apparatus (SCBA). These are commonly used by firefighters and require that you carry a full tank of air. These tanks are heavy and take up a lot of space.

Disposable Masks

Low-cost, disposable N95 masks are an adequate alternative to regular gas masks for the short-term filtering of larger particles, such as anthrax spores. These dust masks are widely available for less than a dollar. They filter down to 0.1 micron and will remove around 99 percent of biological agents, thought they are ineffective against chemicals or gas.

Goggles

These may be useful to protect your eyes from chemicals hazards if your gas mask only covers your nose and mouth.

Protective Clothing

Hazmat suits are labeled A to D. A is the highest level of protection and includes total encapsulation. It protects against mists, vapors, gases, and particles and is used with a SCBA and steel-toe boots. B and C are not vaporproof. B uses a SCBA, while C uses a respirator. D protects only from chemical splashes.

You can improvise protective clothing for biological, chemical, and nuclear attacks from disposable Tyvek coveralls, rubber or latex gloves, rubber overboots, and even rain ponchos. Duct tape can be useful for sealing gaps in clothing.

Cleaners and Disinfectants

Store a supply of chlorine bleach to purify water and to make disinfectant solutions. Mix water with regular bleach in a 10:1 ratio for cleaning skin. Wash off within five minutes and then rinse with soap and water. Use the bleach solution to wipe surfaces. Also store a supply of antibacterial soaps, cleaners, and detergents with bleach.

Antibiotics, Nerve-Agent Antidotes, and Painkillers

Stock up on antibiotics for bacterial biological threats. Even a three-day supply makes it more likely you'll survive until the authorities get additional antibiotics to your area. Atropine and pralidoxime chloride are antidotes for nerve agents. Painkillers are also good to have on hand.

Vaccinations and Immunizations

Be sure to keep vaccinations and immunizations up to date. Experts tell us that smallpox is eradicated and there have been no new naturally occurring cases since 1977. However, the CDC does maintain a stockpile of immunizations in case of a bioterrorism attack.

Creating a Safe Room

If your home is located within the range of a biochemical attack, you can improvise a shelter if you've stored a few items beforehand. If possible, select an upstairs room because many nerve gases and other toxic gases are heavier than air and settle at ground level. Try to include a bathroom or even a kitchen in your safe area if you can.

Store several rolls of plastic packaging tape or duct tape and sheets of clear plastic sheeting. Plastic trash bags could also be used. Use these to seal doors and windows and to close off heating and air-conditioning ducts. Use rolled wet towels pushed against the bottom of doors to seal them.

Preparing for a Nuclear Crisis

Nuclear War Survival Skills

If you're contemplating how to prepare for a nuclear attack, see *Nuclear War Survival Skills* by Cresson H. Kearny (1914–2003), a civil-defense expert at Oak Ridge National Laboratories in the 1960s, at the height of the Cold War. He gives advice on how to survive as best one can when not prepared.

First published in 1978 and updated in 1987 and 2001, it has been released into public domain. It can be purchased inexpensively or is available for free online as a PDF.

Assess the Likelihood

Begin by determining if you live near a likely nuclear target. If you do, there may be nothing short of moving that will give you much of a chance of surviving. In this era of terrorism, targets include population centers where a nuclear attack would wreak as much havoc as possible. And while you might be thinking that heavily populated areas like Los Angeles, New York City, or Houston are the most likely targets, military installations with nuclear-war-related infrastructure and nuclear stockpiles, Minutemen III ICBM bases, and command control centers are actually the prime candidates.

Dirty Bomb

A dirty bomb, or radiological dispersal device (RDA) combines conventional explosive weaponry with radioactive material. It would not release enough radioactivity to kill or cause radiation related illness. Its main result would be to cause injury, panic, and fear.

Evacuation

Evacuating because of an imminent nuclear attack is not the ideal situation to be in. If you think you might have to evacuate, you'll want a copy of Nuclear War Survival Skills to take along with you.

Protecting from Radiation

There are three main types of radiation from a nuclear explosion: alpha, beta, and gamma. Both alpha and beta can be stopped by clothing, but more protection is needed for gamma radiation, which can penetrate much like an X-ray.

The 7:10 Rule of Thumb

Nuclear radiation has a half-life that varies for different types of particles. Most of the radioactive material from a nuclear explosion has a short half-life. The 7:10 Rule of Thumb is a quick way to estimate the dangers of radiation after a nuclear explosion.

The rule states that for every seven-times increase after a detonation, the amount of radiation decreases by ten. What does that mean? Practically speaking, that means that the deadliest exposure is immediately following the blast. The longer you can protect yourself and your family, the less danger.

Table 44.2 7:10 Rule of Thumb Example	
Time Past Detonation of Bomb	Amount of Radiation (hypothetical)
0	1,000 R/hr.
7 hours	100 R/hr.
49 hours (4 days)	10 R/hr.
343 hours (14+ days)	1 R/hr.
2401 hours (100+ days)	0.1 R/hr.

From the hypothetical application shown in the table 44.2, you can see that after two weeks, the radiation level has decreased by one thousand! While it's possible to protect yourself from the deadliest effects of radiation, this is not to say that there will not be other potential long-term problems resulting from nuclear damage.

Nuclear Shelters

In *Nuclear War Survival Skills*, Cresson H. Kearny recommends the use of shelters and describes six types of expedient earth-covered shelters as well as the items needed to stock them. This guide will be useful as you contemplate if and how you'll construct a shelter.

Expedient Shelter

If world conditions threaten nuclear conflict, you may find it necessary to build an expedient shelter, making do with the

Table 44.3 Tools to Build an Expedient Shelter		
· Ax or hatchet · Bleach · Bow saw with extra blade · File · Hammer · Nails, wire, rope · Pick (for hard-soil areas)	· Pillowcases or burlap bags (2 per person) · Pliers · Polyethylene film (4-mil) · Poly trash bags (4 2-gallon)	· Shovel, 1 or more · Siphon or pliable garden hose · Work gloves · KAP* · KFM* · Potassium iodide*

*The Kearny Air Pump (KAP), the Kearny Fallout Meter (KFM), and potassium iodide are explained in following paragraphs.

resources you have available, such as earth, rocks, concrete, etc. Table 44.3 lists useful tools and supplies that will help you build an expedient shelter. Refer to Kearny's Nuclear War Survival Skills for design recommendations. (See page 433 in resource section.)

Permanent Shelter

If you're not located near a probable target, consider constructing a better, more-permanent shelter. The more shielding between you and the radioactive fallout, the better. Based on FEMA recommendations, the minimum protection you'll want is PF-40, or protection factor (PF) of forty. Better to aim for a PF of 1,000.

A PF of 1,000 means only one-thousandth of the outside gamma radiation will penetrate inside. The 1,000 PF only requires about double the thickness of the 40 PF yet keeps out twenty-five times as much radiation.

As shown in table 44.4, 1,000 PF can be provided by two feet of concrete or three feet of dirt. Any material that is about three hundred pounds of material per square foot will be adequate for shelter walls and ceiling.

Table 44.4 Protection Factor of 1,000	
Material	Thickness in Inches
Lead	4
Steel	10
Concrete	24
Packed dirt	36
Water	72
Wood	110

Nuclear Survival Kit

Potassium Iodide (KI)

Nuclear fallout contains radioactive iodine 131. Accidentally inhaled or swallowed, it concentrates in the thyroid gland, destroying its functioning and causing delayed abnormalities and cancer. This is most serious for young children, especially babies. However, you can prevent this problem by storing a supply of stable potassium iodide for use as a blocking agent. You will need one hundred 130-mg tablets or one-half ounce of reagent-grade crystalline or granular potassium iodide per person. Tablets can be bought from several preparedness sources, and the powder can be bought in bulk from a local chemical-supply house without a prescription.

Realize that KI cannot protect you from the immediate effects of the high-level radiation of a nuclear blast and only protects one part of the body from long-term effects.

L-cysteine and Vitamin C may be helpful in preventing radiation damage. L-cysteine tablets can be purchased without a prescription at pharmacies, health-food stores, and chemical-supply companies.

Another drug that is useful in protecting from radiation is 5-androstenediol, an immune-system booster. Amifostine is a chemo-protectant medication currently used to reduce side effects for cancer patients during radiation treatment.

Radiation Meter or KFM

If you've sought shelter during a nuclear attack, you'll need to know when it's safe to come out again. A radiation meter can tell you. There are several radiation detectors you can purchase ranging in price from $20 to over $500. You can also make your own from common supplies for less than $20. Known as the Kearny Fallout Meter (KFM), it was developed by Cresson H. Kearny.

The KAP

Within a few hours, without adequate ventilation or in warm or hot weather, shelters can get so dangerously hot and humid that people will collapse from the heat. The Kearny Air Pump (KAP) is a shelter-ventilating pump that can be made at home with a few inexpensive materials. The plans for the KFM and KAP are found in *Nuclear War Survival Skills*.

Other Shelter Provisions

You also must make provisions for water, food, sanitation, light, and a few other needs.

Provide at least fourteen gallons of water per person and enough food for two weeks. The same types of foods used in the emergency evacuation kit discussed in chapter 33 would be appropriate here. In fact, plan to take your emergency evacuation kit with you into the shelter.

A portable toilet or hose-vented, five-gallon can with heavy-duty plastic-bag liners, toilet paper, and additional smaller bags, plastic buckets or garbage cans will be needed for sanitation. A total waste-storage capacity of at least five to ten gallons per person is necessary.

A supply of small, long-burning candles, matches, and flashlights with extra batteries is good, but additional sustained light is a real advantage. There are several options discussed in chapter 36.

Also, provide a first-aid kit with a tube of antibiotic ointment, blankets or sleeping bags, and a radio with extra batteries.

If you evacuate, take the KAP, KFM, potassium iodide and other supplies with you. Otherwise, keep them in the shelter.

Resources

Fictional and Biographical Books about Survival

Earth Abides by James Stewart (1949)
This novel, first published in 1949, is a classic postapocalyptic story where the main character, Ish, is faced with a world in which most of mankind has died from a fast-moving, airborne disease. Ish connects with other survivors and begins building a community from the ground up. Although he wants to hold on to the vast knowledge accumulated over the centuries, the reality for his children and grandchildren is very different.

Endurance: Shackleton's Incredible Voyage by Alfred Lansing (2015)
This is the story of polar explorer and captain of the ill-fated Endurance, Ernest Shackleton, who led his twenty-seven-member crew to survival and rescue in what is called the "greatest survival story of the twentieth century."

Life as We Knew It by Susan Beth Pfeffer (2008)
This young-adult novel is from the perspective of a teenage girl through her journal entries after an asteroid hits the moon, knocking it closer to earth with cataclysmic results. This is the first in a series that examines the challenges of a postapocalyptic world.

Lucifer's Hammer by Larry Niven and Jerry Pournelle (1979)
Lucifer's Hammer is the name of a comet that slams into earth, resulting in a chain of events that destroy much of civilization on earth. This action-packed novel is about the efforts of the survivors to live on and rebuild civilization. It does a good job of addressing different scenarios that might follow such a cataclysmic event.

One Second After by William R. Forstchen (2009)
This novel describes the aftermath of an EMP attack on America. The book illustrates how ill-equipped twenty-first-century Americans are to deal with life when it suddenly reverts to nineteenth-century conditions. It will give you plenty of things to think about in the event of an off-grid breakdown.

Patriots: A Novel of Survival in the Coming Collapse by James Wesley Rawles (2009)
This is an important novel in the hard-core survivalist world. It is based on the premise that the entire societal framework of America falls apart. A group of sound-minded survivalists use their skills and knowledge to survive the end of American civilization.

Strength in What Remains by Tracy Kidder (2009)
This is the story of Deo, a young medical student who escapes and survives the horrors and atrocities of genocide in Burundi and Rwanda. He makes his way to the United States and through much hardship eventually becomes a physician who establishes a clinic in Burundi. It is worth reading to experience Deo's personal strength and hope in the face of so much struggle.

Unbroken: A World War II Story of Survival, Resilience, and Redemption by Laura Hillenbrand (2010)
This nonfiction account is the story of American Olympian Louis Zamperini, who first survived forty-seven days on a precarious life raft in the Pacific, followed by more than two years of mistreatment in a hellish Japanese prisoner-of-war camp. It describes his struggle with alcoholism and finally his victory as a Christian.

Where the Wind Leads by Vihn Chung (2015)
This is the first-hand account of a Vietnamese family's struggle to survive and eventually flourish after the Vietnam War. When their property is confiscated by the communists and they see no hope for their future, they flee by boat. They are subjected to assaults and robbery by Thai pirates and are brutally traumatized by Malaysian authorities and set adrift to die in the South China Sea. Miraculously they are rescued and then begin the process of assimilating into a new culture.

Chapter 5 – Rule of Three

LAND SHARK INSTANT SURVIVAL SHELTER & STEALTH BAG is an immediate shelter that will help you stay dry, retain heat, and reduce the cooling effects of the wind. Cost is around $70.
WEBSITE: land-shark.com

Chapter 6 – Emergency Water Supply

WATERBOB by Way Safe Florida, Inc. is an emergency drinking-water storage container that fits your bathtub so you can fill it with water before a big storm or hurricane hits. It holds up to one hundred gallons and comes with a pump. Cost is less than $50.
WEBSITE: waterbob.com

RAINXCHANGE by Aquascape offers several rain-collecting options, from a simple decorative barrel to an underground storage system with a water feature.
WEBSITE: rainxchange.com

Chapter 7 – Storing Water

TITAN READY USA's storage system houses two to four 55-gallon water barrels that connect and stack horizontally.
WEBSITE: titanreadyusa.com

SURE WATER makes 260- to 525-gallon elliptical tanks. Their smallest tank is equivalent to about five 55-gallon barrels. The tanks are equipped with two spigots, one at floor level and one about two feet up, for easy access.
WEBSITE: surewatertanks.com

WATERPREPARED makes an elliptical, two-tank, stackable system that holds 160 gallons of water. They also sell the parts for joining tanks if you are a do-it-yourselfer.
WEBSITE: waterprepared.com

EMERGENCY WATER CORP offers the EZ Cycling System—a great way to recycle the water in a 55-gallon drum and maintain a clean water supply.
WEBSITE: emergencywatercorp.com

AQUAMIRA WATER TREATMENT is a chlorine dioxide water treatment used to condition water stored in containers or to purify contaminated water. One ounce treats thirty gallons and costs about fifty cents to treat one gallon.
WEBSITE: aquamira.com

H2ORESQ BY WATER PURE TECHNOLOGIES treats water for the prevention of biofilm growth.
WEBSITE: waterpuretechnologies.com

Chapter 8 – Purifying Water

STERIPEN makes a line of UV water purifiers for individual use. There are several sizes ranging in price from $50 to $100.
WEBSITE: steripen.com

KATADYN MICROPUR contains both chlorine dioxide and NaDCC and costs about fifty cents to purify one liter.
WEBSITE: katadyn.com

POTABLE AQUA manufactures both chlorine-dioxide tablets, called PA Chlorine Dioxide Purification Tablets, and iodine-based tablets, called PA Drinking Water Germicidal Tablets. It costs about fifty cents to purify one liter.
WEBSITE: potableaqua.com

AQUATABS are an effervescent tablet containing NaDCC and are used worldwide in varying sizes. They have a five-year shelf life. A package of thirty costs about ten dollars and will purify sixty quarts.
WEBSITE: aquatabs.com

OASIS WATER PURIFICATION TABLETS by Hydrachem contain NaDCC. They have a five-year shelf life, come in a variety of sizes, and cost about five cents to purify one quart.
WEBSITE: hydrachem.co.uk

POLAR PURE WATER TREATMENT is a lightweight, effective, and economical iodine purification treatment with indefinite shelf life. One bottle treats up to 2,000 gallons for about a penny a cup.
WEBSITE: polarpurewater.com

OPEN by Roving Blue is a pocket water purifier that creates aqueous ozone to kill viruses, bacteria, and protozoa. It comes with rechargeable batteries and a USB cable and costs about $150.
WEBSITE: rovingblue.com

P&G PURIFIER OF WATER is a water purification technology that purifies dirty water using three processes: coagulation, flocculation, and disinfection.
WEBSITE: csdw.org

Chapter 9 – Filtering Water

AQUAMIRA makes several lines of water filters including the Water Basics Red Line Filter that removes viruses, bacteria, and parasites to EPA standards.
WEBSITE: aquamira.com

BERKEY WATER FILTERS makes several freestanding, gravity-fed water filters that eliminate virtually all bacteria and viruses as well as most inorganic substances from water. USA Berkey Filters has been offering these filters for over fifteen years.
WEBSITE: Usaberkeyfilters.com

Chapter 11 – Principles of Successful Food Storage

SHELF RELIANCE makes fee-standing, front-loading can rotation systems. Large free-standing units cost around $500. Smaller units cost less than $50.
WEBSITE: thrivelife.com/all-products/shelving

FIFO CAN STACKER is a can rotation system that sits on existing shelves and accommodates a variety of can sizes. The units cost from $30 to $40.
WEBSITE: foodstorage.com Also available from national retailers.

Chapter 12 – Food Storage Conditions and Containers

JARBOX is a plastic storage container for pint- and quart-size canning jars. Each JarBOX securely holds a dozen jars and can be used for storage or transport. They are stackable to consolidate canning jars. Available from national retailers.

Chapter 13 – Packaging Your Own Dry Food Storage

USA EMERGENCY SUPPLY offers quality Mylar bags manufactured to their specifications. The bags come in a wide range of sizes, from pint to six-gallon. They also sell oxygen absorbers and other food storage buckets.
WEBSITE: usaemergencysupply.com

PACK FRESH USA specializes in oxygen-free storage environments and high-quality barrier film.
WEBSITE: packfreshusa.com

HOUSE OF CANS specializes in containers for a variety of uses and offers a large selection of cans for home processing.
WEBSITE: houseofcans.com

WELLS CAN COMPANY offers metal cans and sealers, including the All-American Senior Flywheel Can Sealer and the Ives-Way Can Sealer. They also offer vacuum sealers and supplies, home pressure canners and jars, and even beekeeping supplies and equipment.
WEBSITE: wellscans.ca

Chapters 14–18 – Food Storage

RAINY DAY FOODS specializes in products for home food storage and emergency preparedness. They have an excellent selection of grains, legumes, dehydrated and freeze-dried foods, and preparedness tools and equipment. You will have to pay for shipping, but definitely check them out. Located in Montpelier, Idaho, they will also sell directly from their warehouse.
WEBSITE: rainydayfoods.com

PLEASANT HILL GRAIN offers a full range of grains, legumes, and other food storage products. They excel in the number of different kinds of food processing equipment they offer, including grain mills and heavy-duty bread mixers.
WEBSITE: pleasanthillgrain.com

HONEYVILLE is an online store specializing in whole grains, seeds, dehydrated and freeze-dried foods, and baking products. They offer a complete line of preparedness and food storage items, including gluten-free alternatives.
WEBSITE: honeyville.com

USA EMERGENCY SUPPLY sells food storage, emergency supplies, survival kits, and dehydrated foods online. This company's information center offers thorough, best-practice advice about self-reliance, preparedness, and food storage topics, including packaging your own food storage.
WEBSITE: usaemergencysupply.com

HOME STORAGE CENTERS, run by The Church of Jesus Christ of Latter-Day Saints, offer wheat, dry beans, rice, powdered milk, and other commodities that have been prepackaged for long-term storage. Though the products are limited to about twenty basic food items, they are very reasonably priced and can be purchased online or in person at a center.
WEBSITE: providentliving.churchofjesuschrist.org

MOUNTAIN HOUSE specializes in freeze-dried foods and offers individual entrées as well as emergency food supplies. Their products are available direct or from sporting goods stores and emergency preparedness retailers.
WEBSITE: mountainhouse.com

EMERGENCY ESSENTIALS is an online store that sells individual dehydrated and freeze-dried foods as well as other preparedness and emergency items. They also have a good selection of blog posts about emergency and home preparedness.
WEBSITE: beprepared.com

OLD WAYS WHOLE GRAIN COUNCIL is an organization that supports the use of whole-grain products and offers a list of places to purchase organic whole grains.
WEBSITE: wholegrainscouncil.org

BLUEGRASS DAIRY AND FOOD specializes in powdered dairy products, including a diverse selection of cheese powders. They are sold in fifty-pound quantities, which would be good for a co-op.
WEBSITE: bluegrassdairy.com

FIREHOUSE PANTRY is an online store selling unique ingredients including gourmet mixes, popcorn seasonings, BBQ rubs, herbs & spices, and cheese powders.
WEBSITE: firehousepantrystore.com

Chapter 19 – Grain Mills, Bread Mixers, and Kitchen Tools

ANKARSRUM is a quality Swedish-made mixer with unique features like self-adjusting speed control and a timer. It has a nice assortment of optional accessories.
WEBSITE: ankarsrumoriginalusa.com

BLENDTEC is known for their powerful blenders, however they also build a high-powered grain mill and heavy-duty mixer.
WEBSITE: blendtec.com

NUTRIMILL offers grain mills, Bosch mixers, and L'Equip food prep appliances. They have two styles of mills: one style grinds using a "microburst" milling heads, and one style uses corundum milling stones.
WEBSITE: nutrimill.com

WONDERMILL makes a high-powered micronizing grain mill and the efficient Wonder Junior handmill. They also manufacture the WonderMix, a heavy-duty kitchen mixer.
WEBSITE: thewondermill.com

LEHMAN'S is a hardware store in Kidron, Ohio. They specialize in practical goods that will help meet many preparedness needs. They carry several brands of hand-cranked grain mills, including Diamant, Country Living, and Leman's Own Hand-Cranked grain mill. They have unique nineteenth-century-style kitchen tools like Lehman's Dazey Butter Churn, and oil lamps.
WEBSITE: lehmans.com

Chapter 20 – Making Basic Food Products

NEW ENGLAND CHEESE MAKING SUPPLY COMPANY offers supplies and equipment for cheese making along with workshops and advice for beginner through expert cheese-makers.
WEBSITE: cheesemaking.com

BREADTOPIA is all about helping folks make great bread. Along with offering recipes, an information-packed blog, and video tutorials, they sell bread-making supplies and specialized tools and equipment.
WEBSITE: breadtopia.com

KING ARTHUR is known for high-quality baking flour. They also offer recipes, tutorials, and baking classes as well as a complete line of baking tools and equipment.
WEBSITE: kingarthurflower.com

MARCATO makes a line of classic manual pasta machines, including the Marga Mulino Oat Flaker and the Atlas 150 Pasta Machine..
WEBSITE: marcato.it

Chapters 21-25 – Growing Plants

TRUE LEAF MARKET offers garden seeds along with seeds specifically for microgreens and sprouting as well as the supplies needed to grow them. They also sell garden seeds packaged for storage.
WEBSITE: trueleafmarket.com

MASONTOPS supplies assorted lids and kits for sprouting and other Mason jar accessories.
WEBSITE: masontops.com

MUMM'S SPROUTING SEEDS specializes in organic, non-GMO sprouting seeds. They are a great resource and offer books and online information about sprouting.
WEBSITE: www.sprouting.com

SPROUT HOUSE is dedicated to sprouts and provides organic sprouting seeds, sprouters, and sprouting supplies.
WEBSITE: www.sprouthouse.com

SPROUT PEOPLE promotes the health benefits of growing sprouts and microgreens and sells a large variety of seeds and unique seed combinations. They also offer sprouters and sprouting equipment.
WEBSITE: sproutpeople.org

Chapters 29-32 – Preserving Food

VKP BRANDS sells the Model 250 Food Strainer (formerly called Victorio Strainer) as well as other canning equipment, including water bath, steam, and pressure canners.
WEBSITE: vkpbrands.com

ALL AMERICAN makes high-quality pressure canners and other canning tools and supplies.
WEBSITE: allamericancanner.com

TATTLER produces a three-part, reusable canning lid. Sold in boxes of 12
WEBSITE: reusablecanninglids.com

FOOD SAVER makes food vacuum sealers and sells vacuum bags, containers, and supplies.
WEBSITE: foodsaver.com

CUISINART FOOD DEHYDRATOR makes a compact vertical dehydrator with five stacking trays.
WEBSITE: cuisinart.com

EXCALIBUR makes several top-rated, horizontal-style food dehydrators. Models come with four to ten drying trays.
WEBSITE: excaliburdehydrator.com

NESCO DEHYDRATORS have a vertical-stacking design, ideal for fruits, vegetables, and jerky. Some models have fruit roll-up trays.
WEBSITE: nesco.com

The *HARVEST RIGHT FREEZE DRYER* preserves both fresh and prepared foods. It comes in three different sizes.
WEBSITE: harvestright.com

Chapter 33-Emergency Evacuation

THORFIRE makes a compact, hand-crank, solar-powered flashlight small enough to fit in a pocket.
WEBSITE: thorfiredirect.com

CYALUME makes emergency light sticks in a variety of colors, manufactured in USA.
WEBSITE: getcyalume.com

UV PAQLITE GLOW LIGHTS are a reusable, reliable light source that recharges in light and has a variety of applications.
WEBSITE: uvpaqlite.com

MPOWERD creates sustainable, thoughtfully designed products including the Luci inflatable solar light and the Luci Core task light.
WEBSITE: mpowerd.com

PRIMUS makes camping and backpacking stoves and other outdoor gear which have crisis preparedness applications.
WEBSITE: primus.us

VARGO STOVES are durable, efficient, simple to use, and ideal for crisis preparedness applications.
WEBSITE: vargooutdoors.com

Chapter 34 – Home Security

WATER-JEL TECHNOLOGIES makes blankets along with burn wraps, burn dressings, and analgesics for burn first-aid.
WEBSITE: waterjel.com

QUICKSAFES offers an in-plain-sight hiding places that look like ordinary shelves or heating vents.
WEBSITE: quicksafes.com

Chapter 35 – Clothing and Bedding

NORTH X NORTH makes comfortable, versatile, protective merino wool scarves.
WEBSITE: northxnorth.co

COOLIBAR makes UPF 50+ clothing, including hats, scarves, tops, bottoms, and jackets.
WEBSITE: coolibar.com

Chapter 36 – Heating, Cooking, Lighting, and Refrigeration

LEHMAN'S, mentioned in an earlier section, also has a wide selection of wood stoves and cook stoves and other non-electric appliances.
WEBSITE: lehmans.com

OBADIAH'S WOODSTOVES sells a large selection of top-quality wood stoves, wood boilers, and wood cook stoves.
WEBSITE: woodstoves.net

UNIQUE OFF GRID APPLIANCES specializes in both propane and solar-powered off-grid appliances.
WEBSITE: uniqueoffgrid.com

CAMP CHEF manufactures heavy-duty outdoor cookers.
WEBSITE:www.campchef.com

SOLO STOVES offers innovative personal and group cooking stoves with efficient use of various fuels— e.g., twigs, pine cones, sticks. They come in a range of sizes, from small personal backpack stoves to large-group firepits.
WEBSITE: solostove.com

SIERRA ZIP STOVES are lightweight titanium and use twigs and small pieces of wood for fuel.
WEBSITE: zzstove.com

The *ALL AMERICAN SUN OVEN* bakes, boils, or steams food and will pasteurize water. These stoves have a sturdy construction and are compact enough to fit into a small suitcase.
WEBSITE: sunoven.com

The *GOSUN SOLAR COOKER* has a unique cylinder design using vacuum tubes. Highly concentrated sun energy cooks foods very quickly. Works in partial sun.
WEBSITE: gosun.co

ALADDIN KEROSENE LAMPS are the quintessential kerosene lamps, come in several styles, and are convertible to electricity.
WEBSITE: aladdinlamps.com

The *IVATION HAND CRANK CAMPING LANTERN* includes a phone charger, radio, SOS siren, and flashlight.
WEBSITE: ivationproducts.com

The *ADVANCED ELEMENTS SUMMER SOLAR SHOWER* heats from three to five gallons of water in the sun for a hot shower.
WEBSITE: advancedelements.com

The *ZODI* portable propane shower is a self-contained unit with a stainless-steel burner and battery-powered water pump.
WEBSITE: zodi.com

Chapter 37 − Managing Energy Sources in a Crisis

The *EXPLOSAFE JERRY CAN*, filled with aluminum mash, stores gasoline and other fuels safely. Sold by various online stores.

SUNJACK makes portable and mountable solar panels, chargers that power USB devices like phones, tablets, lights, etc.
WEBSITE: sunjack.com

GOALZERO SOLAR makes AC and USB power banks, solar chargers, portable and mountable solar panels, and a variety of solar appliances.
WEBSITE: goalzero.com

POINT ZERO ENERGY specializes in solar powered backup systems.
WEBSITE: pointzeroenergy.com

HOME BIOGAS makes a small bio digester that converts bio waste into methane gas that is used to fuel a small cooking stove.
WEBSITE: homebiogas.com

Chapter 38 − Sanitation and Personal Care

RELIANCE SANITATION PRODUCTS makes Double Duty Toilet Waste Bags, Bio-Blue Toilet Deodorant Chemicals, and Bio-Gel Waste Gelation for emergency sanitation situations.
WEBSITE: relianceoutdoors.com

PEEPOOPLE is primarily a humanitarian organization that makes individual disposable toilets.
WEBSITE: peepoople.com

Chapter 41 − Transportation

MORRIS MULE TRAILER COMPANY does its best to make simple, yet strong and functional trailers for both trail and field.
WEBSITE: morrismule.com

The *XVENTURE* line of off-road trailers are built to military standards by Schutt industries.
WEBSITE: schuttindustries.com

RUNGU BIKES makes a two-wheels-in-front, all-terrain e-bike designed for stability and control.
WEBSITE: riderungu.com

POLYMULE is an all-terrain utility handcart with uphill assist and designed for heavy-duty portability.
WEBSITE: polymule.com

Chapter 41 − Firearms

BROWNELLS carries a wide selection of firearms tool kits along with a tremendous parts inventory for firearms.
WEBSITE: www.brownells.com

Chapters 21 to 25 – Growing Fruits and Vegetables

With a quick search you will find countless articles, books, and internet sources about growing vegetable gardens and fruit trees. Those included here focus on the basics and the gardening methods mentioned in the book. Many university extensions offer excellent horticulture information.

Gardening

The Backyard Homestead by Carleen Madigan (2009) is a guide to growing and harvesting a sustainable, self-reliant homestead, including fruits and vegetables, grains, and livestock.

"Specialized Gardening Techniques" Bulletin A3384 (PDF)
This University of Wisconsin publication by Helen C. Harrison describes wide-row planting, square-foot gardening, and raised beds.
WEBPAGE: https://learningstore.uwex.edu/Assets/pdfs/A3384.pdf

All New Square Foot Gardening, Third Edition by Mel Bartholomew (2018), an updated version of the classic *More Food from Grow-Box Gardens* by Jacob R. Mittleider (1982), teaches you how to become a successful DIY square-foot gardener.

How to Grow More Vegetables (And Fruits, Nuts, Berries, Grains and other Products, 9th Edition) by John Jeavons (2017) demonstrates sustainable growing methods for organic and intensive small-scale gardening.

Backyard Winter Gardening: Vegetables Fresh and Simple in Any Climate without Artificial Heat or Electricity, the Way It's Been Done for 2,000 Years by Caleb Warnock (2013) explains how to extend the harvest into winter months to add to your food security.

Fruit Trees

The Fruit Gardener's Bible: A Complete Guide to Growing Fruits and Nuts in the Home Garden by Lewis Hill and Leonard Perry (2011) is an authoritative guide to growing fruit trees.

Field Crops

Homegrown Whole Grains: Grow, Harvest, and Cook Wheat, Barley, Oats, Rice, Corn and More by Sara Pitzer (2009) explains cultivating, harvesting, and using whole grains.

Microgreens and Sprouts

Microgreens: A Guide to Growing Nutrient-Packed Greens by Eric Franks & Jasmine Richardson (2009) is a complete guide for getting started growing microgreens.

Homegrown Sprouts: A Fresh, Healthy, and Delicious Step-by-Step Guide to Sprouting Year Round by Rita Galchus (2013) is about how to grow your own sprouts from the unique perspective of Rita, the Sprout Lady.

Chapter 26 – Raising Animals

Though not comprehensive, the following list gives you a number of well-regarded references and resources to consider when raising animals in general. Reference resources are also listed for specific animals.

Cooperative Extension

Every state has a land-grant university tasked with providing education in agriculture to the citizens of the state. It is a valuable resource for information about raising animals as well as other agricultural and horticulture topics. Many extensions maintain a useful website.

General Guides to Raising Animals

Barnyard in Your Backyard: A Beginner's Guide to Raising Chickens, Ducks, Geese, Rabbits, Goats, Sheep, and Cattle by Gail Damerow (2002) contains basic information to get started raising animals.

The Complete Beginners Guide to Raising Small Animals: Everything You Need to Know about Raising Cows, Sheep, Chickens, Ducks, Rabbits, and More by Carlotta Cooper (2012) offers a good sampling of what is required to raise farm animals.

The Spruce, a website dedicated to all things about your home, has a strand for urban, suburban, and small-sale hobby farms. You will find many good articles to help you make introductory information about raising most small farm animals.
WEBSITE: https://www.thespruce.com/small-farm-4127721

Storey's Guides to Raising Animals
This well-regarded, best-selling series of books is designed to help those who want to raise animals, including rabbits, chickens, ducks, turkeys, goats, llamas, sheep, pigs, beef cattle, horses, and honeybees.

Butchering Poultry, Lamb, Goat, and Pork: The Comprehensive Photographic Guide to Humane Slaughtering and Butchering by Adam Danforth (2014). If you raise animals you will want to know how to slaughter and butcher them.

Merck Veterinary Manual by Susan Aiello and Michael A. Moses. This basic authoritative guide to animal health is used by veterinary professionals. Also available online,
WEBSITE: merckvetmanual.com

Veterinary Guide for Animal Owners: Caring for Cats, Dogs, Chickens, Sheep, Cattle, Rabbits, and More by C. E. Spaulding and Jackie Clay (2015)—an excellent resource for animal owners, this book discusses how to prevent and cure common problems in simple terms.

Rabbits

Raising Rabbits 101 by Aaron G Webster (2015)—a good starting place, covering all aspects of raising rabbits, especially helpful for beginners.

Chickens

The Chicken Chick's Guide to Backyard Chickens: Simple Steps for Healthy, Happy Hens by Kathy Shea Mormino (2017). This book covers it all—coops, chicken care, breed selection, chicken health, etc. The chart "Things I Wish I Had Known Before I Got Chickens" is enlightening!

The Beginner's Guide to Raising Chickens: How to Raise a Happy Backyard Flock by Ann Kuo (2019). The author's experience with raising chickens in an urban setting will help you decide if raising chickens is for you. Good basic information, enthusiastically presented.

"Raising Chickens 101: How to Get Started," *The Old Farmer's Almanac*. This is the first of six articles about raising chickens. Other articles discuss choosing chickens, building a coop, raising chicks, collecting eggs, and what to do when chickens stop laying eggs.
WEBPAGE: https://www.almanac.com/news/home-health/chickens/raising-chickens-101-how-get-started

Goats

The Backyard Goat: An Introductory Guide to Keeping and Enjoying Pet Goats, from Feeding and Housing to Making Your Own Cheese by Sue Weaver (2011). This is a good guide for beginners, giving the basics on getting started raising goats.

Backyard Goats is a website all about raising goats, including breed selection and goat care.
WEBSITE: https://backyardgoats.iamcountryside.com/

Sheep, Pigs, and Cows

The Backyard Sheep: An Introductory Guide to Keeping Productive Pet Sheep by Sue Weaver (2013) is a good guide for beginners that includes information on breed selection, care, feeding, doctoring, breeding, and using sheep for milk and fleece.

"Raising Sheep on a Small Farm" by Lauren Arcuri (2018) is an article from "The Spruce" WEBSITE. This is a good introduction if you want a small flock of sheep.
WEBPAGE: https://www.thespruce.com/how-to-raise-sheep-3016859

"Raising Small Groups of Pigs" by Elizabeth Hines, Penn State Extension, April 20, 2016, is a good place for beginners to get information about how to raise pigs on a small homestead.
WEBPAGE: https://extension.psu.edu/raising-small-groups-of-pigs

"The Beginning Farmer's Guide to Raising Backyard Beef" by Melissa K. Norris (2015). This article contains hints for raising beef cattle. It includes ideas about fencing, breed selection, cattle feed, and butchering.
WEBPAGE: https://www.hobbyfarms.com/the-beginning-farmers-guide-to-raising-backyard-beef-3/

Bees

The Backyard BeeKeeper, Fourth Edition by Kim Flottum (2018). Helpful beginning guide by honeybee researcher.

Fish

Home Aquaponics System is an inclusive website with many articles and tips for aquaculture. They also sell many tools and equipment for home aquaculture.
WEBSITE: aquaponics.com

"Raising Fish for Food: Backyard Fish Farming for Survival," January 27, 2014.
WEBPAGE: https://survivalist101.com/raising-fish/

Edible Wild Plants and Wild Game

The Forager's Harvest: A Guide to Identifying, Harvesting, and Preparing Edible Wild Plants by Samuel Thayer (2006)

Field Guide to Edible Wild Plants of Eastern and Central North America, Third Edition by Steven Foster and James A. DukePeterson (2014)

Edible and Poisonous Plants of the Western/Eastern United States Identification Set. Fifty-two-card set with color pictures and descriptions.

The Bushcraft 101: A Field Guide to the Art of Wilderness Survival by Dave Canterbury (2014) is a good general book, with sections on tools, hunting, trapping, butchering, and cooking.

Hunt, Gather, Cook: Finding the Forgotten Feast by Hank Shaw (2012) is for those just beginning to explore the possibilities of foraging, fishing, and hunting for food.

Chapters 27 to 32 – Preserving Food

National Center for Home Food Preservation is part of the USDA and publishes complete guides for all types of food preservation, including canning, freezing, drying, curing and smoking, fermenting, pickling, and making jams and jellies. Besides factual, accurate information, the publications offer recipes and troubleshooting suggestions. WEBSITE: nchfp.uga.edu

Root Cellars

The Complete Root Cellar Book: Building Plans, Uses, and 100 Recipes by Steve Maxwell and Jennifer MacKenzie (2010) provides complete and detailed information on building root cellars and choosing vegetables to store.

Home Canning

The USDA Complete Guide to Home Canning, 2015 Revision. Thorough factual canning reference. WEBSITE: https://nchfp.uga.edu/publications/publications_usda.html

Ball Blue Book—is published annually and its contents change over time. Thorough instructions for canning, freezing, and dehydrating—includes recipes. *The All New Ball Book of Canning and Preserving* (2016), a recent Ball publication, is larger and expanded.

The Complete Guide to Pressure Canning: Everything You Need to Know to Can Meats, Vegetables, Meals in a Jar, and More Diane Devereaux (2018) A little bit of "how-to" to give you confidence and a lot of recipes for practice.

Putting Food By, authored by Ruth Hertzberg and Janet C. Greene, et al (2010, 5th edition). This classic reference is updated and includes information for all types of food preservation.

A Guide to Canning, Freezing, Curing and Smoking of Meat, Fish and Game by Wilbur F. Eastman, Jr. (2002). This is a no-nonsense resource on time-honored ways of preserving meats.

Extension Bulletins

"Canning Meats in Cans" (2016) FNH-00227 and "Canning Fish in Cans" (2017) FNH-00125, University of Alaska Extension bulletin www.uaf.edu/files/ces/publications-db/catalog/hec/FNH-00227.pdf. They also have a series of YouTube videos with directions for home canning.

"USU Steam Canning: Position Statement" No. 002 (2005) Utah State University Extension bulletin http://extension.usu.edu/files/publications/newsletter/No__002.pdf

"Guidelines for Using an Atmospheric Steam Canner for Home Food Preservation," University of Wisconsin Extension http://winnebago.uwex.edu/files/2015/06/Steam-Canning-PDF1.pdf

Dehydrating Foods

The Dehydrator Bible by Jennifer McKenzie, et al (2009) includes comprehensive information as well as over three hundred recipes for dehydrated food.

The Ultimate Dehydrator Cookbook: The Complete Guide to Drying Food, Plus 398 Recipes, Including Making Jerky, Fruit Leather & Just-Add-Water Meals by Tammy Ganloff. (2014) By a well-regarded dehydrating expert—the title is self-explanatory.

"Home Freeze Drying—The Good, the Bad, and the Ugly" by Laurie Neverman blog: Common Sense Home WEBPAGE: http://commonsensehome.com/home-freeze-drying/

Salt Curing, Smoke Curing, and Pickling

Several university extensions offer excellent information about meat curing. You can be sure the method and recipes are checked for safety.

"The Art and Practice of Sausage Making" by Martin Marchello and Julia Garden-Robinson, University of North Dakota Extension Service

"Basics of Sausage Making," Anand Mohan, PhD, UGA Extension

"Curing and Smoking Poultry Meat," Oregon State University Extension Service, SP 60-593

"Dry Curing Virginia-Style Ham," Virginia Cooperative Extension, Publication 458-233

"Smoking Fish at Home—Safely," Pacific Northwest Extension Publication, Publication, PNW238

Dry-Curing Pork: Make Your Own Salami, Pancetta, Coppa, Prosciutto, and More (Countryman Know How) by Hector Kent (2014)

The Sausage and Jerky Makers' Bible: The Home Processor's Guide to Charcuterie and Sausage and Jerky Handbook (1994) by Eldon Russell Cutlip (2014)

Amazing Ribs by Meathead Goldwyn. This website is about all things meat. Besides information on grilling, there are several well-informed discussions about curing meats.
WEBSITE: amazing ribs.com.

"The Science of Curing Meat Safely" will help you understand the meat curing processes.
WEBPAGE: amazingribs.com/tested-recipes/salting-brining-curing-and-injecting/curing-meats-safely

Chapters 34 to 38 – Securing Your Home for a Crisis

Repair and Home Maintenance

You might want to have books that give a foundation of how your house works and how to maintain and repair it. Although the internet is populated with countless DIY videos, printed material in your hands could be invaluable in a crisis. The following resources will help you with fundamental home maintenance.

How Your House Works: A Visual Guide to Understanding and Maintaining Your Home by Charlie Wing (2012)

Ultimate Guide to Home Repair and Improvement by Editors of Creative Homeowner (2016)

You may also want to keep books on basic plumbing, electrical work, carpentry, auto repair, small-engine repair, blacksmithing, masonry work, and construction on hand. Home Depot and Lowes offer their own repair manuals and specialized instructions, many of which are free.

Home Security

"Be Ember Aware" by University of Nevada Cooperative Extension (2009)
WEBSITE: https://www.yumpu.com/en/document/view/27530896/be-ember-aware-university-of-nevada-cooperative-extension-/6

"Protecting Your Home from Wildfire," California Chaparral Institute
WEBSITE: http://www.californiachaparral.com/fire/protecting-your-home/

Peace of Mind in Earthquake Country: How to Save Your Home, Business and Life by Peter I. Yanev and Andrew C. T. Thompson (2009)

DIY Secret Hiding Places:90 Places to Hide What You Don't Want Found by Steve Plant (2015).

The High Security Shelter: How to Implement a Multi-Purpose Safe Room in the Home by Joel Skousen and Andrew Skousen (2017)

Wood Fuel

Norwegian Wood: Chopping, Stacking, and Drying Wood the Scandinavian Way by Lars Mytting (2015) is both entertaining and factual. If you plan on heating primarily with wood, this book will teach you the tricks of the trade.

"Wood Heating Appliances for Homes and Businesses" by Scott A. Sanford and David S. Liebl, University of Wisconsin Extension (2014).
WEBSITE: https://learningstore.uwex.edu/Assets/pdfs/GWQ066.pdf

Solar Energy

Wholesale Solar "Wholesale Solar Load Evaluation Calculator." This website will also help you calculate your off-grid system size.
WEBSITE: wholesalesolar.com/solar-information/start-here/offgrid-calculator

Sanitation

Manual of Composting Toilet and Greywater Practice (2016). The British Columbia government has published a very helpful 23-page manual in pdf format. You can get an updated website with a simple Google search.

Chapter 39 – Medical and Dental Library

Emergency Medicine

The Survival Medicine Handbook: A Guide for When Help is NOT on the Way, revised (2013) by Joseph Alton, MD, and Amy Alton, ARNP.
WEBSITE: doomandbloom.net offers useful information and emergency medical products.

Alton's Antibiotics and Infectious Disease by Joseph Alton, MD and Amy Alton, ARNP (2018)

Where There is No Doctor: A Village Healthcare Handbook by David Werner, et al (updated 2015)

Medical Dictionaries

Medical Terminology Made Easy by Eva Regan (2016)

Stedman's Medical Dictionary (2005 or current edition)

Anatomy Texts

Atlas of Human Anatomy by Frank H Netter, M.D. (2014)

Gray's Anatomy for Students (2014)

Medical Handbooks

Current Medical Diagnosis and Treatment by Maxine Papadakis, Stephen McPhee, Michael Rabow (2020) updated annually. Used by NP and PA students, practitioners.

The Merck Manual, Robert S. Porter, editor (updated periodically)

Physicians' Desk Reference, updated annually and includes a pharmaceutical reference.

Childbirth Resources

Varney's *Midwifery* by Tekoa L King and Mary C. Brucker (2018)) is a basic midwifery guide.

Chapter 40 – Communications

SWLing Post blog is dedicated to teaching about shortwave-radio communication and offers a forum for the shortwave community.
WEBSITE: swling.com

Passport to World Band Radio is a reference for shortwave communication. Issues from 1986 to 2009 may be found in print from secondary sources.

American Radio Relay League (ARRL) is the United States' amateur radio organization.
WEBSITE: arrl.org

Chapter 42 – Survival Firearms

Chapter 42 is only a very basic overview of survival firearms. A simple internet search will give you ample information about firearms. This list includes basic information and a few classic resources.

Gun Laws

For up-to-date gun laws, refer to the following two websites:
opencarry.org
handgunlaw.us

Basic Firearms References

Gun Digest, Jerry Lee, editor (published annually)
Includes gun reviews and gun articles, also free downloads.
WEBSITE: gundigest.com

Shooter's Bible: The World's Best-Selling Firearms Reference by Jay Cassell (published annually)
This is the most comprehensive reference for new guns and those currently in production.
WEBSITE: skyhorsepublishing.com/9781510748125/shooters-bible-111th-edition/

Chapter 44 – Surviving Terrorism and War

Nuclear War Survival Skills by Cresson H. Kearny (1979, 1987, 2001)
The classic nuclear war survival manual by an Oakridge National Laboratory scientist is still relevant today, forty years after it was written. Available in paperback or downloadable as a PDF.

Index